Boris Shapiro
Polynomial Solutions of Linear Differential Equations

De Gruyter Expositions in Mathematics

Volume 80

Boris Shapiro

Polynomial Solutions of Linear Differential Equations

Spectral Problems, Asymptotic and WKB-Analysis

DE GRUYTER

Mathematics Subject Classification 2020
Primary: 34M03, 30C15, 34M46; Secondary: 34M05, 34M40

Author
Boris Shapiro
Department of Mathematics
Stockholm University
Albanovavägen 28
11419 Stockholm
Sweden
shapiro@math.su.se

ISSN 0938-6572
ISBN 978-3-11-914710-1
ISBN 978-3-11-220774-1 (PDF)
ISBN 978-3-11-220781-9 (E-PUB)
DOI https://doi.org/10.1515/9783112207741

Library of Congress Control Number: 2026935824

Bibliographic information published by the Deutsche Nationalbibliothek
The Deutsche Nationalbibliothek lists this publication in the Deutsche Nationalbibliografie; detailed bibliographic data are available on the Internet at https://dnb.dnb.de.

De Gruyter and Walter de Gruyter GmbH are part of De Gruyter Brill.
www.degruyterbrill.com

Questions about General Product Safety Regulation: productsafety@degruyterbrill.com

Printing: CPI books GmbH, Leck

Contents

Preface

> The mathematical facts worthy of being studied are those which, by their analogy with other facts, are capable of leading us to the knowledge of a physical law. They reveal the kinship between other facts, long known, but wrongly believed to be strangers to one another.
>
> Henri Poincaré, *Science and Hypothesis*

Frankly speaking, I never expected that I would be able to write a book. Now and then I had some thoughts about a research monograph, but I was always waiting for this or that part of the next project to be completely finished. Due to a heavy teaching load I never had enough time to concentrate on putting the already obtained results in the area together.

However, when I came to Beijing Institute for Mathematical Sciences and Applications (BIMSA) for a semester and was supposed to give a lecture course, I decided, in spite of the fact that many pieces of the overall project were (and still are) unfinished, to summarize the material scattered throughout many different publications. To compensate for the incompleteness of the currently available information, I included a number of open problems at the end of each chapter hoping that the aesthetic appeal of the subject together with the abundance of concrete examples will attract other mathematicians to its study. (I strongly believe that a good mathematical theory should have a clear motivation, many examples, and connect to many previously known results, see the above epigraph by H. Poincaré.)

I put some effort in trying not to overload the text with complicated definitions and make it as understandable as I possibly could, in particular, by providing many illustrations of the root configurations for the considered polynomials. I really hope that many portions of this book are accessible to master students having some knowledge of complex analysis and potential theory.

I gladly acknowledge that this book would never have seen the light of day without the help of several of my colleagues, collaborators, and students whose collective efforts are presented below.

I want to thank my former doctoral students Tanja Bergkvist, Thomas Holst, and Hans Rullgård for their break-through contributions. Although all three have since moved on from mathematics, their impact on this work is lasting. I am also deeply grateful to my colleague Rikard Bøgvad and especially to the late Professor Jan-Erik Björk—a true Renaissance man, whose vast knowledge and interests spanned every imaginable area of human activity.

I have been fortunate to collaborate with many talented researchers across the globe, including Jorge Borrego-Morell, Emil Horozov, Andrei Martínez-Finkelshtein, Gisli Mason, Guillaume Tahar, Kouichi Takemura, Milos Tater, and many others. My sincere thanks also go to my friend and collaborator Christian Hägg, whose amazing skills and assistance with the figures were invaluable.

https://doi.org/10.1515/9783112207741-203

I am grateful to Stockholm University for granting and funding the sabbatical semester that allowed me the time to complete this project.

Last but not least, I want to thank my wife, Karin Peters, for her wholehearted support during my time in China and my children Sonya, Jacob, Elias, Benjamin, and Rebecka for being wonderful and (almost) grown-up kids.

Stora Benhamra, Sweden, January 2026 Boris Shapiro

1 Introducing the main protagonists

1.1 Motivation and short historical account

Below we discuss polynomial solutions of linear differential equations with the main emphasis on polynomial eigenfunctions of (multi-parameter) spectral problems for univariate linear ordinary differential operators with polynomial coefficients as well as their asymptotics.

This area of mathematics is by no means new. It goes back at least to the second half of the 19th century. For example, the algebraic form of the classical Lamé equation (see [197, Chapter 23]) was introduced by Gabriel Lamé in 1830s in connection with the separation of variables in the Laplace equation in $\mathbb{R}^l$ with respect to elliptic coordinates. It has the form

$$Q(z)\frac{d^2S}{dz^2} + \frac{1}{2}Q'(z)\frac{dS}{dz} + V(z)S = 0, \tag{1.1}$$

where $Q_l(z)$ is a real polynomial of degree l with all real and distinct roots, and $V(z)$ is a polynomial of degree at most $l-2$ whose choice depends on what type of solution to (1.1) we are looking for. In the second half of the 19th century, several famous mathematicians including M. Bôcher, E. Heine, F. Klein, and T. Stieltjes studied the number as well as properties of the Lamé polynomials (sometimes also referred as Lamé solutions of a given degree and a given kind).

Such solutions to (1.1) exist for some choices of $V(z)$ and are characterized by the property that their logarithmic derivative is a rational function. For a given $Q(z)$ of degree $l \geq 2$ with simple roots, there exist 2^l different kinds of Lamé polynomials depending on whether these solutions are smooth or have a square root singularity at each given root of $Q(z)$ (for details, see [153] and [197]). An excellent modern study of these questions can be found in [121].

Existence of a polynomial solution of a given linear differential equation imposes natural restrictions on the indices of its singular points and implies that its monodromy group has an invariant 1-dimensional eigenspace. Existence of a sequence of polynomial solutions for a spectral problem of a given linear differential operator imposes further restrictions on the operator. In many cases, such sequences of polynomial solutions can be studied by various classical methods of analysis and potential theory, and one can obtain detailed exact and asymptotic results about their roots.

But in spite of the relative abundance of classical results, the field is still surprisingly modern and its asymptotic aspects are closely related to several nowadays popular areas such as, for example, the WKB-expansion for linear differential equations of order exceeding 2 and the theory of differentials of high order.

It is common knowledge that very few spectral problems involving linear ordinary differential operators allow explicit calculation of their spectra and eigenfunctions. There are, however, special but important exceptions from this general rule;

https://doi.org/10.1515/9783112207741-001

the most well-known of them being the Schrödinger equation on the real line with a real quadratic potential and boundary conditions given by vanishing of eigenfunctions at $\pm\infty$.

Such potentials for which similar explicit calculations can be carried out are usually referred to as *exactly solvable*. A classical example of this kind is the Pöschl–Teller potential, see [152]. Additionally there is an interesting class of quantum mechanical systems/potentials discovered relatively recently for which only a finite part of the spectrum and the respective eigenfunctions can be explicitly obtained. Such systems are called *quasi-exactly solvable*, and they are also quite rare, see, e. g., [189].

At the same time, if we abandon the physically motivated class of second order differential equations, there exists a sufficiently large and easily characterized class of linear differential operators that possess sequences of polynomial eigenfunctions (one in each degree) and whose spectra can be explicitly found. In what follows we shall call such operators *exactly solvable* (shorthand ES-operators) and refer to their polynomial eigenfunctions as *eigenpolynomials*. Besides being closely related to the theory of orthogonal polynomials, they are much easier to analyze from the point of view of asymptotic analysis, and they are connected to interesting geometric objects such as high order differentials. Their study constitutes Part I of this book.

Another source of polynomial solutions of linear differential equations with polynomial coefficients is related to the studies of Lamé, Heine, and Stieltjes briefly mentioned above. In their approach, instead of just adding a constant term to a given operator, one is allowed to add a polynomial term of a certain degree with arbitrary coefficients, which under certain conditions guarantees the existence of finitely many polynomial solutions occurring for different choices of the added polynomial. In the classical approach one only considered operators of second order. Part II of this book is devoted to similar studies and extension of several classical results to operators of arbitrary order.

1.2 Exactly solvable operators, basic notions

The fundamental (and quite simple) definition of exactly solvable operators going back to Salomon Bochner [35], which we will use throughout this book, is as follows.

Definition 1.1. A univariate linear differential operator

$$T = \sum_{j=1}^{k} Q_j(z)\frac{d^j}{dz^j} \tag{1.2}$$

with polynomial coefficients is called *exactly solvable* if the following two conditions hold:

(i) $\deg Q_j \leq j, \forall j = 1, \ldots, k$;
(ii) $\exists j_0$ such that $\deg Q_{j_0} = j_0$.

Since $\deg Q_j \leq j, j = 1, \ldots, k$, we will use the notation $Q_j(z) = \sum_{i=0}^{j} q_{ji} z^i$ throughout the book.

Remark 1.1. To explain why our summation index j starts from 1 and not from 0, observe that the above condition (i) implies that the term Q_0 of any ES-operator must be constant. Since we shall mainly consider spectral problems for ES-operators, it is natural to get rid of Q_0 adding it to the spectral parameter. Another reason for our choice is that it allows us to avoid certain pathologies occurring in examples like $T = 1 + \frac{d}{dz}$, which are not ES-operators according to our definition, and although they are easy to study on their own, they have properties drastically different from those of "our" ES-operators.

Figure 1.1: Salomon Bochner, 1899–1982.

Remark 1.2. Naturally S. Bochner has not been using the term "exactly solvable" for the class of operators he introduced. This terminology appeared much later and originates from physics.

As we will show in Chapter 2, exact solvability of T is equivalent to the fact that T preserves the infinite complete flag of linear subspaces in $\mathbb{C}[z]$ given by

$$\mathbb{C}_0[z] \subset \mathbb{C}_1[z] \subset \cdots \subset \mathbb{C}_n[z] \subset \cdots$$

where $\mathbb{C}_n[z]$ is the standard $(n+1)$-dimensional linear subspace of $\mathbb{C}[z]$ containing all polynomials of degree at most n.

Choosing the standard monomial basis $\{1, z, z^2, \ldots,\}$ of $\mathbb{C}[z]$, we immediately observe that, by definition, any ES-operator T acts in this basis as an infinite upper triangular matrix. Namely, for any non-negative integer n, using (1.2) we obtain

$$T(z^n) = \lambda_n z^n + \text{terms of degree less than } n$$

where the coefficients $\lambda_0, \lambda_1, \lambda_2, \ldots$ (with $\lambda_0 = 0$) are given by the formula

$$\lambda_n = \sum_{j=1}^{k} (n)_j q_{j,j}, \quad n = 0, 1, 2, \dots \tag{1.3}$$

As above $q_{j,j}$ is the coefficient of $Q_j(z)$ of (the maximal possible) degree j and

$$(n)_j = n(n-1)\dots(n-j+1)$$

is the Pochhammer symbol. (Observe that by condition (ii) in Definition 1.1 at least one $q_{j,j}$ is non-vanishing.)

For generic ES-operators T, all λ_n, $n = 0, 1, 2, \dots$ are distinct, and therefore $\mathbb{C}[x]$ has a basis consisting of monic eigenpolynomials $p_n(z)$ of T, $\deg p_n(z) = n$ defined for all integers $n \geq 0$.

For an arbitrary ES-operator T, we can only conclude that there always exists a minimal non-negative integer n_T such that λ_n will be distinct for all integer $n \geq n_T$. Indeed, formula (1.3) immediately implies that $|\lambda_n|$ is monotone increasing for all sufficiently large n, which guarantees that for n large, all λ_n are distinct. (For this fact to be true, we need our summation in (1.2) to start from 1 and not from 0; it does not hold for $T = 1 + \frac{d}{dz}$.) This monotonicity implies that n_T always exists. On the other hand, the restriction of T to the linear subspace $\mathbb{C}_n[z]$ for n small compared to the order k of T can be a (more or less) arbitrary triangular matrix and, in particular, can have non-trivial Jordan blocks.

Notation 1.1. For any ES-operator T, we call the latter number n_T the *least uniqueness degree* of T. Further, let us denote by $\{p_n\}_{n=n_T}^{\infty}$ the sequence of uniquely defined monic eigenpolynomials of T in degrees $n_T, n_{T+1}, n_{T+2}, \dots$.

To build up the intuition, let us describe ES-operators of orders 1 and 2. The following facts are easy to prove.

Example 1.1. (i) Up to a scaling and affine change of z, the only ES-operator of order 1 is $T = z\frac{d}{dz}$ whose monic eigenpolynomials are $1, z, z^2, \dots, z^n, \dots$ and $\lambda_n = n$.

(ii) The following operators are exactly solvable of order 2:

$$T_1 = \frac{d^2}{dz^2} - 2z\frac{d}{dz}, \quad p_n(z) = H_n(z), \quad \lambda_n = -2n;$$

$$T_2 = z\frac{d^2}{dz^2} + (\alpha + 1 - z)\frac{d}{dz}, \quad p_n(z) = L_n^{\alpha}(z), \quad \lambda_n = -n;$$

$$T_3 = z^2\frac{d^2}{dz^2} + 2(z+1)\frac{d}{dz}, \quad p_n(z) = Y_n(z), \quad \lambda_n = n(n+1);$$

$$T_4 = (z^2 - 1)\frac{d^2}{dz^2} + (\beta - \alpha - (\alpha + \beta + 2)z)\frac{d}{dz}, \quad p_n(z) = P_n^{\alpha,\beta}(z), \quad \lambda_n = -n(n + \alpha + \beta + 1).$$

Here $H_n(z)$ stands for the n-th Hermite polynomial, $L_n^{\alpha}(z)$ for the n-th Laguerre polynomial with parameter α, $Y_n(z)$ for the n-th Bessel polynomial, and $P_n^{\alpha,\beta}(z)$ for the n-th

Jacobi polynomial with parameters (α, β). These four families describe all possible interesting ES-operators of order 2 up to an affine transformation and scaling. (Notice that we allow α and β to be complex.)

1.3 ES-operators and Bochner–Krall problem

Based on the observations that every ES-operator has a sequence of eigenpolynomials and that for ES-operators of order 2 one typically obtains the four classical families of orthogonal polynomials with complex-valued parameters, Salomon Bochner posed the following general problem, see [35].

Problem 1.1. Describe ES-operators whose sequences of eigenpolynomials are orthogonal with respect to some positive measure supported on $\mathbb{R}$.

In loc. cit. he solved this problem for operators of order 2 showing that the above four cases of operators T_1, T_2, T_3, T_4 with the classical values of parameters (α, β) provide the complete answer. He also made a number of important general observations. In [105] his student H. L. Krall solved Problem 1.1 for operators of order 4 obtaining a number of additional families. These families are various classical systems such as the Jacobi-type, the Laguerre-type, the Legendre-type, the Bessel, and the Hermite polynomials, see [66, Theorem 3.1].

An assortment of families corresponding to order six operators has been found in the ensuing decades, see, e. g., [106, Chapter XVI]. W. Hahn showed that the four classical families are the only orthogonal polynomial sequences $\{P_n(z)\}_{n=0}^{\infty}$ for which $\{\frac{d}{dz} P_n(z)\}_{n=0}^{\infty}$ is also an orthogonal polynomial sequence, see [81]. An analog of this was proved for the order 4 case by K. H. Kwon, L. L. Littlejohn, J. K. Lee, and B. H. Yoo [113]. The most general form of Problem 1.1 is still open for operators of order six or higher, but K. H. Kwon and J. K. Lee have found a satisfactory solution if the polynomials are required to be orthogonal with respect to a compactly supported positive measure, see [114]. (A weaker result in the same direction was somewhat earlier obtained in [28].)

Problem 1.1 is now known as the *Bochner–Krall* problem. The respective ES-operators are called *Bochner–Krall* operators, and their sequences of eigenpolynomials are called *Bochner–Krall* systems (BKS, for short). Classification of Bochner–Krall operators has attracted considerable attention, see, e. g., [66] with its 100 references and [115, 116, 117] and the references therein. But currently it is still far from being solved.

The Bochner–Krall problem and its natural generalizations provide the initial motivation for the study of ES-operators. Another motivation comes from the possibility to study in detail the asymptotic behavior of their eigenpolynomials, which leads to several interesting results similar to those obtained by the classical WKB-method for Schrödinger equations. An intriguing circumstance is that in this asymptotic study naturally appears a family of rational higher order differentials.

Figure 1.2: Eduard Heine, 1821–1881, and Thomas Stieltjes, 1856–1894.

1.4 Heine–Stieltjes theory

Obviously, any linear ordinary operator with polynomial coefficients that is not exactly solvable can have at most finitely many eigenpolynomials. However, there is a natural way to "improve the situation" by varying polynomial coefficients of the operator instead of just adding to it a constant term. In other words, instead of considering the standard spectral problem for a given linear differential operator, one should consider an appropriate multi-parameter spectral problem in which infinitely many polynomial solutions will naturally appear.

The classical example of this approach is usually referred to as the *Heine–Stieltjes theory* after E. Heine and T. J. Stieltjes who were first to develop it for operators of order 2 in the second half of the 19th century. Some old and new results in this direction are presented in Part II of our book.

Definition 1.2. Consider an arbitrary linear ordinary differential operator

$$T = \sum_{j=1}^{k} Q_j(z) \frac{d^j}{dz^j} \tag{1.4}$$

with polynomial coefficients. (Here compared to the case of (1.2), we put no restrictions on the degrees of $Q_j(z)$.) The number

$$r = \max_{i=1,\dots,k} (\deg Q_i(z) - i)$$

is called the *Fuchs index* of T. The operator T is called a *(higher) Lamé operator* if its Fuchs index r is non-negative. T is called *non-degenerate* if $\deg Q_k(z) = k + r$ and *degenerate* otherwise.

Remark 1.3. Notice that the classical Lamé operator corresponds to $k = 2$. Observe that if $r = 0$, then T is exactly solvable. Further, the non-degeneracy of T is a natural condition equivalent to the requirement that T has either a regular or a regular singular point at ∞.

Given a (higher) Lamé operator T, consider the following multi-parameter spectral problem.

Problem 1.2. For each non-negative integer n, find all polynomials $V(z)$ of degree at most r such that the equation

$$TS(z) + V(z)S(z) = 0 \tag{1.5}$$

has a polynomial solution $S(z)$ of degree n.

Following the classical terminology, we call (1.5) a *(higher) Heine–Stieltjes spectral problem*, $V(z)$ is called a *(higher) Van Vleck polynomial*, and the corresponding polynomial $S(z)$ is called a *(higher) Stieltjes polynomial*. Below, we will often skip mentioning "higher".

In Part II of the book we discuss the asymptotic root-counting measures for sequences of Heine–Stieltjes polynomials of increasing degrees when the respective sequence of Van Vleck polynomials (which all are of the same degree) has a limit. The answer is expressed in terms of this limit and the leading coefficient of the Lamé operator.

One can also ask about the asymptotic root-counting measures for the product of all Van Vleck polynomials that correspond to Stieltjes polynomials of degree $n \to \infty$. This problem seems to be considerably more difficult. Below we present its solution only in the case of Lamé operator of second order with Fuchs index 1. The answer is given in quite non-trivial terms related to the modulus of the symbol curve, which in this case is elliptic. So far our attempts to solve a similar problem for high order Lamé operators with Fuchs index 1 failed.

A promising aspect of both asymptotic theory of ES-operators and Heine–Stieltjes theory is the appearance of an interesting family of rational differentials whose order equals the order of the differential operator and which have certain properties similar to those of the famous Strebel quadratic differentials. One can hope that the material of this book will stimulate further development of the theory of higher differentials on surfaces of any genus.

1.5 Short historical notes

Heinrich Eduard Heine (Figure 1.2, left) made fundamental contributions to several areas of mathematics, mainly in real analysis, but also to special functions and mathematical logic/formalism. In particular, he introduced the notion of uniform convergence, which everybody studies in Calculus 1 classes and his name appears in Heine–Borel theorem, Heine–Cantor theorem, Heine–Stieltjes polynomials (considered below), Mahler–Heine formula, etc.

In spite of his untimely death at the age of 38, **Thomas Joannes Stieltjes** (Figure 1.2, right) has worked on almost all branches of analysis, continued fractions, and number theory. Stieltjes' work is also seen as an important first step towards the theory of Hilbert spaces. Stieltjes also contributed to the theory of divergent series and discontinuous functions, to ordinary and partial differential equations, the gamma function, interpolation, and elliptic functions.

Salomon Bochner (Figure 1.1) made substantial contributions to harmonic analysis, probability theory, differential geometry as well as history of mathematics. Several notions and results such as the Bochner integral, Bochner theorem on Fourier transforms, Bochner–Riesz means, Bochner–Martinelli formula bear at present his name. His mathematical production consists of 6 books on various subjects including history of mathematics and about 140 research papers. Among these papers 32 were published in Proc. Nat. Acad. USA, 46 in Annals of Mathematics, 3 in Acta Math., and 4 in Duke Math. J. He belonged to a sizeable group of European mathematicians of Jewish origin who moved to the USA before or during the WWII and contributed to an enormous development of mathematics in their new motherland. The list of his doctoral students looks almost scary. It contains: Richard Askey, Eugenio Calabi, Jeff Cheeger, M. T. Cheng, Charles L. Dolph, Hillel Furstenberg, Robert Gunning, Israel Halperin, Sigurdur Helgason, Carl Herz, Gilbert Hunt, Samuel Karlin, Anthony Knapp, Paco Lagerstrom, Lynn Loomis, Harry Rauch, Joseph H. Sampson, Herbert Scarf, William A. Veech, Gerard Washnitzer, Bernard Russell. The majority of these names are well known to every professional mathematician.

Part I: **Exactly solvable operators**

2 Non-degenerate exactly solvable operators

(The material of this chapter is based on [122] and [27].)

2.1 Preliminaries

We start with a description of several general properties of non-degenerate ES-operators and their eigenpolynomials, which were initially formulated as conjectures in [122] after extensive computer experiments. Their validity will be proved in the second part of this chapter and in Chapter 4.

Notation 2.1. Given a univariate polynomial P, denote by μ_P its *root-counting measure*, i. e., the discrete probability measure supported on the zero locus of P and obtained by placing the point mass $\frac{\mathrm{mult}_i}{\deg P}$ at each of its geometrically distinct roots z_i. Here mult_i is the multiplicity of the root z_i.

Consider an ES-operator T given by (1.2) and assume that T is non-degenerate, i. e., $\deg Q_k(z) = k$. As we mentioned in § 1.2, T has a unique monic eigenpolynomial denoted by $p_n(z)$ in every degree n exceeding its least uniqueness degree n_T.

Let μ_n denote the *root-counting measure of* p_n.

Consider the limit

$$\mu = \lim_{n\to\infty} \mu_n$$

if it exists. (Here convergence is understood as the weak convergence of measures.) Later in this chapter we show that μ exists for any non-degenerate T, and we call it *the asymptotic root-counting measure* of the sequence $\{p_n\}_{n=n_T}^{\infty}$.

Denote by $\mathcal{S} = supp(\mu)$ the support of the asymptotic measure μ and call it the *(root) accumulation curve* for the sequence $\{p_n\}$.

Before we describe the support $\mathcal{S}$ of μ and its density, we need to recall another notion.

Definition 2.1. A real univariate polynomial is called *hyperbolic* if all its zeros (counted with multiplicities) are real.

Different properties of hyperbolic polynomials and criteria of hyperbolicity were extensively studied in the beginning of the last century, see, e. g., [151, Chapters 5–6]. In 60s and 70s the interest to hyperbolic polynomials (mostly in the case of several variables) was revitalized due to the fundamental contributions of I. G. Petrovsky and L. Hörmander to the theory of linear partial differential equations with constant coefficients. But some new results were obtained even in the case of one variable, see, e. g., [143]. In 80s V. Arnold and his students wrote a number of papers on hyperbolic polynomials motivated by their application to potential theory, see [12, 13, 103, 104].

https://doi.org/10.1515/9783112207741-003

A typical context in which hyperbolic polynomials appear is the spectral theory of Sturm–Liouville problems or, more specifically, the theory of orthogonal polynomials, see, e. g., [50]. It is well known that every polynomial in a sequence of polynomials orthogonal with respect to some positive measure supported on $\mathbb{R}$ is hyperbolic and all its zeros lie in the convex hull of its support. Such a measure is called the *weight of orthogonality*.

The classical polynomial families such as Hermite, Laguerre, Jacobi polynomials (and their special cases such as Tchebyschev, Legendre, and Gegenbauer polynomials) arise in this way.

Let us consider a sequence of monic polynomials $\{p_n(x)\}_{n=0}^{\infty}$ orthogonal with respect to some probability measure ν supported on $\mathbb{R}$ and satisfying the so-called Nevai condition, see [141]. We will not need the precise definition of this class, but an interested reader can consult loc. cit. for the precise definition of the *Nevai class*, denoted there by $M(a, b)$. In particular, if the density of ν is a function ρ and $\log \rho$ is integrable on the smallest interval containing $supp(\nu)$, then ν is of the Nevai class, but this condition is not necessary.

Besides hyperbolicity, i. e., the real-rootedness of orthogonal polynomials, their main properties are as follows:

I) All the zeros of $p_n(z)$ are simple.

II) The zeros of any pair of consecutive orthogonal polynomials $(p_n(z), p_{n+1}(z))$ are interlacing, i. e., $\tilde{z}_1 < z_1 < \tilde{z}_2 < z_2 < \cdots < \tilde{z}_{n-1} < z_{n-1} < \tilde{z}_n$ where z_j (resp. $\tilde{z}_j$) is the jth smallest zero of p_n (resp. of p_{n+1}).

III) If the orthogonality measure ν is supported on a finite interval $[a, b]$ and satisfies the Nevai condition, then the density of the asymptotic root-counting measure of $\{p_n(x)\}$ when $n \to \infty$ on the interval $[a, b]$ is independent of ν and is given by $\frac{b-a}{2\pi\sqrt{(b-z)(z-a)}}$.

2.2 Properties of non-degenerate ES-operators

Let us now state a number of properties of the asymptotic measure μ for an arbitrary non-degenerate ES-operator (1.2).

2.2.1 Hyperbolic leading coefficient $Q_k(z)$

Assume that the leading coefficient $Q_k(z)$ of a non-degenerate ES-operator T is hyperbolic and has at least two distinct real roots; we denote by α the minimal and by β the maximal of the roots of $Q_k(z)$. Numerical studies of the root distribution of the eigenpolynomials $p_n(z)$ in this case led us to the following guess, see Figure 2.1.

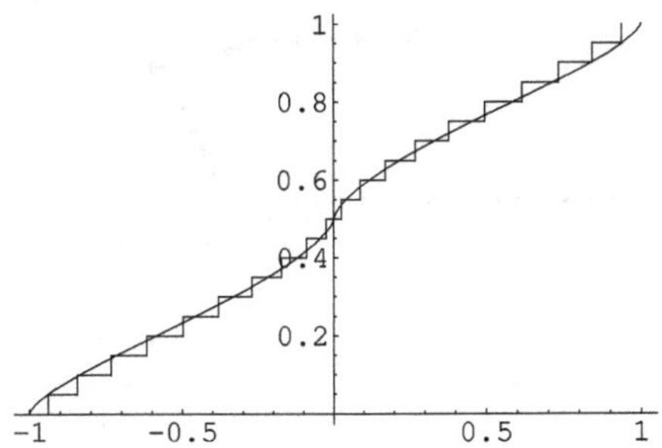

Figure 2.1: Distribution of zeros of $p_{20}(z)$ for $Q_3(z) = z^3 - z$ and the asymptotic distribution function $\frac{\Gamma(\frac{4}{3})}{3\Gamma(\frac{2}{3})}\int_{-1}^{x}\frac{dx}{\sqrt[3]{|x^3-x|}}$, $x \in [-1, 1]$.

Property 2.1. If the leading coefficient $Q_k(z)$ is a hyperbolic polynomial of degree k and has at least two distinct (real) roots, then μ is supported on the interval $[\alpha, \beta]$ with the density proportional to $\frac{1}{\sqrt[k]{|Q_k(z)|}}$, $z \in [\alpha, \beta]$.

Important observation

Property 2.1 implies that for any $Q_k(z)$ different from an integer power of a quadratic polynomial, the sequence $\{p_n(z)\}$ is neither a sequence of orthogonal polynomials with a weight function $w(x)$ supported on a real interval $[\alpha, \beta]$ satisfying Nevai's condition nor a polynomial sequence obtained by taking derivatives of such since for such families the asymptotic density of the zeros coincides with $\frac{\beta-\alpha}{2\pi\sqrt{|(x-\alpha)(x-\beta)|}}$, $x \in [\alpha, \beta]$, see property III) of orthogonal polynomials.

2.2.2 General complex-valued $Q_k(z)$

Denote by $\mathrm{Conv}_{Q_k} \subset \mathbb{C}$ the convex hull of the set of all roots of the leading coefficient $Q_k(z)$ of the operator T.

Property 2.2. For any non-degenerate ES-operator T as in (1.2), the asymptotic root-counting measure μ and the accumulation curve $\mathcal{S}$ enjoy the following properties:

a) The measure μ (and therefore the accumulation curve $\mathcal{S}$) depends only on the leading coefficient $Q_k(z)$ of T.
b) $\mathcal{S}$ is a planar tree imbedded in $\mathrm{Conv}_{Q_k} \subset \mathbb{C}$ whose leaves are all the roots of $Q_k(z)$.
c) For a generic $Q_k(z)$, the total number of vertices of $\mathcal{S}$ equals $2k - 2$, i. e., the number of internal vertices of $\mathcal{S}$ equals $k - 2$.

Remark 2.1. By a "planar tree" we mean an embedded planar graph without cycles, see, e. g., Figure 2.2. Property 2.2 a) implies that for any non-degenerate operator T, one can interchangeably denote $\mu = \mu_{Q_k}$ and $\mathcal{S} = \mathcal{S}_{Q_k}$, which we will sometimes use later on without special mentioning. (Notice that for a general degenerate ES-operator T, its asymptotic root-counting measure μ is no longer determined only by its leading coefficient, see Chapter 3.)

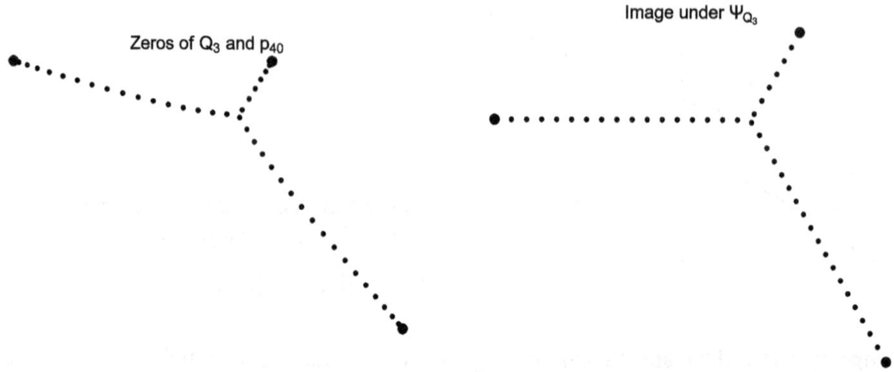

Figure 2.2: Zeros of the polynomial $Q_3 = (z-3+i)(z-i)(z-2-i)$ (larger dots), approximation to the accumulation curve $\mathcal{S}$ for $T = Q_3(z)\frac{d^3}{dz^3}$ (left), and its image under the mapping Ψ_{Q_3} (right). (The figures show the roots of eigenpolynomial $p_{40}(z)$ and their Ψ_{Q_3}-images presenting embedded trees in $\mathbb{C}$.)

Assuming that Property 2.2 holds, select a branch cut BC of $\mathbb{C} \setminus \mathcal{Z}(Q_k)$ where $\mathcal{Z}(Q_k)$ stands for the zero locus of Q_k such that BC consists of simple Jordan curves connecting each distinct root of $Q_k(z)$ to infinity and which do not intersect each other and the accumulation curve $\mathcal{S}$. (Such choice is possible due to item b) of Property 2.2.)

Further select the unique branch of $1/\sqrt[k]{Q_k(z)}$ in the simply connected domain $\Omega_{Q_k} := \mathbb{C} \setminus BC$, which asymptotically coincides with $1/z$ near infinity. (Recall that $Q_k(z)$ is monic.) Finally choose any point z_0 in Ω_{Q_k} and consider the holomorphic mapping $\Psi_{Q_k} : \Omega_{Q_k} \to \mathbb{C}$ defined by the integral

$$\Psi_{Q_k}(z) = \int_{z_0}^{z} \frac{d\xi}{\sqrt[k]{Q_k(\xi)}}.$$

Obviously, Ψ_{Q_k} is a (multi-valued) analytic function.

Property 2.3. The map Ψ_{Q_k} sends $\mathcal{S}$ onto a planar tree with *straight* edges, see Figure 2.2. Furthermore, the angles between the edges of $\mathcal{S}$ and the densities of the asymptotic measure μ along these edges are uniquely determined by the combinatorics of the tree, see Properties 2.4–2.5 below.

Note that for a generic Q_k, the image $\Psi_{Q_k}(\mathcal{S})$ is independent of the choice of a branch cut with the above properties, and a choice of a base point z_0 will result in a translation of the image.

To illustrate, we show in Figure 2.2 the zeros of the eigenpolynomial p_{40} for the operator T of order 3 with $Q_3 = (z-3+i)(z-i)(z-2-i)$ and $Q_0 = Q_1 = Q_2 = 0$. Large dots represent the zeros of Q_3 (which are, in this case, also zeros of p_n) and small dots represent (the remaining) zeros of p_{40}. It is remarkable how well the zeros of the eigenpolynomial line up along the curves predicted by our results. Notice also how these curves are straightened out by the mapping Ψ_{Q_3}.

Denote by $\Psi\mathcal{S}$ the image $\Psi_{Q_k}(\mathcal{S})$ of the accumulation curve $\mathcal{S}$ under the mapping Ψ_{Q_k}.

Property 2.4. The angles between the neighboring edges at the vertices of $\Psi\mathcal{S}$ (and therefore of $\mathcal{S}$) are defined as follows. (Observe that by Property 2.3 the edges of $\Psi\mathcal{S}$ are straight segments and $\Psi\mathcal{S}$ is an embedded tree.)

Take a vertex $v \in \Psi\mathcal{S}$. Consider the graph obtained from $\Psi\mathcal{S}$ by removing all the edges adjacent to v. This graph is a forest whose subtrees are in 1–1 correspondence with the set of the removed edges. Namely, a subtree in the complement corresponds to the edge that attaches it to v. Set the weight $\sharp(e, v)$ of each edge e adjacent to the vertex v equal to the number of leaves in the subtree corresponding to e. Let e_1 and e_2 be any two neighboring edges of $\Psi\mathcal{S}$ (considered as a planar graph) that are adjacent to v. Then the angle $\epsilon(e_1, e_2)$ between them equals

$$\epsilon(e_1, e_2) = \frac{\pi}{k}(\sharp(e_1, v) + \sharp(e_2, v)).$$

See illustration in Figure 2.3.

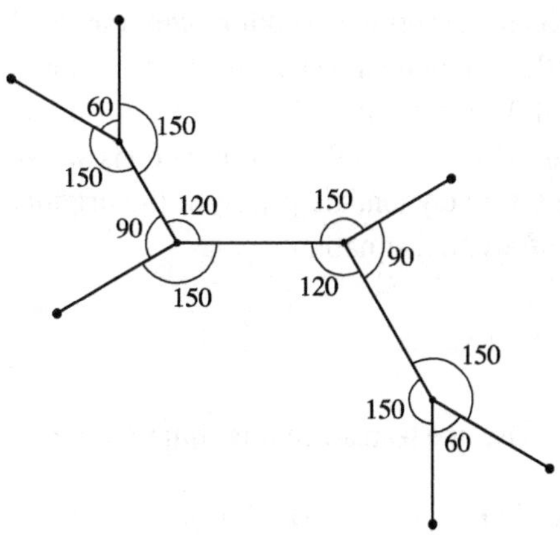

Figure 2.3: How to determine the angles for an embedded tree in $\mathbb{C}$.

To simplify the notation, set $Q := Q_k$. Let $\Psi\mu_Q$ denote the image of the asymptotic root-counting measure μ_Q under the transformation Ψ_Q. Its support obviously coincides with $\Psi\mathcal{S}$.

The image measure $\Psi\mu_Q$ (and therefore the original measure μ_Q) is uniquely determined by the following condition.

Property 2.5. The image $\Psi\mu_Q$ of the asymptotic root-counting measure μ_Q enjoys the following properties.

a) The density $\nu(e)$ of $\Psi\mu_Q$ is constant on every edge e of $\Psi\mathcal{S}$ and therefore equals $\nu(e) = \frac{\mu(e)}{|e|}$ where $\mu(e)$ is the total mass of the edge e w. r. t. the measure $\Psi\mu_Q$ and $|e|$ is its usual length.
b) For any pair (v, e) where v is a vertex of $\Psi\mathcal{S}$ and e is an edge of $\Psi\mathcal{S}$ adjacent to v, let $\bar{U}(v, e)$ denote the unit vector parallel to e and pointing towards v. Then, for every vertex v of $\Psi\mathcal{S}$, one has the following balancing condition:

$$\sum_{e\in E(v)} \nu(e)\bar{U}(v,e) = 0 \tag{2.1}$$

where $E(v)$ is the set of all edges adjacent to v.

Note that for a generic $Q(x)$, system (2.1) consists of $2(k-2)$ linear equations in $2k-3$ variables $\nu(e)$, which together with the condition $\sum_{e\in E} \nu(e)|e| = 1$ allow us to determine the densities $\nu(e)$ uniquely.

Moreover, with some straightforward modifications, the same recipe works for all polynomials Q of degree k and not just for generic ones.

To move further, let us make the following heuristic remark. Notice that assuming that Properties 2.1–2.5 hold, one gets in many respects the same asymptotic behavior of $\{p_n(z)\}$ if the sequence of eigenpolynomials for $Q(z)\frac{d^k}{dz^k} + \cdots$ is replaced by the sequence of eigenpolynomials for $Q^r(z)\frac{d^{rk}}{dz^{rk}} + \cdots$ where r is any positive integer. Therefore it makes sense to study the solutions of the equation, which one gets when taking even fractional power like $r = 1/k$. In that case the "degree" of the polynomial is equal to 1 and the corresponding differential operator is thus of order 1. Moreover, since $\lambda_{k,n} = (n+1)\cdots(n+k)$ for $k \geq 1$, therefore for $k = 1$, the "correct" eigenvalue for the nth eigenfunction is $(n+1)$.

In other words, we want to compare the eigenpolynomials $\{p_n(z)\}$ of the original operator to the solutions y_n of the first order differential equation

$$\frac{d}{dz}(\sqrt[k]{Q}y_n) = (n+1)y_n.$$

Multiplying it through by $\sqrt[k]{Q}$ and setting $f_n = \sqrt[k]{Q}y_n$, we transform this equation into

$$f_n' = (n+1)Q^{-1/k}f_n$$

or

$$d\log f_n := \frac{f_n'}{f_n} = (n+1)Q^{-1/k}.$$

Now $f_n = Q^{1/k}y_n$, so $d\log f_n = d\log y_n + (1/k)\, d\log Q$, and therefore we get

$$\frac{1}{n}\, d\log y_n = \frac{n+1}{n}Q^{-1/k} - \frac{1}{kn}\, d\log Q.$$

This implies that

$$\lim_{n\to\infty}\frac{1}{n}\,dlog\,y_n = Q^{-1/k}.$$

Since y_n corresponds to p_n, we can formulate the following guess supported by computer experiments.

Property 2.6. In $\mathbb{C}\setminus\mathcal{S}$ we have that

$$\frac{1}{n}\lim_{n\to\infty}\frac{p_n'(z)}{p_n(z)} = \frac{1}{\sqrt[k]{Q(z)}}$$

uniformly on compact sets.

By solving the above first order equation, one gets that

$$f_n(x) = e^{(n+1)\Psi_Q(z)}$$

or, equivalently,

$$y_n(z) = \frac{e^{(n+1)\Psi_Q(z)}}{\sqrt[k]{Q(z)}}$$

which must be an asymptotic solution of the equation $Ty = \lambda y$. Obviously, it is not reasonable to expect that $y_n(z)$ approaches $p_n(z)$ in any sense (note that $y_n(z)$ has no roots!). However, the ratio between two consecutive eigenfunctions can asymptotically stay the same.

Property 2.7. In $\mathbb{C}\setminus\mathcal{S}$ we have that

$$\lim_{n\to\infty}\frac{p_{n+1}(z)}{p_n(z)} = e^{\Psi_Q(z)}$$

uniformly on compact sets.

An even stronger form of this conjecture reminding of a first order *WKB*-approximation is as follows.

Property 2.8. In $\mathbb{C}\setminus\mathcal{S}$ we have that

$$\lim_{n\to\infty}\frac{p_n(z)}{(e^{\Psi_Q(z)})^n} = \left(\frac{e^{\Psi_Q(z)}}{\sqrt[k]{Q(z)}}\right)^{(k+1)/2}$$

uniformly on compact sets.

Remark 2.2. We shall prove Properties 2.1 and 2.2–2.8 later in this chapter and in Chapter 4, see also [27] and [40].

Remark 2.3. (i) The accumulation curve $\mathcal{S}$ introduced above has a strong resemblance with the Stokes curves occurring in the theory of differential equations with small parameter and asymptotic expansions, see [53] and [195]. The role of a small parameter in our case is played by $\frac{1}{\lambda_n}$ when $n \to \infty$.

(ii) As was mentioned by Prof. M. Kontsevich around 2000, the integral $\Psi_Q(z) = \int_{z_0}^{z} \frac{d\xi}{\sqrt[k]{Q(\xi)}}$ considered on the plane curve $y^k = Q(z)$ apparently has properties similar to the properties of Strebel differentials and might be considered as a Strebel differential of higher order. We discuss this aspect in Chapter 12.

2.3 Warm-up example

In this section we consider a special type of non-degenerate ES-operators of the form

$$T_Q(p) = \frac{d^k}{dz^k}(Q(z)p)$$

where $Q(z)$ is a polynomial of degree k. The general case will be treated later in this chapter. This operator can be thought of as a generalization of the famous hypergeometric operator

$$(z^2 - 1)f'' + (az + b)f' + c.$$

Note that most of the above-mentioned families of orthogonal polynomials are among its polynomial solutions.

The intriguing detail is that, in general, T_Q is *neither* a positive-definite *nor* a self-adjoint operator in which case many natural properties of its eigenfunctions are expected. (Recall that Q is always a monic polynomial of degree k.)

Our nearest goal is to settle the following statements.

Theorem 2.1. *a) For any monic $Q(z)$ of degree k, the operator T_Q has one polynomial eigenfunction $p_n(z)$ in each degree $n \geq 0$ and its eigenvalue $\lambda_{n,k}$ equals $\frac{(n+k)!}{n!}$.*

b) For any $n \geq 0$, all the zeros of $p_n(z)$ belong to the convex hull Conv_Q of the set of zeros of $Q(z)$.

Corollary 2.1. *If $Q(z)$ is hyperbolic, then each $p_n(z)$ is also hyperbolic and its zeros lie between the minimal root α and the maximal root β of $Q(z)$.*

Theorem 2.2. *If $Q(z)$ is hyperbolic and has at least two pairwise different zeros (i. e., $\alpha < \beta$), then besides the property that they belong to (α, β) all the zeros of $p_n(z)$ are simple.*

We start by providing explicit determinantal formulas for the coefficients of the eigenpolynomial $p_n(x)$. Set $Q(z) = z^k + q_{k-1}z^{k-1} \cdots + q_1 z + q_0$ and $p_n(z) = z^n + a_{n,n-1}z^{n-1} + \cdots + a_{n,1}z + a_{n,0}$, where

$$\frac{d^k}{dz^k}[Q(z)p_n(z)] = \lambda_{n,k}p_n(z), \quad \lambda_{n,k} = \frac{(n+k)!}{n!}. \tag{2.2}$$

Consider the upper triangular $(n \times n)$-matrix

$$M_n = \begin{pmatrix} 1-\frac{\lambda_{n,k}}{\lambda_{0,k}} & q_{k-1} & q_{k-2} & \cdots & q_0 & 0 & & \cdots & 0 \\ 0 & 1-\frac{\lambda_{n,k}}{\lambda_{1,k}} & q_{k-1} & q_{k-2} & \cdots & q_0 & 0 & \cdots & 0 \\ \vdots & \ddots & \ddots & \ddots & \ddots & & \ddots & \ddots & \vdots \\ & & & & & & & \ddots & 0 \\ & & & & & & & & q_0 \\ & & & & & & & \ddots & \vdots \\ & & & & & & & \ddots & q_{k-2} \\ \vdots & & & & & & \ddots & \ddots & q_{k-1} \\ 0 & \cdots & & & & & \cdots & 0 & 1-\frac{\lambda_{n,k}}{\lambda_{n-1,k}} \end{pmatrix}.$$

Theorem 2.3. *For every $n \geq 1$, the coefficients of $p_n(z)$ satisfy the linear system*

$$M_n \cdot A = B \tag{2.3}$$

where

$$A = \begin{pmatrix} a_{n,0} \\ a_{n,1} \\ \vdots \\ \vdots \\ \vdots \\ a_{n,n-1} \end{pmatrix}, \quad B = -\begin{pmatrix} 0 \\ \vdots \\ 0 \\ q_0 \\ q_1 \\ \vdots \\ q_{k-1} \end{pmatrix}.$$

Proposition 2.1. *If $Q = (z^2-1)^l$ then the family $\{p_n(z)\}$ of eigenpolynomials of the operator T_Q coincides (up to constant factors) with the family of Gegenbauer polynomials with the parameter value $\lambda = l + \frac{1}{2}$ (or, equivalently, with the Jacobi (l, l)-polynomials).*

Basic information about Gegenbauer polynomials can be found in, e. g., [50].

2.4 Proving results for $T_Q(p) = \frac{d^k}{dz^k}(Q(z)p)$

Proof of Theorem 2.1. Let $\mathbb{C}_m[x]$ be the linear space of all polynomials with complex coefficients whose degree is less than or equal to m. In order to prove a), note that the action of T_Q in the natural monomial basis $(z^m, z^{m-1}, \dots, 1)$ of $\mathbb{C}_m[x]$ is given by an upper triangular matrix with non-vanishing diagonal entries $\lambda_{n,k} = (n+k)\cdots(n+1) =$ for any $n \le m$. Thus, for each $n \ge 0$, there exists a unique monic polynomial $p_n(z)$, which is the eigenfunction of T_Q corresponding to the eigenvalue $\lambda_{n,k} = \frac{(n+k)!}{n!}$.

In order to prove b), recall that for any univariate polynomial ϕ with complex coefficients, by the well-known Gauss–Lucas theorem, the zeros of ϕ' lie in the convex hull Conv_ϕ of zeros to ϕ. Moreover, any zero of ϕ' lying on the boundary of Conv_ϕ is a multiple zero of ϕ. Let us now assume that $p_n(z)$ contains a zero not lying in Conv_Q. Then we can always choose such a zero κ of $p_n(z)$ that lies on the boundary of the convex hull Conv_{Qp_n} of the zeros of the product $Q(z)p_n(z)$. Since $p_n(z)$ satisfies (2.2), one gets that κ should have the same multiplicity both in $p_n(z)$ and $\frac{d^k}{dz^k}(Q(z)p_n(z))$, which is impossible by the assumption that $\kappa \notin \mathrm{Conv}_Q$. □

Proof of Corollary 2.1. If $Q(z)$ is hyperbolic with the minimal root α and the maximal root β, then $\mathrm{Conv}_Q = [\alpha, \beta]$. Therefore by Theorem 2.1 all zeros of $p_n(z)$ belong to $[\alpha, \beta]$. □

In order to prove Theorem 2.2, we need a number of additional statements.

Recall that $p_n(z)$ denotes the monic eigenpolynomial of degree n for the differential operator $T_Q = \frac{d^k}{dz^k} Q(z)$ where $Q(z)$ is a monic polynomial of degree k. This means that the polynomials Q and p_n satisfy the relation

$$\frac{d^k}{dz^k}(Qp_n) = \lambda_{k,n} p_n$$

where $\lambda_{n,k} := (n+k)!/n!$.

Lemma 2.1. *Write $p_n = \bar{p}_n z^{n_0}$ and $Q = \bar{Q} z^{k_0}$ where $\bar{p}_n$ and $\bar{Q}$ have non-zero constant terms. If $k_0 < k$, then $k_0 + n_0 < k$ and the polynomial Qp_n is of the form*

$$Qp_n = z^{k+n} + c_{k+n-1} z^{k+n-1} + \dots + c_{k+n_0} z^{k+n_0} + c_{k-1} z^{k-1} + \dots + c_{k_0+n_0} z^{k_0+n_0}$$

where $c_{k_0+n_0} \neq 0$.

Proof. If $n_0 = 0$, there is nothing to prove. We may therefore assume throughout the proof that $n_0 > 0$. Write

$$Qp_n = \bar{Q}\,\bar{p}_n z^{k_0+n_0} = c_{k_0+n_0} z^{k_0+n_0} + \text{higher order terms}.$$

By construction we have that $c_{k_0+n_0} \neq 0$.

Assume that $k \le k_0 + n_0$. Then

$$\lambda_{n,k} \bar{p}_n z^{n_0} = \frac{d^k}{dz^k}(Qp_n)$$

$$= \frac{d^k}{dz^k}(c_{k_0+n_0} z^{k_0+n_0} + \text{higher order terms})$$
$$= c_{k_0+n_0} \frac{(k_0+n_0)!}{(k_0+n_0-k)!} z^{k_0+n_0-k} + \text{higher order terms}$$
$$= c_{k_0+n_0} \frac{(k_0+n_0)!}{(k_0+n_0-k)!} z^{n_0-(k-k_0)} + \text{higher order terms},$$

which is impossible by the assumption $k > k_0$. Therefore we must have $k > k_0 + n_0$.

Choose an integer i with $0 \le i < n_0$ and write $Qp_n = cz^{k+i}$ + other terms. Then

$$\lambda_{n,k}\bar{p}_n z^{n_0} = \frac{d^k}{dz^k} Q(z)(Qp_n)$$
$$= \frac{d^k}{dz^k} Q(z)(cz^{k+i} + \text{other terms})$$
$$= c\frac{(k+i)!}{i!} z^i + \text{other terms}.$$

Since $i < n_0$ while the left-hand side is a multiple of z^{n_0}, which implies that $c = 0$. This proves the second claim. □

Proposition 2.2. *If Q has at least two different roots, then all the roots of p_n have multiplicity at most $k - 1$.*

Proof. Select a root of p_n. Changing the independent variable $z \to z + y$, we may assume that this root is equal to zero. Since Q has at least two different roots, $Q \neq z^k$. Thus Lemma 2.1 applies and we have $n_0 \le n_0 + k_0 < k$. □

Remark. Note that we get an even lower bound on the multiplicity in the case when the root of p_n is also a root of Q.

The following lemma is well known, but we prove it here for the convenience of the reader.

Lemma 2.2. *Let $s = a_{n+m}z^{n+m} + \cdots + a_m z^m$ be a real polynomial with $a_{n+m} \neq 0 \neq a_m$. If $a_{m+i} = a_{m+i+1} = 0$ for some $i = 1, \ldots, n-2$, then s is not hyperbolic.*

Proof. Without loss of generality we may assume that $m = 0$. First we treat the special case $s = a_n z^n + a_{n-r}z^{n-r} + \cdots + a_0$ where $r > 2$ and $a_{n-r} \neq 0$. Then $s^{(n-r)}$ is not hyperbolic, and consequently s is not hyperbolic (since the derivative of a hyperbolic polynomial is hyperbolic).

The special case $s = a_n z^n + \cdots + a_r z^r + a_0$ where $r > 2$ and $a_r \neq 0$ can be reduced to the first case by looking at the polynomial $z^n s(1/z)$.

Finally, the general case may be reduced to the second case by a suitable number of derivations. □

Proof of Theorem 2.2. By Proposition 2.2 above we have that all the roots of p_n are real. Assume that p_n has a multiple root. Changing $z \to z+y$ we may without loss of generality assume that this root equals zero. Lemma 2.1 thus implies that the polynomial Qp_n has at least two consecutive vanishing terms away from the ends. But since all the roots of Qp_n are real, this is impossible by Lemma 2.2. □

Proof of Theorem 2.3. For every polynomial $p(z)$, its Laplace transform is given by

$$L[p](t) = \int_0^{+\infty} e^{-tz} p(z)dz, \quad t > 0.$$

If a polynomial p_n satisfies equation (2.2), then

$$L\left[\frac{d^k}{dz^k}(Qp_n)\right] = L[\lambda_n p_n].$$

By the standard properties of the Laplace transform (see, e. g., [102]), we have

$$\begin{aligned}
L[\lambda_{n,k}p_n] &= \lambda_{n,k}L[p_n],\\
L\left[\frac{d^k}{dz^k}(Qp_n)\right] &= t^k L[Qp_n] - \sum_{i=0}^{k-1} t^i \left.\frac{d^{(k-i-1)}}{dz^{(k-i-1)}}\right|_{z=0}(Qp_n),\\
L[Qp_n] &= L\left[\sum_{i=0}^{k} q_i z^i p_n\right] = \sum_{i=0}^{k} q_i(-1)^i \frac{d^i}{dt^i}L[p_n] = Q\left(-\frac{d}{dt}\right)[L[p_n]],\\
L[p_n] &= L\left[\sum_{i=0}^{n} a_{n,i}z^i\right] = \sum_{i=0}^{n} a_{n,i}\frac{i!}{t^{i+1}},\\
Q\left(-\frac{d}{dt}\right)[L[p_n]] &= \sum_{i=0}^{n} a_{n,i}i!Q\left(-\frac{d}{dt}\right)\left[\frac{1}{t^{i+1}}\right] = \sum_{i=0}^{n} a_{n,i}\sum_{j=0}^{k} q_j\frac{(i+j)!}{t^{i+j+1}},\\
L\left[\frac{d^k}{dz^k}(Qp_n)\right] &= t^k \sum_{\substack{0\leqslant i\leqslant n, 0\leqslant j\leqslant k\\ i+j\geqslant k}} a_{n,i}q_j\frac{(i+j)!}{t^{i+j+1}}.
\end{aligned}$$

Hence, the coefficients of the polynomial p_n satisfy the identity

$$\sum_{m=0}^{n}\frac{(m+k)!}{t^{m+1}}\sum_{\substack{m\leqslant i\leqslant n\\ i\leqslant m+k}} a_{n,i}q_{m+k-i} = \lambda_{n,k}\sum_{m=0}^{n} a_{n,m}\frac{m!}{t^{m+1}}.$$

It implies the equalities

$$\sum_{\substack{m\leqslant i\leqslant n\\ i\leqslant m+k}} a_{n,i}q_{m+k-i} = \frac{\lambda_{n,k}}{\lambda_{m,k}}a_{n,m}, \quad m = 0,\dots,n.$$

The equality for $m = n$ is trivial since $a_{n,n} = q_k = 1$. The other equalities define the linear system (2.3). Since $\det M_n = \prod_{j=0}^{n-1}(1 - \frac{\lambda_{n,k}}{\lambda_{j,k}}) \neq 0$, this system has a unique solution. □

Applying Cramer's formulas to the above linear system, we get the following.

Corollary 2.2. *The coefficients of the polynomial $p_n(z)$ equal to*

$$a_{n,i} = -\frac{D_{n,i}}{\prod_{j=i}^{n-1}(1 - \frac{\lambda_{n,k}}{\lambda_{j,k}})}, \qquad i = 0, \dots, n-1$$

where $D_{n,i}$ is the determinant of the matrix obtained from the upper triangular $(n-i) \times (n-i)$-matrix $M_{n,i}$ given below substituting its first column by the numbers $(q_{k-1}, q_{k-2}, \dots, q_0, 0, \dots, 0)$ in the reverse order, i. e., starting from the last entry

$$M_{n,i} = \begin{pmatrix} 1-\frac{\lambda_{n,k}}{\lambda_{i,k}} & q_{k-1} & q_{k-2} & \cdots & q_0 & 0 & & \cdots & 0 \\ 0 & 1-\frac{\lambda_{n,k}}{\lambda_{i+1,k}} & q_{k-1} & q_{k-2} & \cdots & q_0 & 0 & \cdots & 0 \\ \vdots & \ddots & \ddots & \ddots & \ddots & & \ddots & \ddots & \vdots \\ & & & & & & & \ddots & 0 \\ & & & & & & & & q_0 \\ & & & & & & & \ddots & \vdots \\ & & & & & & & \ddots & q_{k-2} \\ \vdots & & & & & & \ddots & \ddots & q_{k-1} \\ 0 & \cdots & & & & & \cdots & 0 & 1-\frac{\lambda_{n,k}}{\lambda_{n-1,k}} \end{pmatrix}.$$

Corollary 2.3. *If the sum of roots of a polynomial $Q(z)$ vanishes, then the same property holds for every $p_n(z)$ with $n > 0$.*

Proof. According to Lemma 2.2,

$$a_{n,n-1} = -\frac{q_{k-1}}{1 - \frac{\lambda_{n,k}}{\lambda_{n-1,k}}}.$$

Hence, $a_{n,n-1} = 0$ if $q_{k-1} = 0$. □

Corollary 2.4. *If a polynomial $Q(z)$ is even or odd, then the polynomials $p_n(z)$ with even (odd) n are even (odd, respectively).*

Proof. According to Lemma 2.2, we have to prove that $D_{n,i} = 0$ for all $i = n-1 \mod 2$ if $q_j = 0$ for all $j = k-1 \mod 2$. But for such polynomial $Q(z)$ and odd $n-i$, the determinant $D_{n,i}$ is equal to

$$\prod_{\substack{i<j<n-1 \\ j \equiv i+1 \mod 2}} \left(1 - \frac{\lambda_{n,k}}{\lambda_{j,k}}\right) \det \widetilde{M}_{n,i}$$

where the matrix $\widetilde{M}_{n,i}$ is obtained from $M_{n,i}$ by deletion of all even columns and rows. It remains to notice that the first column of the matrix $\widetilde{M}_{n,i}$ vanishes. □

Example 2.1. It is easy to calculate the coefficients of the polynomials $p_n(z)$ in the case when $Q(z)$ has only two terms. Indeed, let $Q(z) = z^k - az^{k-m}$ where $1 \le m \le k$. Then system (2.3) can be easily solved by the Gauss elimination method

$$a_{n,i} = 0 \quad \text{if } i \ne n \mod m,$$
$$a_{n,n-ml} = -\frac{a^l}{\prod_{j=1}^{l}(1 - \frac{\lambda_{n,k}}{\lambda_{n-mj,k}})}$$

for all natural l such that $ml < n$.

Proof of Proposition 2.1. We start with the following statement.

Lemma 2.3. *For $Q = (z^2 - 1)^l$, the eigenpolynomials $p_n(z)$ and $p_m(z)$, $n \ne m$ satisfy the orthogonality condition*

$$\int_{-1}^{1} (z^2 - 1)^l p_n(z) p_m(z) dz = 0.$$

Consider the polynomial family $\{q_n\}$ where $q_n = (z^2 - 1)^{2l} p_n$. One checks directly that $p_n = q_n^{(2l)}/\lambda_{n,2l}$ and that q_n satisfies the equation $\lambda_{n,2l} q_n(z) = (z^2 - 1)^l q_n^{(2l)}$. Take

$$I_{n,m} = \int_{-1}^{1} (z^2 - 1)^l p_n(z) p_m(z) dz.$$

In terms of q_is the integral $I_{n,m}$ can be rewritten as

$$I_{n,m} = \lambda_{n,2l} \int_{-1}^{1} q_n^{(2l)}(z) q_m(z) dz.$$

Notice that both $q_n(z)$ and $q_m(z)$ have zeros of multiplicity at least l at ± 1, and therefore their derivatives of order at most $l-1$ vanish at these points. Applying integration by parts to the right-hand side of the above integral $2l$ times and using the previous remark, we arrive at

$$I_{n,m} = \lambda_{n,2l} \int_{-1}^{1} q_n^{(2l)}(z) q_m(z) dz = \lambda_{n,2l} \int_{-1}^{1} q_n(z) q_m^{(2l)}(z) dz = \frac{\lambda_{n,2l}}{\lambda_{m,2l}} I_{n,m}.$$

Since $0 \ne \lambda_{m,2l} \ne \lambda_{n,2l} \ne 0$ for $m \ne n$, one has $I_{n,m} = 0$. Therefore the family $\{p_n(x)\}$ is orthogonal on the interval $[-1, 1]$ w. r. t. the non-negative weight function $(x^2 - 1)^l$.

Proposition 2.1 follows since a system of polynomials orthogonal on $[-1,1]$ with the weight function $Q = (z^2-1)^l$ coincide (up to constant factors) with the Gegenbauer polynomials with the value $\lambda = l + \frac{1}{2}$ of its parameter. □

By Corollary 2.1 if $Q(z)$ is hyperbolic and α (resp. β) is its minimal (resp. maximal) roots, then all zeros of $p_n(z)$ belong to (α,β). Moreover the following stronger statement holds.

Proposition 2.3. *If $Q(z)$ is hyperbolic and has at least two distinct zeros, then the zeros of any pair of consecutive polynomial eigenfunctions $p_i(z)$ and $p_{i+1}(z)$ are interlacing.*

The proof consists of the following three simple lemmas. (Lemma 2.6 is usually called Obreschkov's theorem, see [144, 56].)

Lemma 2.4. *If R_n and R_{n+1} are strictly hyperbolic polynomials of degrees n and $n+1$ resp., then $R_n + \epsilon R_{n+1}$ is hyperbolic for any sufficiently small ϵ.*

Proof. For any sufficiently small ϵ, the n real zeros of $R_n + \epsilon R_{n+1}$ are located in some small neighborhoods of the n simple real zeros of R_n and the $(n+1)$th real zero has a very big absolute value. □

Lemma 2.5. *If Q has only real zeros, then any linear combination $\alpha p_n + \beta p_{n+1}$ with real coefficients of the polynomial eigenfunctions of T_Q is a hyperbolic polynomial.*

Proof. Indeed, applying to $\alpha p_n + \beta p_{n+1}$ some high power T_Q^{-N} of the inverse operator, one gets

$$T_Q^{-N}(\alpha p_n + \beta p_{n+1}) = \frac{\alpha}{\lambda_{n,k}^N} p_n + \frac{\beta}{\lambda_{n+1,k}^N} p_{n+1} = \frac{\alpha}{\lambda_{n,k}^N}(p_n + \epsilon p_{n+1}),$$

where ϵ is arbitrarily small for the appropriate choice of N (since $0 < \lambda_{n,k} < \lambda_{n+1,k}$). Thus by Lemma 2.4 the polynomial $T_Q^{-N}(\alpha p_n + \beta p_{n+1})$ is hyperbolic for sufficiently big N. Assume that $\alpha p_n + \beta p_{n+1}$ is non-hyperbolic and take the largest N_0 for which $R_{N_0} = T_Q^{-N_0}(\alpha p_n + \beta p_{n+1})$ is still non-hyperbolic. Then, obviously, $R_{N_0} = T_Q(R_{N_0+1})$ where $R_{N_0+1} = T_Q^{-N_0+1}(\alpha p_n + \beta p_{n+1})$. Note that R_{N_0+1} is hyperbolic and that if Q has only real zeros then the application of T_Q to any hyperbolic polynomial results in a hyperbolic polynomial. Contradiction. □

Lemma 2.6. *If R_n and R_{n+1} are any real polynomials of degrees n and $n+1$ resp., then their arbitrary linear combination $\alpha R_n + \beta R_{n+1}$ with real coefficients is hyperbolic if and only if*

i) *Both R_n and R_{n+1} are hyperbolic.*
ii) *Their roots are interlacing.*

Proof. The condition i) is obviously necessary since in the opposite case one can choose $\alpha = 0$ or $\beta = 0$ to get a non-hyperbolic polynomial. Assume that both R_n and R_{n+1} are hyperbolic and $x_i < x_{i+1}$ are two consecutive zeros of, say, R_n not separated by a zero of R_{n+1}. Then, considering $R_n + tR_{n+1}$ where $|t|$ is increasing and the sign of t is chosen so that tR_{n+1} and R_n have different signs on (x_i, x_{i+1}), we get that for sufficiently big $|t|$ the polynomial $R_n + tR_{n+1}$ looses its hyperbolicity.

On the other hand, i) and ii) are sufficient. (We consider only the case of simple zeros of R_{n+1}, which implies the general case as well.) Indeed, any linear combination $\alpha R_n + \beta R_{n+1}$ with $\alpha \neq 0$ changes its sign at any two consecutive zeros of R_{n+1}. Therefore $\alpha R_n + \beta R_{n+1}$ has at least n real zeros and thus all its $n + 1$ zeros are real. □

2.5 Proving the fundamental properties

In the remaining part of this chapter we concentrate on general non-degenerate ES-operators T given by (1.2) and prove the majority of the properties presented in § 2.2. (This material is based on [27].)

More precisely, we shall show that in the limit when $n \to \infty$, the zeros of p_n are distributed according to a certain probability measure. This probability measure depends only on the leading coefficient Q_k of T and its description is provided by Theorem 2.4 below. We start with the following notion.

Definition 2.2. Given a positive measure μ supported in $\mathbb{C}$, we define its *Cauchy transform* as

$$C_\mu(z) = \int \frac{d\mu(\zeta)}{z - \zeta} \tag{2.4}$$

and its *logarithmic potential* as

$$u(z) = \int \log|z - \zeta|\, d\mu(\zeta). \tag{2.5}$$

Detailed description of properties of the Cauchy transform of a measure can be found in, e. g., [155]. Let *supp*(μ) denote the support of a measure μ.

Theorem 2.4. *Let Q_k be a monic polynomial of degree k. Then there exists a unique probability measure μ_{Q_k} with compact support whose Cauchy transform $C := C_{Q_k}$ satisfies the algebraic equation*

$$C^k(z) = 1/Q_k(z) \tag{2.6}$$

for almost all $z \in \mathbb{C}$.

We record some properties of the measure μ_{Q_k}, which will be encountered in the proof of Theorem 2.4. Consider a primitive function of $Q_k^{-1/k}(z)$

$$\Psi(z) := \Psi_{Q_k}(z) = \int Q_k^{-1/k}(\xi)\, d\xi.$$

At this point, we think of Ψ as a locally defined function in any simply connected domain where Q_k does not vanish. The choice of a branch of $Q_k^{1/k}(z)$ and an integration constant is of no importance here. As need arises, specifications will be made concerning these choices.

The next result settles Properties 2.2–2.3 of § 2.2.2.

Theorem 2.5. *Let Q_k and μ_{Q_k} be as in Theorem 2.4. Then the support $supp(\mu_{Q_k})$ of the measure μ_{Q_k} (which coincides with the accumulation curve $\mathcal{S}$ of the operator T) is the union of finitely many smooth curve segments, and each of these curves is mapped to a straight line by the mapping Ψ. Moreover, $supp(\mu_{Q_k})$ contains all the zeros of Q_k and is contained in the convex hull of the zeros of Q_k. It is connected and has connected complement. In other words, $\mathcal{S}$ is an embedded tree in $\mathbb{C}$.*

Given a monic polynomial $Q_k(z)$ of degree k, let $\mathbf{Q} = (Q_k(z), Q_{k-1}(z), \ldots, Q_1(z))$ be a collection of polynomials satisfying $\deg Q_j \le j, j = 1, \ldots, k-1$. Let $T_{\mathbf{Q}}$ be the ES-operator of order k given by

$$T_{\mathbf{Q}} = Q_k(z)\frac{d^k}{dz^k} + Q_{k-1}(z)\frac{d^{k-1}}{dz^{k-1}} + \cdots + Q_1(z)\frac{d}{dz}. \tag{2.7}$$

Our main result (which settles Property 2.6 of Subsection 2.2.2) is as follows.

Theorem 2.6. *Let p_n be the monic degree n eigenpolynomial of the operator $T_{\mathbf{Q}}$ and let μ_n be the root-counting measure of p_n. Then μ_n converges weakly to μ_{Q_k} when $n \to \infty$.*

It is not difficult to deduce various other features of the measure μ_{Q_k} from the statement of Theorem 2.5 and the defining property (2.6).

Let us now prove Property 2.4, in other words, a recipe for computing the angles between the different curve segments suggested in the beginning of Chapter 2, see [122]. The key observation is the following. Suppose z_0 is a (smooth) point on one of the curve segments in $supp(\mu_{Q_k})$, and let $\mathcal{C}_1$ and $\mathcal{C}_2$ be the limiting values of $\mathcal{C}(z)$ as z approaches z_0 from different sides of the curve. Then $\mathcal{C}_1$ and $\mathcal{C}_2$ are kth roots of $1/Q_k(z_0)$, and their actual values among all possible kth roots are easily found if the combinatorics of $supp(\mu_{Q_k})$ is known (which was assumed in the considered recipe). Namely, from the fact that

$$\pi\mu_{Q_k} = \frac{\partial \mathcal{C}}{\partial \bar{z}} \ge 0,$$

it follows that the curve segment in $supp(\mu_{Q_k})$ containing z_0 must be perpendicular to $\overline{\mathcal{C}}_1 - \overline{\mathcal{C}}_2$ at z_0 and that the density of μ_{Q_k} at z_0 is proportional to $|\mathcal{C}_1 - \mathcal{C}_2|$. Using this observation where several curves meet, it is possible to deduce the angles between them as illustrated in Figure 2.3.

In particular, it is easy to compute the measure μ_{Q_k} when Q_k is hyperbolic, i. e., has only real zeros thus proving Property 2.1. Namely, denote the zeros by $\alpha = z_1 \le \cdots \le$

$z_k = \beta$ in increasing order. From Theorem 2.5 we know that $supp(\mu_{Q_k}) = [\alpha, \beta]$. A direct computation shows that on the subinterval $[z_j, z_{j+1}]$ the measure is given by

$$\mu_{Q_k} = \frac{1}{\pi}\frac{\partial C}{\partial \bar{x}} = \frac{2}{\pi |Q_k|^{1/k}} \sin\left(\frac{\pi j}{k}\right) dx$$

where dx denotes Lebesgue measure on the real line. This remains true even if Q_k has multiple zeros, except if all the zeros coincide. In this case, μ_{Q_k} reduces, of course, to a point mass at this multiple zero.

The rest of the chapter is organized as follows. In § 2.6 we compute the matrix for the operator T_{Q_k} with respect to the basis of monomials $1, z, z^2, \ldots$. § 2.7 contains a proof of the uniqueness part of Theorem 2.4. Along the way, we also prove essentially all the statements in Theorem 2.5. In § 2.8 we recall some basic facts on the weak topology of measures in the complex plane and on logarithmic potentials and Cauchy transforms. We also outline the connection of these concepts to root-counting measures of polynomials and prove a general result on the relation between the zeros of a polynomial and those of its derivative. In the final § 2.9 we apply the ideas from the previous section to give a proof of Theorem 2.6. The existence part of Theorem 2.4 is also a consequence of this proof.

2.6 Calculation of the matrix

Here we generalize Theorem 2.3 to the case of a general non-degenerate ES-operator T_Q given by (2.7). Let $p_n(z) = \sum_{i=0}^{n} a_{n,i} z^i$ be a monic polynomial of degree n and let $Q_j(z) = \sum_{\ell=0}^{j} q_{j,\ell} z^\ell$. Using these notations, we get

$$\begin{aligned} T_Q(p_n) &= \sum_{j=1}^{k} Q_j \cdot \frac{d^j}{dz^j} p_n = \sum_{j=1}^{k} \left[\sum_{\ell=0}^{j} q_{j,\ell} z^\ell \right] \cdot \left[\sum_{i \geq j}^{n} a_{n,i} \cdot \frac{i!}{(i-j)!} z^{i-j} \right] \\ &= \sum_{j=1}^{k} \sum_{s=0}^{n} \left[\sum_{\substack{s=\ell+i-j \\ j \leq i \leq n \\ 0 \leq \ell \leq j}} q_{j,\ell} \cdot a_{n,i} \cdot \frac{i!}{(i-j)!} \right] z^s \\ &= \sum_{s=0}^{n} \left[\sum_{j=1}^{k} \sum_{\substack{s=\ell+i-j \\ j \leq i \leq n \\ 0 \leq \ell \leq j}} q_{j,\ell} \cdot a_{n,i} \cdot \frac{i!}{(i-j)!} \right] z^s. \end{aligned}$$

Lemma 2.7. *If p_n is monic and $T_Q(p_n) = \lambda_n \cdot p_n$, then*

$$\lambda_n = \sum_{j=1}^{k} q_{j,j} \cdot \frac{n!}{(n-j)!}.$$

(This formula coincides with that in Chapter 1, see (1.3).)

Proof. With p_n monic and $T_Q(p_n) = \lambda_n \cdot p_n = \lambda_n z^n + \lambda_n \cdot a_{n,n-1} z^{n-1} + \cdots + \lambda_n \cdot a_{n,0}$, finding the eigenvalue λ_n amounts to finding the coefficient at z^n in $T_Q(p_n)$. Note that $\deg(Q_j(z)\frac{d^j}{dz^j}p_n) \le j + n - j = n$ with equality if and only if $\deg Q_j(z) = j$. Thus we can assume that $p_n = z^n$ since any lower degree terms of p_n will result in terms of degree lower than n in $T_Q(p_n)$. Then

$$
\begin{aligned}
T_Q(z^n) &= \sum_{j=1}^{k} Q_j(z) \cdot \frac{d^j}{dz^j} z^n = \sum_{j=1}^{k} Q_j(z) \cdot \frac{n!}{(n-j)!} z^{n-j} \\
&= \sum_{j=1}^{k} \left[\left(\sum_{\ell=0}^{j} q_{j,\ell} z^{\ell} \right) \cdot \frac{n!}{(n-j)!} z^{n-j} \right] \\
&= \sum_{j=1}^{k} \left[\sum_{\ell=0}^{j} q_{j,\ell} \cdot \frac{n!}{(n-j)!} z^{\ell+n-j} \right].
\end{aligned}
$$

Setting $\ell = j$ we get

$$
\lambda_n z^n = \sum_{j=1}^{k} q_{j,j} \cdot \frac{n!}{(n-j)!} z^n \quad \Longrightarrow \quad \lambda_n = \sum_{j=1}^{k} q_{j,j} \cdot \frac{n!}{(n-j)!}. \qquad \square
$$

Lemma 2.8. *For $n \ge 1$, the coefficient vector $X = (a_{n,0}, \dots, a_{n,n-1})$ of p_n satisfies the linear system $MX = Y$, where Y is a vector and M is an upper triangular matrix, both having entries expressible in the coefficients $q_{m,j}$ (see below).*

Proof. The relation

$$
T_Q(p_n) = \lambda_n \cdot p_n
$$

is equivalent to

$$
\sum_{s=0}^{n} \left[\sum_{j=1}^{k} \sum_{\substack{s=\ell+i-j \\ j\le i\le n \\ 0\le \ell\le j}} q_{j,\ell} \cdot a_{n,i} \cdot \frac{i!}{(i-j)!} \right] z^s = \lambda_n \sum_{s=0}^{n} a_{n,s} z^s.
$$

With $\ell = j + s - i$, the condition $\ell \le j$ gives $i \ge s$ and the condition $\ell \ge 0$ results in $j \ge i - s$. Therefore the above system is equivalent to

$$
\sum_{s=0}^{n} \left[\sum_{s\le i\le n} \quad \sum_{i-s\le j\le \min(i,k)} q_{j,j+s-i} \cdot \frac{i!}{(i-j)!} \cdot a_{n,i} - \lambda_n \cdot a_{n,s} \right] z^s = 0.
$$

Thus for each s we have

$$
\sum_{s\le i\le n} \quad \sum_{i-s\le j\le \min(i,k)} q_{j,j+s-i} \cdot \frac{i!}{(i-j)!} \cdot a_{n,i} - \lambda_n \cdot a_{n,s} = 0
$$

or, equivalently,

$$\sum_{s\le i\le n-1}\ \sum_{i-s\le j\le \min(i,k)} q_{j,j+s-i}\cdot\frac{i!}{(i-j)!}\cdot a_{n,i}-\lambda_n\cdot a_{n,s}=\sum_{n-s\le m\le \min(n,k)} q_{j,j+s-n}\cdot\frac{n!}{(n-j)!}\cdot a_{n,n}$$

where $a_{n,n}=1$. The $n\times n$ matrix M is thus constructed for $0\le s\le n-1$ and $0\le i\le n-1$. The left-hand side of the above equation corresponds to the $(s+1)$th row in M multiplied by the column vector X, and the right-hand side represents the $(s+1)$th row in Y. Thus the entries of M are given by

$$M_{s+1,i+1}=\sum_{i-s\le j\le \min(i,k)} q_{j,j+s-i}\cdot\frac{i!}{(i-j)!}-\lambda_n\cdot\delta_{i,s} \tag{2.8}$$

where δ denotes the Kronecker delta. The condition $i\ge s$ implies that M is upper triangular. □

We can now prove an important claim about the spectral asymptotics. Using Lemma 2.7 we get

$$\begin{aligned}
&\frac{\lambda_n}{n(n-1)\cdots(n-k+1)}\\
&=\frac{\sum_{j=1}^{k} q_{j,j}\cdot\frac{n!}{(n-j)!}}{n(n-1)\cdots(n-k+1)}\\
&=\left(q_{1,1}\frac{n!}{(n-1)!}+q_{2,2}\frac{n!}{(n-2)!}+\cdots+q_{k-1,k-1}\frac{n!}{(n-k+1)!}+q_{k,k}\frac{n!}{(n-k)!}\right)\frac{1}{(n)_k}\\
&=\frac{q_{1,1}}{(n-1)\cdots(n-k+1)}+\cdots+\frac{q_{k-1,k-1}}{(n-k+1)}+q_{k,k}.
\end{aligned}$$

Thus

$$\lim_{n\to\infty}\frac{\lambda_n}{(n)_k}=q_{k,k}=1.$$

Let us show that the determinant of the matrix M constructed in Lemma 2.8 is non-zero for sufficiently large values of n. Since the matrix is upper triangular, its determinant equals the product of the diagonal elements. Thus it suffices to prove that for sufficiently large n, every diagonal element is non-zero.

The diagonal element $M_{i+1,i+1}$ of M is obtained by letting $i=s$ in (2.8), and so

$$M_{i+1,i+1}=\sum_{0\le j\le \min(i,k)} q_{j,j}\cdot\frac{i!}{(i-j)!}-\lambda_n$$

for $i=0,\dots,n-1$. But the last expression equals $\lambda_i-\lambda_n$. Indeed, if $i\ge k$ then

$$\sum_{0\le m\le \min(i,k)} q_{j,j}\cdot\frac{i!}{(i-j)!}=\sum_{0\le j\le k} q_{j,j}\cdot\frac{i!}{(i-j)!}=\lambda_i.$$

If $i < k$ then this is again true since by definition $i!/(i-j)! = 0$ for $i < j \le k$. Thus we have to show that $\lambda_i - \lambda_n \ne 0, \forall i < n$ as $n \to \infty$. For small values of i (for example, $i < k$), we have $\lambda_i < \infty$ and $\lambda_n \to \infty$ as $n \to \infty$, implying $\lambda_i - \lambda_n \ne 0$. For larger values of i $(0 < j < k \le i)$, we get

$$\begin{aligned}\lambda_n - \lambda_i &= \sum_{j=1}^{k} q_{j,j}\frac{n!}{(n-j)!} - \sum_{j=1}^{k} q_{j,j}\frac{i!}{(i-j)!}\\ &= \sum_{j=1}^{k} q_{j,j}\left[\frac{n!}{(n-j)!} - \frac{i!}{(i-mj)!}\right].\end{aligned}$$

Dividing the last expression by $\frac{n!}{(n-k)!} - \frac{i!}{(i-k)!}$ we obtain

$$\frac{\lambda_n - \lambda_i}{\frac{n!}{(n-k)!} - \frac{i!}{(i-k)!}} = q_{k,k} + \sum_{j=1}^{k-1} q_{j,j}\frac{\frac{n!}{(n-j)!} - \frac{i!}{(i-j)!}}{\frac{n!}{(n-k)!} - \frac{i!}{(i-k)!}},$$

which tends to $q_{k,k} = 1$ as $n \to \infty$, since for each $j \le k-1$ we get

$$\begin{aligned}\lim_{n\to\infty}\frac{\frac{n!}{(n-j)!} - \frac{i!}{(i-j)!}}{\frac{n!}{(n-k)!} - \frac{i!}{(i-k)!}} &= \lim_{n\to\infty}\frac{\frac{i!}{(n-j)!}\left(\frac{n!}{i!} - \frac{(n-j)!}{(i-j)!}\right)}{\frac{i!}{(n-k)!}\left(\frac{n!}{i!} - \frac{(n-k)!}{(i-k)!}\right)}\\ &= \lim_{n\to\infty}\frac{(n-k)!}{(n-j)!}\cdot\frac{\left(\frac{n!}{i!} - \frac{(n-j)!}{(i-j)!}\right)}{\left(\frac{n!}{i!} - \frac{(n-k)!}{(i-k)!}\right)}\\ &= 0.\end{aligned}$$

Therefore $\lambda_n - \lambda_i \ne 0$. Thus, as $n \to \infty$, every diagonal element of M becomes non-zero. Thus M is invertible and the system $MX = Y$ has a unique solution.

2.7 Probability measures whose Cauchy transform satisfies an algebraic equation

In this section we will prove the uniqueness part of Theorem 2.4 and show that the measure μ_{Q_k}, if it exists, has the properties stated in Theorem 2.5. The proof relies heavily on the following lemma.

Lemma 2.9. *Let $A \subset \mathbb{C}$ be a finite set, $U \subset \mathbb{C}$ be a convex domain, and $\chi : U \to A$ be a measurable function such that $\partial\chi/\partial\bar{z} \ge 0$ (in the sense of distributions). Let $a \in A$, $z_0 \in U$ and assume that $\chi^{-1}(a) \cap \{|z - z_0| < r\}$ has positive Lebesgue measure for every $r > 0$. Then $\chi(z) = a$ almost everywhere in $U \cap (z_0 + \Gamma_a)$ where*

$$\Gamma_a = \{z \in \mathbb{C}; \Re(az) \ge \Re(bz), \forall b \in A\}. \tag{2.9}$$

Note that if $\chi^{-1}(a) \cap \{|z - z_0| < r\}$ has positive Lebesgue measure for every $a \in A$ and all $r > 0$, then χ is determined completely (outside a set of measure 0) since the cones Γ_a cover the whole complex plane.

Proof. Let χ_a denote the characteristic function of the set $\chi^{-1}(a)$. We will show that if $z_1, z_2 \in U$ with $z_2 - z_1 \in \Gamma_a$, and ϕ is a positive test function such that $z_1 + supp(\phi)$ and $z_2 + supp(\phi)$ are both contained in U, then

$$(\phi * \chi_a)(z_1) \leq (\phi * \chi_a)(z_2). \tag{2.10}$$

The desired conclusion follows from this. Indeed, let ϕ_j be a sequence of positive test functions such that $supp(\phi_j) \to 0$ and $\int \phi_j \, d\lambda = 1$, where λ denotes planar Lebesgue measure. We know then that $\phi_j * \chi_a$ converges in L^1_{loc} to χ_a. Hence, for any $\epsilon, r > 0$ we can find for all sufficiently large j a point z_1 with $|z_1 - z_0| < r$ such that $(\phi_j * \chi_a)(z_1) > 1 - \epsilon$. It follows from (2.10) that $(\phi_j * \chi_a)(z_2) > 1 - \epsilon$, and hence

$$|(\phi_j * \chi)(z_2) - a| = \left| \int \phi_j(z_2 - \zeta)(\chi(\zeta) - a) \, d\lambda(\zeta) \right| < \epsilon \max_{b \in A} |b - a|$$

for all $z_2 \in z_1 + \Gamma_a$. Letting ϵ and r tend to 0 and $j \to \infty$, it follows that $\chi(z) = \lim_{j \to \infty} (\phi_j * \chi)(z) = a$ for almost all z in $z_0 + \Gamma_a$.

We now prove inequality (2.10). Without loss of generality we may assume that $z_2 - z_1 > 0$ and that $a = 0$ for the general case can be reduced to this situation by replacing χ with the function $e^{i\theta}(\chi(e^{i\theta} z) - a)$ where $\theta = \arg(z_2 - z_1)$. The assumption that $z_2 - z_1 \in \Gamma_a$ then implies that A is contained in the closed left half plane $\{\Re z \leq 0\}$.

For any $\epsilon > 0$, let $\tilde{\chi}_\epsilon = \log(\chi - \epsilon)$ where we have chosen a branch of the logarithm function that is continuous in the left half plane. Let ψ be a positive test function and note that $\partial(\psi * \chi)/\partial \bar{z} \geq 0$ and $\Re \psi * \chi \leq 0$. It follows that

$$\Re \frac{\partial}{\partial \bar{z}} \log(\psi * \chi - \epsilon) = \Re \left(\frac{1}{\psi * \chi - \epsilon} \cdot \frac{\partial(\psi * \chi)}{\partial \bar{z}} \right) \leq 0.$$

When $supp(\psi) \to 0$ with $\int \psi \, d\lambda = 1$, we have that $\log(\psi * \chi - \epsilon) \to \tilde{\chi}_\epsilon$ in L^1_{loc}, and hence as a distribution. By passing to the limit, it follows that

$$\Re \frac{\partial \tilde{\chi}_\epsilon}{\partial \bar{z}} \leq 0.$$

If we write $\tilde{\chi}_\epsilon = \sigma_\epsilon + i\tau_\epsilon$, this means that

$$\frac{\partial \sigma_\epsilon}{\partial x} \leq \frac{\partial \tau_\epsilon}{\partial y}. \tag{2.11}$$

Fix a positive test function ϕ such that $z_j + supp(\phi) \subset U$ for $j = 1, 2$ and consider the function $(\phi * \sigma_\epsilon)(z_1 + \xi)$ of the real variable ξ. It follows from (2.11) and the fact that τ_ϵ is uniformly bounded for all ϵ that

$$\begin{aligned}\frac{\partial}{\partial \xi}(\phi * \sigma_\epsilon)(z_1+\xi) &= \int \frac{\partial \phi}{\partial x}(z_1+\xi-\zeta)\sigma_\epsilon(\zeta)\,d\lambda(\zeta)\\ &\le \int \frac{\partial \phi}{\partial y}(z_1+\xi-\zeta)\tau_\epsilon(\zeta)\,d\lambda(\zeta)\\ &\le M\end{aligned}$$

where the constant M does not depend on ϵ. In particular,

$$(\phi * \sigma_\epsilon)(z_2) - (\phi * \sigma_\epsilon)(z_1) \le M|z_2 - z_1|. \tag{2.12}$$

On the other hand, it is clear that

$$(\phi * \sigma_\epsilon)(z) = \log \epsilon \cdot (\phi * \chi_a)(z) + O(1). \tag{2.13}$$

Now (2.10) follows from (2.12) and (2.13) when $\epsilon \to 0$. □

We deduce two corollaries of Lemma 2.9.

Corollary 2.5. *Let $U \subset \mathbb{C}$ be a convex domain and $A \subset \mathbb{C}$ be a finite set. If v is a subharmonic function defined in U such that $2\partial v/\partial z \in A$ almost everywhere, then v is convex.*

Recall that a subharmonic function can locally be written as the sum of a harmonic function and a logarithmic potential. It follows that the distribution $\partial v/\partial z$ is represented by a locally integrable function. The condition $2\partial v/\partial z \in A$ should be interpreted by saying that $2\partial v/\partial z$ is represented by a measurable function with values in A.

Proof. Let $\chi = 2\partial v/\partial z$. Since v is subharmonic, $\partial \chi/\partial \bar{z} \ge 0$. Take any point $z_0 \in U$ and let A_0 be the set of all $a \in A$ such that $\chi^{-1}(a)$ has positive measure in every neighborhood of z_0. Let U_0 be a convex neighborhood of z_0 such that $\chi(z) \in A_0$ almost everywhere in U_0. By Lemma 2.9, $\chi(z) = a$ almost everywhere in $U_0 \cap (z_0 + \Gamma_a)$ where Γ_a is defined by (2.9) but with A_0 in place of A. This implies that $v(z) = v(z_0) + \mathbb{R}a(z - z_0)$ for all $z \in U_0 \cap (z_0 + \Gamma_a)$, so that

$$v(z) = v(z_0) + \max_{a \in A_0} \mathbb{R}a(z - z_0), \quad z \in U_0.$$

We have shown that in a neighborhood of z_0, v is the maximum of certain linear functions, hence it is convex there. Since z_0 was arbitrary, it follows that v is convex. □

Corollary 2.6. *Let $A \subset \mathbb{C}$ be a finite set, $U \subset \mathbb{C}$ be a convex domain, and let $\chi : U \to A$ be a measurable function. Then $\partial \chi/\partial \bar{z} \ge 0$ if and only if there exist real numbers c_a (possibly equal to $-\infty$) such that $\chi(z) = a$ almost everywhere in G_a where*

$$G_a = \{z \in U; c_a + \mathbb{R}(az) \ge c_b + \mathbb{R}(bz), \forall b \in A\}.$$

Proof. Suppose that c_a are real numbers such that $\chi(z) = a$ almost everywhere in G_a. Let $v(z) = \max_{a\in A}(c_a + \mathbb{R}(az))$. Then v is subharmonic and $\chi = 2\partial v/\partial z$, hence

$$\frac{\partial \chi}{\partial \bar{z}} = 2\frac{\partial^2 v}{\partial z \partial \bar{z}} \geq 0.$$

Suppose conversely that $\partial \chi / \partial \bar{z} \geq 0$. Since $\partial \chi / \partial \bar{z}$ is real, there exists a real valued function v defined in U with $2\partial v / \partial z = \chi$. It follows from Corollary 2.5 that v is convex. Moreover, we see from the proof that

$$v(z) = \max_{a \in A} (c_a + \Re(az))$$

where

$$c_a = \inf_{z \in U} (v(z) - \Re(az)).$$

If we define G_a using these constants c_a, it follows that $v(z) = c_a + \Re(az)$ for $z \in G_a$, hence $\chi(z) = 2\partial v / \partial z = a$ in G_a. □

Fix a monic polynomial Q_k of degree k and suppose that μ is a probability measure compactly supported in $\mathbb{C}$ whose Cauchy transform $\mathcal{C}(z)$ satisfies

$$\mathcal{C}^k(z) = 1/Q_k(z). \tag{2.14}$$

We will first show that μ has the properties asserted in Theorem 2.5, except that *supp*(μ) is contained in the convex hull of the zeros of Q_k, which will be proved in § 2.9.

Lemma 2.10. *If the Cauchy transform of μ satisfies* (2.14), *then the support of μ is the union of finitely many smooth curve segments. These curves are mapped to straight segments by Ψ.*

Proof. It is sufficient to prove that *supp*(μ) has these properties in a neighborhood of any given point z_0. Assume first that $Q_k(z_0) \neq 0$. Choose a branch of $Q_k^{-1/k}(z)$ defined in a simply connected neighborhood of z_0 and let Ψ be a primitive function of $Q_k^{-1/k}(z)$. Let U be a convex neighborhood of $\Psi(z_0)$ so small that Ψ maps a neighborhood of z_0 bijectively onto U. By (2.14) we can write $\mathcal{C}(z) = \chi(\Psi(z))Q_k^{-1/k}(z)$ for $z \in \Psi^{-1}(U)$, where χ has values in the set of kth roots of unity. If we write $w = \Psi(z)$, then

$$\pi\mu = \frac{\partial \mathcal{C}}{\partial \bar{z}} = \frac{\partial \chi(\Psi(z))}{\partial \bar{z}} \cdot Q_k^{-1/k} = \Psi^*\left(\frac{\partial \chi}{\partial \bar{w}}\right) \cdot \overline{\frac{\partial \Psi}{\partial z}} \cdot Q_k^{-1/k} = \Psi^*\left(\frac{\partial \chi}{\partial \bar{w}}\right) \cdot |Q_k|^{-2/k}$$

where Ψ^* denotes the pullback of distributions in U by Ψ. Since μ is positive, it follows that

$$\frac{\partial \chi}{\partial \bar{w}} \geq 0.$$

By Corollary 2.6, U is the union of sets G_a whose boundaries are finite unions of line segments, such that χ is constant in each G_a. It follows that *supp*(μ) $\cap \Psi^{-1}(U)$ =

$\Psi^{-1}(supp(\partial\chi/\partial\bar{z}))$ is the union of finitely many curve segments that are mapped to straight lines by Ψ.

If z_0 is a zero of Q_k, we take a disc D centered at z_0 that does not contain any other zeros of Q_k. If γ is any ray emanating at z_0, we can define single valued branches of $Q^{-1/k}(z)$ and Ψ in $D \setminus \gamma$. Notice that Ψ is continuous up to z_0. Let U be any half disc centered at $\Psi(z_0)$ and contained in $\Psi(D \setminus \gamma)$. As in the first part of the proof, it follows that $supp(\mu)$ has the required properties in $\Psi^{-1}(U)$. By varying γ and U, we see that the same holds in a full neighborhood of z_0. □

Hence $supp(\mu)$ can be thought of as a graph whose edges are smooth curve segments connecting certain vertices. The statement that $supp(\mu)$ is connected and has connected complement then means that it is a connected graph without cycles, that is a tree. Recall that a connected graph is a tree precisely if the number of vertices exceeds the number of edges by exactly one.

Lemma 2.11. *If the Cauchy transform of μ satisfies* (2.14), *then the support of μ is a tree.*

Proof. We will first prove that $supp(\mu)$ is connected. To do this we will show that if U is a bounded domain that is connected and simply connected, and the boundary of U does not intersect $supp(\mu)$, then either $supp(\mu) \subset U$ or $supp(\mu) \subset \mathbb{C} \setminus U$. From this it easily follows that $supp(\mu)$ is connected. Now it is clear that all the zeros of Q_k are either contained in U or in the complement of U since $\mathcal{C}(z)$ defines a continuous branch of $Q_k^{-1/k}(z)$ along ∂U. Observe also that

$$\frac{1}{2\pi i}\int_{\partial U} \mathcal{C}(z)\,dz = \frac{1}{2\pi i}\int_{\mathbb{C}}\int_{\partial U} \frac{dz}{z-\zeta}\,d\mu(\zeta) = \int_{U} d\mu(\zeta). \tag{2.15}$$

Now if all the zeros of Q_k are contained in the complement of U, there is an analytic continuation of $\mathcal{C}(z)$ across U, hence the left-hand side of (2.15) vanishes. It follows that $supp(\mu) \subset \mathbb{C} \setminus U$. If, on the other hand, all the zeros of Q_k are contained in U, then $\mathcal{C}(z)$ has an analytic continuation in $\mathbb{C} \setminus U$, which is asymptotically equal to a/z for some kth root a of unity when $z \to \infty$. Thus the left-hand side of (2.15) is equal to a. Since the right-hand side is positive, a must be 1, which means that all the mass of μ is in U. Hence we have proved that $supp(\mu)$ is connected.

Now let E be the set of all curve segments in $supp(\mu)$ and let V be the set of vertices that are endpoints of the edges in E. We may assume that V contains all the zeros of Q_k. To every pair $e \in E$, $v \in V$ such that v is an endpoint of e, we assign a number $\nu(e, v)$ by the following rule. Let γ be a small loop winding once around v in the clockwise direction, and let $\nu(e, v)$ be the jump of $(2\pi i)^{-1}\log \mathcal{C}(z)$ when z crosses e moving along γ. This number, which is defined modulo $\mathbb{Z}$, will be uniquely determined if we require that $0 < \nu(e, v) < 1$. Assume now that v is not a zero of Q_k and let $e_1, \dots, e_r$ be the curves in E having v as one endpoint. (If some curve has both its endpoints in v, it will be counted twice.) Select a branch of $Q_k^{1/k}(z)$ near v and observe that by Lemma 2.9 and the proof of

Lemma 2.10, $Q_k^{1/k}(z)C(z)$ is a kth root of unity, which moves once around the unit circle in the counterclockwise direction as z moves along γ. It follows that $\nu(e_1, v)+\cdots+\nu(e_r, v) = 1$. If instead v is a zero of Q_k of multiplicity m, a slight modification of the argument shows that $\nu(e_1, v)+\cdots+\nu(e_r, v) = 1-m/k$. On the other hand, it is clear that $\nu(e, v_1)+\nu(e, v_2) = 1$ where v_1, v_2 are the endpoints of $e \in E$. Hence the sum of all the $\nu(e, v)$ is equal both to $\sharp V - 1$ and to $\sharp E$. Since *supp*(μ) is a connected graph, this implies that it is a tree. □

We are now ready to prove the uniqueness part of Theorem 2.4. This is done by means of the following two lemmas.

Lemma 2.12. *Suppose the Cauchy transform of μ satisfies* (2.14) *and let u be the logarithmic potential of μ. If Ψ^{-1} is a (locally defined) inverse of a primitive function of $Q_k(z)^{-1/k}$, then $u \circ \Psi^{-1}$ is convex.*

Proof. Let χ be as in the proof of Lemma 2.10. Since $2\partial u/\partial z = C(z)$ we have

$$\begin{aligned}2\frac{\partial}{\partial w}u(\Psi^{-1}(w)) &= 2\frac{\partial u}{\partial z}(\Psi^{-1}(w)) \cdot Q_k(\Psi^{-1}(w))^{1/k}\\ &= C(\Psi^{-1}(w)) \cdot Q_k(\Psi^{-1}(w))^{1/k}\\ &= \chi(w).\end{aligned}$$

It follows from Corollary 2.5, that $u \circ \Psi^{-1}$ is convex. □

Lemma 2.13. *Let μ be a measure whose Cauchy transform satisfies* (2.14). *Let $\Omega = \mathbb{C} \setminus$ supp(μ) and let $\Psi(z)$ be the* complex potential *of μ defined in Ω by*

$$\Psi(z) = \int \log(z - \zeta)\, d\mu(\zeta).$$

Notice Ψ is a multi-valued function mapping Ω onto a domain $H = \{w; \Re w > h(\Im w)\}$ where h is a continuous function, and $\Psi^{-1} : H \to \Omega$ is a single valued function.

Proof. It is clear that Ψ is a holomorphic function in Ω defined up to multiples of $2\pi i$ and that $\Psi'(z) = C(z)$. Let γ be a curve segment of *supp*(μ) and let U be a one-sided neighborhood of γ in Ω on which Ψ has a single valued branch. Now the restriction of Ψ to U has an analytic continuation across γ, and by Lemma 2.10, Ψ maps γ to a line segment. Moreover, since in the notation of the proof of Lemma 2.10, $\chi = 1$ in $\Psi(U)$ and $\Re\chi \le 1$ everywhere, it follows that $\Psi(\gamma)$ is not horizontal and that $\Psi(U)$ is on the right-hand side of $\Psi(\gamma)$. Putting the segments $\Psi(\gamma)$ together as U moves around *supp*(μ), we obtain a curve of the form $\{\Re w = h(\Im w)\}$ bounding a domain $H = \{\Re w > h(\Im w)\}$. It is clear that Ψ maps Ω into H and the boundary of Ω to the boundary of H. Now $\psi(z) = \exp(-\Psi(z))$ is a single valued proper mapping from $\Omega \cup \{\infty\}$ to $D = \{\zeta; \log|\zeta| < -h(-\arg\zeta)\}$ that does not vanish in Ω and has a simple zero at ∞. It follows that $\psi : \Omega \cup \{\infty\} \to D$ is a bijection, hence $\Psi^{-1}(w) = \psi^{-1}(e^{-w})$ is a single valued holomorphic mapping. □

Corollary 2.7. *If μ_1 and μ_2 are two probability measures whose Cauchy transforms satisfy* (2.14), *then* $\mu_1 = \mu_2$.

Proof. Let Ψ be defined as in Lemma 2.13 with μ_1 in place of μ, and let u_1 and u_2 be the logarithmic potentials of μ_1 and μ_2. Then $u_1(\Psi^{-1}(w)) = \Re w$ for all $w \in H$ and $u_2(\Psi^{-1}(w)) = \Re w$ when $\Re w$ is sufficiently large. Since $u_2 \circ \Psi^{-1}$ is convex by Lemma 2.12, it follows that $u_2(\Psi^{-1}(w)) \geq \Re w$ for all $w \in H$, hence $u_1(z) \leq u_2(z)$ for almost all z. Similarly, $u_2(z) \leq u_1(z)$ for almost all z, and it follows that $\mu_1 = \Delta u_1/2\pi = \Delta u_2/2\pi = \mu_2$. □

2.8 On measures and their Cauchy transforms

In this section we describe the basic connections between root-counting measures and the Cauchy transform, which will be used to prove Theorem 2.6.

Let μ_n be a sequence of measures in the complex plane. The sequence is said to converge weakly to a measure μ if

$$\int \phi(z)\, d\mu_n(z) \to \int \phi(z)\, d\mu(z)$$

for every continuous function ϕ with compact support. If in addition there exists a compact set K such that $supp(\mu_n) \subset K$ for every n, we will say that μ_n converges weakly with compact support to μ and write $\mu_n \to \mu$ (w. c. s.).

If $K \subset \mathbb{C}$ is a compact set and $M(K)$ denotes the space of all probability measures with support in K, equipped with the weak topology, it is known that $M(K)$ is a sequentially compact Hausdorff space. We will use this fact to select a convergent subsequence from a sequence of measures as a first step in the proof of Theorem 2.6.

If ϕ is a locally integrable function and μ is a compactly supported measure, the convolution

$$(\phi * \mu)(z) = \int \phi(z - \zeta)\, d\mu(\zeta)$$

is a locally integrable function defined almost everywhere in the complex plane. If $\mu_n \to \mu$ (w. c. s.), it is easy to show that $\phi * \mu_n \to \phi * \mu$ in L^1_{loc}.

We will be particularly interested in the cases where $\phi(z) = \log|z|$ or $\phi(z) = 1/z$. Convolution with these functions defines the logarithmic potential

$$u(z) = \int \log|z - \zeta|\, d\mu(\zeta)$$

and the Cauchy transform

$$\mathcal{C}(z) = \int \frac{d\mu(\zeta)}{z - \zeta}$$

of μ. It is well known that the measure μ can be reconstructed from either u or $\mathcal{C}$ by the formula

$$\mu = \frac{1}{2\pi} \cdot \Delta u = \frac{1}{\pi} \cdot \frac{\partial \mathcal{C}}{\partial \bar{z}}$$

where $\Delta = (\partial/\partial x)^2 + (\partial/\partial y)^2$ is the Laplace operator and $\partial/\partial\bar{z} = (\partial/\partial x + i\partial/\partial y)/2$. These identities should be understood in the sense of distribution theory.

Let p be a polynomial of degree n and let μ be the root-counting measure of p, as defined in the introduction to this chapter. If p is monic, the logarithmic potential of μ is given by

$$\frac{1}{n}\log|p(z)| = \int \log|z-\zeta|\,d\mu(\zeta), \tag{2.16}$$

and for any p, the Cauchy transform of μ is

$$\frac{p'(z)}{np(z)} = \int \frac{d\mu(\zeta)}{z-\zeta}. \tag{2.17}$$

These two identities, which can easily be verified, are among the main ingredients in the proof of Theorem 2.6. We will here use them to prove a general lemma, which will be needed later.

Lemma 2.14. *Let p_m be a sequence of polynomials such that $n_m := \deg p_m \to \infty$ and let μ_m and μ'_m be the root-counting measures of p_m and p'_m respectively. If $\mu_m \to \mu$, $\mu'_m \to \mu'$ (w. c. s.) and u and u' are the logarithmic potentials of μ and μ', then $u' \leq u$ with equality in the unbounded component of $\mathbb{C} \setminus \operatorname{supp}(\mu)$.*

Proof. Assume with no loss of generality that p_m are monic. Let K be a compact set containing the zeros of every p_m. By (2.16) we then have

$$u(z) = \lim_{m\to\infty} \frac{1}{n_m}\log|p_m(z)|$$

and

$$u'(z) = \lim_{m\to\infty} \frac{1}{n_m - 1}\log\left|\frac{p'_m(z)}{n_m}\right| = \lim_{m\to\infty} \frac{1}{n_m}\log\left|\frac{p'_m(z)}{n_m}\right|$$

with convergence in L^1_{loc}. Hence by (2.17),

$$u'(z) - u(z) = \lim_{m\to\infty} \frac{1}{n_m}\log\left|\frac{p'_m(z)}{n_m p_m(z)}\right| = \lim_{m\to\infty} \frac{1}{n_m}\log\left|\int \frac{d\mu_m(\zeta)}{z-\zeta}\right|. \tag{2.18}$$

Now, if ϕ is a positive test function, it follows that

$$\begin{aligned}\int \phi(z)(u'(z) - u(z))\, d\lambda(z) &= \lim_{m\to\infty} \frac{1}{n_m} \int \phi(z) \log\left|\int \frac{d\mu_m(\zeta)}{z-\zeta}\right| d\lambda(z)\\ &\le \lim_{m\to\infty} \frac{1}{n_m} \int \phi(z) \int \frac{d\mu_m(\zeta)}{|z-\zeta|}\, d\lambda(z)\\ &= \lim_{m\to\infty} \frac{1}{n_m} \iint \frac{\phi(z)\, d\lambda(z)}{|z-\zeta|}\, d\mu_m(\zeta) \end{aligned} \tag{2.19}$$

where λ denotes Lebesgue measure in the complex plane. Since $1/|z|$ is locally integrable, the function $\int \phi(z)|z-\zeta|^{-1}\, d\lambda(z)$ is continuous, and hence bounded by a constant M for all z in K. Since $supp(\mu_m) \subset K$, the last expression in (2.19) is bounded by M/n_m, hence the limit when $m \to \infty$ is 0. This proves that $u' \le u$.

In the complement of $supp(\mu)$, u is harmonic and u' is subharmonic, hence $u' - u$ is a negative subharmonic function. Moreover, in the complement of K, $p_m'/(n_m p_m)$ converges uniformly on compact sets to the Cauchy transform $\mathcal{C}(z)$ of μ. Since $\mathcal{C}(z)$ is a non-constant holomorphic function in the unbounded component of $\mathbb{C} \setminus K$, it follows from (2.18) that $u' - u = 0$ there. By the maximum principle for subharmonic functions it follows then that $u' - u = 0$ in the unbounded component of $\mathbb{C} \setminus supp(\mu)$. The proof is complete. □

2.9 Root-counting measures of eigenpolynomials

We now turn to the proof of Theorem 2.6. The plan is to show that μ_n converges to a measure whose Cauchy transform satisfies (2.14). This will prove Theorem 2.6 and the existence part of Theorem 2.4. Let μ_n be the root-counting measure of p_n as in the statement of Theorem 2.6. Also let $\mu_n^{(i)}$ be the root-counting measure of the ith derivative $p_n^{(i)}$. We begin by showing that there is a compact set K containing the supports of all the measures $\mu_n^{(i)}$.

Lemma 2.15. *Let* $\mathbf{Q} = (Q_1, \dots, Q_k)$ *be fixed and let* p_n *be an eigenpolynomial of degree* n *of the operator* $T_{\mathbf{Q}}$*. Then there exists a compact set* K *such that all the zeros of every* $p_n^{(i)}$ *lie in* K *for every* n *and every* $i \ge 0$*. If* $Q_1 = \dots = Q_{k-1} = 0$*,* K *may be taken as the convex hull of the zeros of* Q_k*.*

Proof. The case with $Q_1 = \dots = Q_{k-1} = 0$ was treated in § 2.3 and [122]. In the general case it suffices to check the roots of p_n since by the Gauss–Lucas theorem the roots of any derivative $p_n^{(i)}$ are contained in the convex hull of the roots of p_n. Furthermore, it suffices to show that there exists a compact set containing the zeros of p_n for large values of n, since for any finite value of n we have finitely many roots of the polynomial p_n, and these are clearly contained in some compact set.

Let z be a root of p_n. Then

$$T_{\mathbf{Q}}(p_n)(z) = \sum_{j=1}^{k} Q_j(z) \cdot p_n^{(j)}(z) = \lambda_n \cdot p_n(z) = 0$$

or, equivalently,

$$Q_k(z)\cdot p_n^{(k)}(z)+Q_{k-1}(z)\cdot p_n^{(k-1)}(z)+\cdots+Q_1(z)\cdot p_n^{(1)}(z)=0. \tag{2.20}$$

We will show that for sufficiently large choices of $|z|$ and n, this equation will not hold. It is possible to find some r_0 and some n_0 such that if $|z|\geq r_0$ and $n>n_0$ then z cannot be a root of p_n. Using formula (2.17) we have

$$\frac{p_n^{(i+1)}(z)}{(n-i)\cdot p_n^{(i)}(z)}=\int\frac{d\mu_n^{(i)}(\zeta)}{z-\zeta}=:b_i.$$

Thus

$$p_n^{(k-1)}(z)=\frac{p_n^{(k)}(z)}{(n-k+1)\cdot b_{k-1}},$$
$$p_n^{(k-2)}(z)=\frac{p_n^{(k-1)}(z)}{(n-k+2)\cdot b_{k-2}}=\frac{p_n^{(k)}(z)}{(n-k+1)(n-k+2)\cdot b_{k-1}\cdot b_{k-2}},$$

and so on. Generally we have

$$p_n^{(i)}(z)=\frac{p_n^{(k)}(z)}{(n-k+1)\cdots(n-i)\cdot\prod_{j=i}^{k-1}b_j}.$$

Now assume that z is the root of p_n with the largest modulus and let $|z|=r$. With ζ being a root of some $p_n^{(i)}$, we have $|\zeta|\leq|z|$ by the Gauss–Lucas theorem. We will estimate $b_i=\int\frac{d\mu_n^{(i)}(\zeta)}{z-\zeta}$ so that $|b_i|\geq 1/2r,\forall i\leq k$. We have

$$\frac{1}{z-\zeta}=\frac{1}{z}\cdot\frac{1}{1-\zeta/z}=\frac{1}{z}\cdot\frac{1}{1-\theta}$$

and $|\theta|=|\zeta/z|\leq 1$.

With $w=1/(1-\theta)$ we obtain

$$|w-1|=\left|\frac{1}{1-\theta}-\frac{(1-\theta)}{(1-\theta)}\right|=\frac{|\theta|}{|1-\theta|}=|\theta||w|\leq|w|$$
$$\Leftrightarrow$$
$$|w-1|\leq|w|$$
$$\Leftrightarrow$$
$$\Re(w)\geq\frac{1}{2}.$$

Using this result we get

$$\begin{aligned}|b_i| &= \left|\int \frac{d\mu_n^{(i)}(\zeta)}{z-\zeta}\right| = \frac{1}{r}\left|\int \frac{d\mu_n^{(i)}(\zeta)}{1-\theta}\right| \\ &= \frac{1}{r}\left|\int w d\mu_n^{(i)}(\zeta)\right| \geq \frac{1}{r}\left|\int \Re(w) d\mu_n^{(i)}(\zeta)\right| \\ &\geq \frac{1}{2r}\int d\mu_n^{(i)}(\zeta) = \frac{1}{2r}.\end{aligned}$$

Now we choose r_0 in such a way that $|Q_k(w)| \geq r^k/2$ as $|w| \geq r_0$ and then a constant C such that $|Q_i(w)| \leq C \cdot r^i$ for every $i = 1, \ldots, k-1$. Finally, we choose n_0 such that $\frac{C \cdot 2^{k-i+1}}{(n-i)\cdots(n-k+1)} < \frac{1}{k-1}$ as $n > n_0$ for every $i = 1, \ldots, k-1$. Then, as $|z| = r \geq r_0$ and $n > n_0$, we get

$$\begin{aligned}\left|\frac{Q_i(z) \cdot p_n^{(i)}(z)}{Q_k(z) \cdot p_n^{(k)}(z)}\right| &= \frac{|Q_i(z)|}{|Q_k(z)|} \cdot \frac{(n-k)!}{(n-i)!} \cdot \frac{1}{\prod_{j=i}^{k-1} |b_j|} \\ &\leq \frac{|Q_i(z)|}{|Q_k(z)|} \cdot \frac{(n-k)!}{(n-i)!} \cdot 2^{k-i} \cdot r^{k-i} \\ &\leq \frac{C \cdot r^i}{r^k/2} \cdot \frac{(n-k)!}{(n-i)!} \cdot 2^{k-i} \cdot r^{k-i} \\ &= \frac{C \cdot 2^{k-i+1}}{(n-i)\cdots(n-k+1)} < \frac{1}{k-1}.\end{aligned}$$

Dividing (2.20) by $Q_k(z) \cdot p_n^{(k)}(z)$, we obtain

$$1 + \sum_{i=1}^{k-1} \frac{Q_i(z) \cdot p_n^{(i)}(z)}{Q_k(z) \cdot p_n^{(k)}(z)} = 0,$$

but with $r \geq r_0$ and $n > n_0$ we get

$$\left|\sum_{i=1}^{k-1} \frac{Q_i(z) \cdot p_n^{(i)}(z)}{Q_k(z) \cdot p_n^{(k)}(z)}\right| \leq \sum_{i=1}^{k-1} \left|\frac{Q_i(z) \cdot p_n^{(i)}(z)}{Q_k(z) \cdot p_n^{(k)}(z)}\right| < \sum_{i=1}^{k-1} \frac{1}{k-1} = 1,$$

and so (2.20) cannot be fulfilled with such choices of r and n. □

Assume that N is a subsequence of the natural numbers such that

$$\mu^{(j)} = \lim_{n\to\infty, n\in N} \mu_n^{(j)} \tag{2.21}$$

exists for $j = 0, \ldots, k$. The following lemma shows that the Cauchy transform of $\mu = \mu^{(0)}$ satisfies (2.14).

Lemma 2.16. *The measures $\mu^{(j)}$ are all equal and the Cauchy transform $\mathcal{C}(z)$ of this common limit satisfies $\mathcal{C}^k(z) = 1/Q_k(z)$ for almost every z.*

Proof. By (2.17) we have that

$$\frac{p_n^{(j+1)}(z)}{(n-j)p_n^{(j)}(z)} \to \int \frac{d\mu^{(j)}(\zeta)}{z-\zeta} \tag{2.22}$$

with convergence in L^1_{loc}, and by passing to a subsequence once again we can assume that we have pointwise convergence almost everywhere. From the relation $T_Q p_n = \lambda_n p_n$ it follows that

$$Q_k \frac{p_n^{(k)}}{n\cdots(n-k+1)p_n} = \frac{\lambda_n}{n\cdots(n-k+1)} - \sum_{l=0}^{k-1} \frac{Q_l}{(n-l)\cdots(n-k+1)} \prod_{j=0}^{l-1} \frac{p_n^{(j+1)}}{(n-j)p_n^{(j)}}. \tag{2.23}$$

Now $\lambda_n / n\cdots(n-k+1) \to 1$ while the sum converges pointwise to 0 almost everywhere by virtue of the factors $(n-l)\cdots(n-k+1)$ in the denominators. It follows that

$$\frac{p_n^{(k)}(z)}{n\cdots(n-k+1)p_n(z)} \to \frac{1}{Q_k(z)} \tag{2.24}$$

when $n \to \infty$ through the sequence N for almost every z. If $u^{(j)}$ denotes the logarithmic potential of $\mu^{(j)}$, then it follows from (2.16) and (2.24) that

$$u^{(k)} - u^{(0)} = \lim_{n\to\infty} \frac{1}{n} \log\left|\frac{p_n^{(k)}}{n\cdots(n-k+1)p_n}\right| = -\lim_{n\to\infty} \frac{1}{n} \log|Q_k| = 0.$$

On the other hand we have from Lemma 2.14 that $u^{(0)} \geq u^{(1)} \geq \cdots \geq u^{(k)}$, hence the potentials $u^{(j)}$ are all equal, and it follows that $\mu^{(j)} = \Delta u^{(j)}/2\pi$ are all equal. Finally we have from (2.22) and (2.24) that

$$\mathcal{C}^k(z) = \lim_{n\to\infty} \prod_{j=0}^{k-1} \frac{p_n^{(j+1)}(z)}{(n-j)p_n^{(j)}(z)} = \lim_{n\to\infty} \frac{p_n^{(k)}(z)}{(n)_k \cdot p_n(z)} = \frac{1}{Q_k(z)}$$

for almost every z. This completes the proof. □

Corollary 2.8. *There exists a unique measure μ_{Q_k} satisfying the requirements in Theorem 2.4. The sequence μ_n converges weakly to μ_{Q_k}. Moreover, supp(μ_{Q_k}) is contained in the convex hull of the zeros of Q_k.*

Proof. The operator T_Q has an eigenpolynomial p_n of degree n for all sufficiently large n. By Lemma 2.15, there exists a compact set K such that $supp(\mu_n^{(j)}) \subset K$ for all n. By compactness, there exists a subsequence N such that the limit (2.21) exists for $j = 0, \ldots, k$. By Lemma 2.16, $\mu_{Q_k} = \mu^{(0)}$ has the required properties, so existence is proved. Uniqueness was established in § 2.7. Since we may take $Q_0 = \cdots = Q_{k-1} = 0$, and in this case $supp(\mu_n^{(j)}) \subset K$ where K is the convex hull of the zeros of Q_k by Lemma 2.15, it follows that $supp(\mu_{Q_k})$ is also contained in K.

Assume that μ_n does not converge to μ_{Q_k} Then we can find a subsequence N' of the natural numbers such that μ_n stays away from a fixed neighborhood of μ_{Q_k} in the weak topology for all $n \in N'$. Again by compactness, we can find a subsequence N of N' such that the limit (2.21) exists for $j = 0, \dots, k$. By Lemma 2.16 and the uniqueness of μ_{Q_k}, it follows that $\mu^{(0)} = \mu_{Q_k}$, contradicting the assumption that μ_n stays away from μ_{Q_k} for all n in N' and hence all n in N. The proof is complete. □

2.10 Open problems in Chapter 2

2.1. Let us call a monic polynomial Q of degree k *special* if its accumulation tree $\mathcal{S}_Q$ is not a three-valent graph with $2k - 2$ distinct vertices. The set of all special polynomials form a real hypersurface Deg_k in the space $\mathrm{Pol}_k^{\mathbb{C}}$ of monic degree k polynomials. Deg_k contains the usual discriminant Disc_k consisting of all polynomials with multiple zeros (which is a complex hypersurface). In the simplest case $k = 3$ the discriminant Deg_3 consists of all 3rd degree polynomials whose zeros lie on some real affine line in $\mathbb{C}$.

Problem 2.1. Describe (find the equation for) Deg_k.

In the first non-trivial case $k = 4$ our computer experiments led us to the following guess.

Conjecture 2.1. *Deg_4 is the analytic continuation of the set of all convex 4-tuples of points on $\mathbb{C}$ such that their convex hull admits an inscribed circle.*

2.2. For fixed k and every n, define the discriminantal hypersurface $D_n \subset \mathrm{Pol}_k^{\mathbb{C}}$ consisting of polynomial Q for which the corresponding eigenpolynomial $p_n(x)$ has multiple zeros. Here $\mathrm{Pol}_k^{\mathbb{C}}$ (resp. $\mathrm{Pol}_k^{\mathbb{R}}$) is the space of all monic degree k polynomials with complex (resp. real) coefficients. Theorem 2.2 says that if $Q(z)$ is hyperbolic, then every $p_n(z)$ is hyperbolic and has only simple zeros. Therefore we know that all the discriminants D_n do not intersect the domain of strictly hyperbolic polynomials in $\mathrm{Pol}_k^{\mathbb{R}}$. An example of these discriminants for the family $Q_3(z) = z^3 + az + b$ is shown on Figure 2.4.

Problem 2.2. Find equations for D_n and study their topological properties.

Conjecture 2.2. *The domain of hyperbolic polynomials is the limit of the intersection of the connected components in the complement to all D_n containing it.*

Example 2.2. The explicit equations for the discriminants D_2, D_3, D_4 for the family $Q_3(z) = z^3 + az + b$ are as follows:

$$D_2 = \frac{4a}{9}; \quad D_3 = \frac{a^3}{16} + \frac{27b^2}{361}; \quad D_4 = \frac{36864a^6}{15353125} + \frac{14336a^3b^2}{2042125} - \frac{6912b^4}{923521}.$$

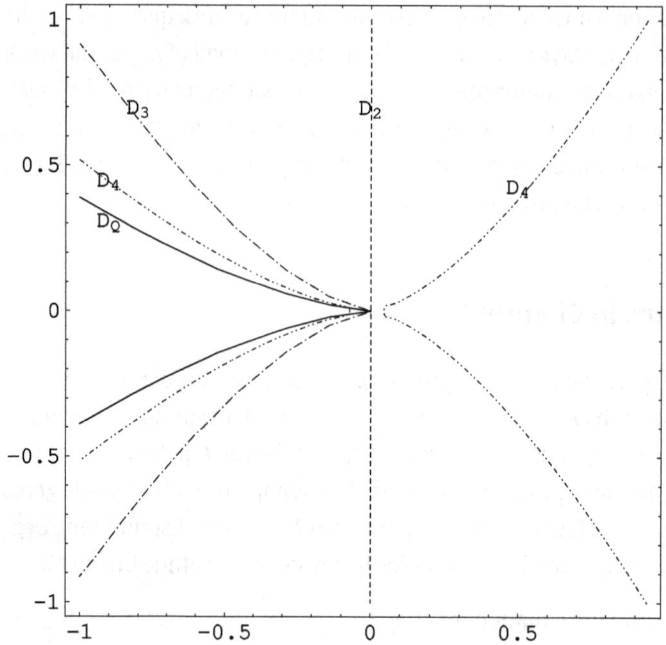

Figure 2.4: Discriminantal curves for $n \le 4$ in the family $Q_3(z) = z^3 + az + b$.

Note that the standard discriminant D_Q is given by

$$D_Q = 4a^3 + 27b^2.$$

2.3. Is it true that all types of embedded trees are obtained by the above construction? It is true that the type of tree determines the connected component in the space of generic polynomials?

2.4. Find some topological information about the connected components in the space of generic polynomials. For example, is it possible to calculate their fundamental groups?

2.5. Does the accumulation tree of this chapter appear in other mathematical situations?

3 Degenerate exactly solvable operators

(The material of this chapter is based on [26].)

3.1 Preliminaries

In this chapter we study root asymptotics for sequences of eigenpolynomials of degenerate ES-operators, see § 1.2. In other words, we consider a linear differential operator $T = \sum_{j=1}^{k} Q_j \frac{d^j}{dz^j}$, where Q_j are complex polynomials in z satisfying the condition $\deg Q_j \leq j$ for all j, $\deg Q_k < k$, and $\deg Q_{j_0} = j_0$ for some $j_0 < k$.

We concentrate on the polynomial eigenfunctions of such operators, that is, polynomials satisfying

$$T(p_n) = \lambda_n p_n \tag{3.1}$$

for some value of the spectral parameter λ_n, where n is a positive integer and $\deg p_n = n$.

One of the major differences between non-degenerate and degenerate operators is that in the non-degenerate case the union of all roots of all eigenpolynomials of T is contained in a compact set (see Chapter 2 and [27]), contrary to the degenerate case, which we, in particular, prove below.

The importance of studying eigenpolynomials of ES-operators is, among other things, motivated by numerous examples coming from classical orthogonal polynomials. Our results can be considered as natural generalizations of the theorems about the asymptotic behavior of the maximal root in classical orthogonal polynomial families such as Laguerre and Hermite polynomials, which appear as solutions to the eigenvalue problem (3.1) for certain degenerate ES-operators of second order. Namely, the Laguerre polynomials are solutions to the differential equation

$$zy''(z) + (1 - z)y'(z) + ny(z) = 0,$$

while the Hermite polynomials are solutions to the differential equation of

$$y''(z) - 2zy'(z) + 2ny(z) = 0$$

where n is a non-negative integer.

Recent studies and interesting results on the asymptotic zero behavior for these polynomials and the corresponding generalized polynomials can be found in, e. g., [57, 72, 111, 130, 128, 108, 109] and the references therein. Paper [108] provides bounds on the spacing of zeros of certain functions belonging to the Laguerre–Pólya class satisfying a second order differential equation, and as their corollary new sharp inequalities for the extreme zeros of the Hermite, Laguerre, and Jacobi polynomials are established.

https://doi.org/10.1515/9783112207741-004

Below we partially extend the results of Chapter 2 to the case of degenerate ES-operators. Numerical evidence shows that when $n \to \infty$, the roots of the eigenpolynomial of degree n are distributed along a certain curvilinear tree in the complex plane in this case as well, but only after an appropriate scaling, which depends on n. We will assume wlog that p_n is monic.

In what follows we denote by r_n the largest modulus of all roots of the unique monic eigenpolynomial p_n of T of degree n where $n \geq n_T$, see Chapter 1.2, i. e.,

$$r_n = \max\{|z| : p_n(z) = 0\}.$$

The main results of this chapter are as follows.

Theorem 3.1. *Let T be a degenerate ES-operator of order k. Then $r_n \to \infty$ when $n \to \infty$.*

Next we establish a lower bound for r_n when $n \to \infty$.

Theorem 3.2. *Let $T = \sum_{j=1}^{k} Q_j(z)\frac{d^j}{dz^j} = \sum_{j=1}^{k}(\sum_{i=0}^{\deg Q_j} q_{j,i}z^i)\frac{d^j}{dz^j}$ be a degenerate ES-operator of order k. Then, for any $\gamma < b$,*

$$\lim_{n\to\infty} \frac{r_n}{n^\gamma} = \infty,$$

where

$$b := \min_{j\in[1,k-1]}^{+} \left(\frac{k-j}{k-j+\deg Q_j - \deg Q_k} \right).$$

The notation $\min^+$ *means that the minimum is taken only over positive terms* $(k - j + \deg Q_j - \deg Q_k)$.

Corollary 3.1. *Let $T = \sum_{j=1}^{k} Q_j\frac{d^j}{dz^j}$ be a degenerate ES-operator of order k such that $\deg Q_j \leq j_0$ for all $j > j_0$, and in particular, $\deg Q_k = j_0$, where j_0 is the largest j such that $\deg Q_j = j$. Then $\lim_{n\to\infty} \frac{r_n}{n^\gamma} = \infty$ for any $\gamma < 1$.*

Corollary 3.2. *Let $T = \sum_{j=1}^{k} Q_j\frac{d^j}{dz^j}$ be a degenerate ES-operator of order k such that $\deg Q_j = 0$ for all $j > j_0$, where j_0 is the largest j for which $\deg Q_j = j$. Then $\lim_{n\to\infty} \frac{r_n}{n^\gamma} = \infty$ for any $\gamma < \frac{k-j_0}{k}$.*

3.2 Growth rate conjecture

In fact our extensive numerical experiments and natural heuristic arguments (see § 3.4) support the following conjecture:

Conjecture 3.1 (Growth rate conjecture). *Let $T = \sum_{j=1}^{k} Q_j\frac{d^j}{dz^j}$ be a degenerate ES-operator of order k and denote by j_0 the largest j for which $\deg Q_j = j$. Then*

$$\lim_{n\to\infty} \frac{r_n}{n^d} = c_T,$$

where $c_T > 0$ is a positive constant and

$$d := \max_{j\in[j_0+1,k]}\left(\frac{j-j_0}{j-\deg Q_j}\right).$$

Remark 3.1. Note that the growth rate conjecture implies Theorem 3.2 since $b \le d$. The next two theorems support the growth rate conjecture.

Theorem 3.3. *Let T be a degenerate ES-operator of order k consisting of precisely two terms: $T = Q_{j_0}\frac{d^{j_0}}{dz^{j_0}} + Q_k\frac{d^k}{dz^k}$. Then there exists a positive constant c such that*

$$\lim_{n\to\infty}\inf\frac{r_n}{n^d} \ge c,$$

where $d := \max_{j\in[j_0+1,k]}(\frac{j-j_0}{j-\deg Q_j}) = \frac{k-j_0}{k-\deg Q_k}$.

This result can be generalized to operators consisting of any number of terms, but with certain conditions on the degrees of the polynomial coefficients Q_j for which $j > j_0$, where j_0 is the largest j for which $\deg Q_j = j$. Namely,

Theorem 3.4. *Let T be a degenerate ES-operator of order k. Denote by j_0 the largest j such that $\deg Q_j = j$ and let $(j-\deg Q_j) \ge (k-\deg Q_k)$ for every $j > j_0$. Then there exists a positive constant $c > 0$ such that*

$$\lim_{n\to\infty}\inf\frac{r_n}{n^d} \ge c,$$

where $d := \max_{j\in[j_0+1,k]}(\frac{j-j_0}{j-\deg Q_j}) = \frac{k-j_0}{k-\deg Q_k}$.

Numerical experiments show that roots of eigenpolynomials scaled according to the growth rate conjecture fill certain intriguing compact curves in $\mathbb{C}$. To illustrate this phenomenon, let us present three typical pictures. Below p_n denotes the unique and monic eigenpolynomial of the given operator T of degree n, and $s_n(z) = p_n(n^d z)$ denotes the corresponding appropriately scaled polynomial, where d is as in the growth rate conjecture. Conjecturally the union of all roots of all $s_n(z)$ is contained in a compact set, see Figure 3.1.

In § 3.4 we will empirically derive the conjectural algebraic equation satisfied by the Cauchy transform $\mathcal{C}(z)$ of the asymptotic root-counting measure of the scaled eigenpolynomial s_n for an arbitrary degenerate ES-operator. From this equation one can obtain detailed information about the above curves and also conclude which terms of the operator are relevant for the asymptotic zero distribution of its eigenpolynomials.[1]

1 As was mentioned in Chapter 2, in the non-degenerate case, which we have treated previously, the asymptotic zero distribution of the eigenpolynomials depends *only* on the leading coefficient Q_k. For the degenerate operators, however, the situation is more complicated.

T_1 | T_2 | T_3

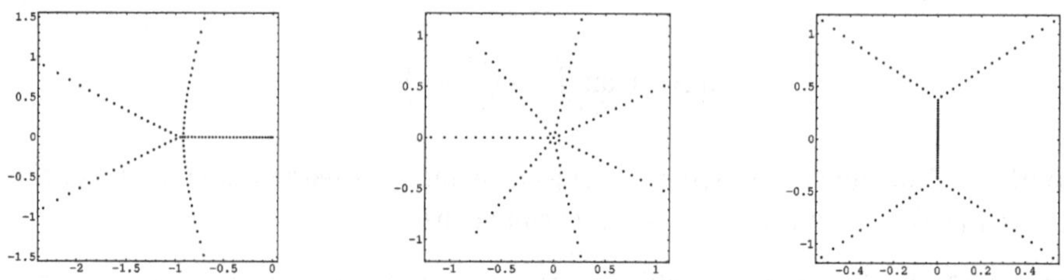

roots of $s_{100}(z) = p_{100}(100z)$ | roots of $s_{100}(z) = p_{100}(100z)$ | roots of $s_{100}(z) = p_{100}(100z)$

Figure 3.1: Roots of the scaled eigenpolynomial $s_{100}(z)$ for $T_1 = z\frac{d}{dz} + z\frac{d^2}{dz^2} + z\frac{d^3}{dz^3} + z\frac{d^4}{dz^4} + z\frac{d^5}{dz^5}$; $T_2 = z^2\frac{d^2}{dz^2} + \frac{d^7}{dz^7}$; $T_3 = z^3\frac{d^3}{dz^3} + z^2\frac{d^4}{dz^4} + z\frac{d^5}{dz^5}$.

Numerical evidence clearly illustrates that distinct operators whose scaled eigenpolynomials satisfy the same Cauchy transform equation when $n \to \infty$ will yield identical asymptotic zero distributions, see Figure 3.2. For further details, see § 3.5.3.

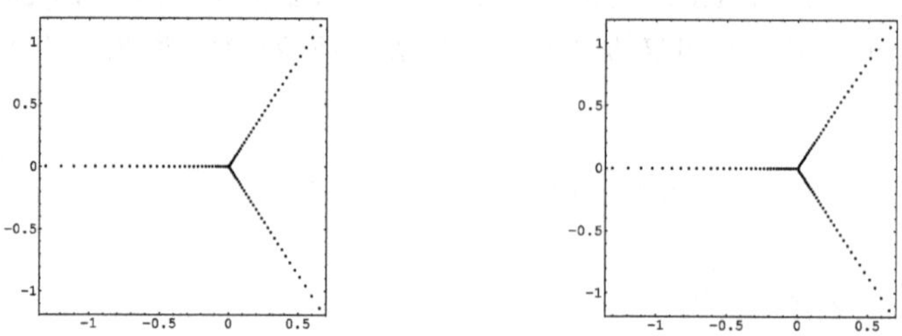

T_4, roots of $s_{100}(z) = p_{100}(100^{2/3}z)$ | $\tilde{T}_4$, roots of $s_{100}(z) = p_{100}(100^{2/3}z)$

Figure 3.2: Roots of the scaled eigenpolynomial $s_{100}(z)$ for $T_4 = z^3\frac{d^3}{dz^3} + z^2\frac{d^5}{dz^5}$ and $\tilde{T}_4 = z^2\frac{d^2}{dz^2} + z^3\frac{d^3}{dz^3} + z\frac{d^4}{dz^4} + z^2\frac{d^5}{dz^5} + \frac{d^6}{dz^6}$.

The rest of this chapter is organized as follows. In § 3.3 we give the proofs of the lemma, the theorems, and the corollaries stated in this section. In § 3.4 we explain how we arrived at the growth rate conjecture and how we obtained as its corollary the algebraic equation for the Cauchy transform of the asymptotic root-counting measure. In § 3.5 we give numerical evidence supporting the growth rate conjecture and its corollary. In § 3.6 we give the detailed calculations that led to the corollary of the growth rate conjecture. We also prove that for a class of operators of the considered type, the conjectured upper bound for r_n implies the conjectured lower bound. Finally in § 3.7 we discuss some open problems and directions for further research.

3.3 Proofs of the main results for degenerate ES-operators

To prove Theorem 3.1, we need the following. Given a sequence of polynomials $\{s_n\}$, where $\deg s_n = n$, we denote by μ_n the root-counting measure of s_n. Further, for any polynomial s_n of degree n and an integer $j = 0, 1, \ldots, n-1$, introduce the Cauchy transform $\mathcal{C}_{n,j}$ of the root-counting measure $\mu_n^{(j)}$ for the jth derivative $s_n^{(j)}$ as defined by

$$\mathcal{C}_{n,j}(z) := \frac{s_n^{(j+1)}(z)}{(n-j)s_n^{(j)}(z)} = \int \frac{d\mu_n^{(j)}(\zeta)}{z-\zeta}.$$

It is well known that the measure $\mu_n^{(j)}$ can be reconstructed from $\mathcal{C}_{n,j}$ by the formula $\mu_n^{(j)} = \frac{1}{\pi} \cdot \frac{\partial \mathcal{C}_{n,j}}{\partial \bar{z}}$, where $\partial/\partial\bar{z} = \frac{1}{2}(\partial/\partial x + i\partial/\partial y)$, see, e. g., [155].

Proof of Theorem 3.1. For $T = \sum_{j=1}^{k} Q_j \frac{d^j}{dz^j}$, denote by j_0 the largest j for which $\deg Q_j = j$. Note that since T is degenerate we have $j_0 < k$. Let us apply the above notation to the sequence $\{p_n\}$ of monic eigenpolynomials of the operator T. (By our previous results we know that p_n exists and is unique for any ES-operator and all sufficiently large n.) By definition, $\mathcal{C}_{n,j}(z) = \frac{p_n^{(j+1)}(z)}{(n-j)p_n^{(j)}(z)}$ is the Cauchy transform of $p_n^{(j)}(z)$, and we get

$$\frac{p_n^{(j)}(z)}{p_n(z)} = \mathcal{C}_{n,0}(z)\mathcal{C}_{n,1}(z)\cdots\mathcal{C}_{n,j-1}(z)\cdot n(n-1)\cdots(n-j+1) = \frac{n!}{(n-j)!}\prod_{m=0}^{j-1}\mathcal{C}_{n,m}(z).$$

Expanding $Q_j(z) = \sum_{i=0}^{\deg Q_j} q_{j,i}z^i$ for $j = 1, \ldots, k$, we have the expression

$$\lambda_n = \sum_{j=1}^{j_0} q_{j,j}(n)_j$$

for the eigenvalue λ_n, which we already mentioned in the previous chapters. Now, dividing the equation $T(p_n) = \lambda_n p_n$ by p_n, we obtain

$$Q_k(z)\frac{p_n^{(k)}(z)}{p_n(z)} + Q_{k-1}(z)\frac{p_n^{(k-1)}(z)}{p_n(z)} + \cdots + Q_1(z)\frac{p_n'(z)}{p_n(z)} = \sum_{j=1}^{j_0} q_{j,j}\frac{n!}{(n-j)!}$$

$$\Leftrightarrow$$

$$Q_k(z)\frac{n!}{(n-k)!}\prod_{m=0}^{k-1}\mathcal{C}_{n,m}(z) + Q_{k-1}(z)\frac{n!}{(n-k+1)!}\prod_{m=0}^{k-2}\mathcal{C}_{n,m}(z) + \cdots + Q_1(z)\frac{n!}{(n-1)!}\mathcal{C}_{n,0}(z)$$
$$= \sum_{j=1}^{j_0} q_{j,j}\frac{n!}{(n-j)!}. \tag{3.2}$$

Dividing both sides of this equation by $\frac{n!}{(n-k)!}$, we get

$$Q_k(z)\prod_{m=0}^{k-1} C_{n,m}(z)\Big[1+\frac{(n-k)!}{(n-k+1)!}\frac{1}{C_{n,k-1}(z)}\frac{Q_{k-1}(z)}{Q_k(z)} + \frac{(n-k)!}{(n-k+2)!}\frac{1}{C_{n,k-1}(z)C_{n,k-2}(z)}\frac{Q_{k-2}(z)}{Q_k(z)}+\cdots + \frac{(n-k)!}{(n-1)!}\frac{1}{\prod_{m=1}^{k-1} C_{n,m}(z)}\frac{Q_1(z)}{Q_k(z)}\Big] = \sum_{j=1}^{j_0} q_{j,j}\frac{(n-k)!}{(n-j)!}. \tag{3.3}$$

Now assume that the statement of Theorem 3.1 is false, i. e., all zeros of all p_n are uniformly bounded. Then we can take a subsequence $\{p_{n_i}\}$ such that all the corresponding root-counting measures μ_{n_i} are weakly convergent to a compactly supported probability measure. Then all Cauchy transforms $C_{n_i,m}$ will be uniformly convergent to a nonvanishing function outside some large disc, which in particular contains all the roots of $Q_k(z)$. Since $j_0 < k$, the right-hand side of (3.3) tends to zero when $n \to \infty$. On the other hand, in the left-hand side of (3.3), all terms in the bracket except for the constant term 1 tend to zero when $n \to \infty$, and thus the limit of the left-hand side equals $\lim_{n\to\infty} Q_k(z)\prod_{m=0}^{k-1} C_{n,m}(z) = K \neq 0$, and we obtain a contradiction when $n \to \infty$. ☐

In order to prove Theorem 3.2, we need the following two lemmas, where Lemma 3.1 is used to prove Lemma 3.2.

Lemma 3.1. *Let z_n be a root of p_n with the largest modulus r_n. Then, for any complex number z_0 such that $|z_0| = r_0 \geq r_n$, we have $|C_{n,j}(z_0)| \geq \frac{1}{2r_0}$ for all $j \geq 0$.*

Proof. Recall that $C_{n,j}(z) := \int \frac{d\mu_n^{(j)}(\zeta)}{z-\zeta} = \frac{p_n^{(j+1)}(z)}{(n-j)p_n^{(j)}(z)}$. With ζ being some root of $p_n^{(j)}(z)$, we have $|\zeta| \leq |z_0|$ by the Gauss–Lucas theorem. Thus $\frac{1}{z_0-\zeta} = \frac{1}{z_0}\cdot\frac{1}{1-\zeta/z_0} = \frac{1}{z_0}\cdot\frac{1}{1-\theta}$ where $|\theta| = |\zeta/z_0| \leq 1$. With $w = \frac{1}{1-\theta}$ we obtain

$$|w-1| = \frac{|\theta|}{|1-\theta|} = |\theta||w| \leq |w| \quad \Leftrightarrow \quad |w-1| \leq |w| \quad \Rightarrow \quad \mathrm{Re}(w) \geq 1/2,$$

and thus

$$\begin{aligned}|C_{n,j}(z_0)| &= \left|\int \frac{d\mu_n^{(j)}(\zeta)}{z_0-\zeta}\right| = \frac{1}{r_0}\left|\int \frac{d\mu_n^{(j)}(\zeta)}{1-\theta}\right| = \frac{1}{r_0}\left|\int w\,d\mu_n^{(j)}(\zeta)\right| \\ &\geq \frac{1}{r_0}\left|\int \mathrm{Re}(w)\,d\mu_n^{(j)}(\zeta)\right| \geq \frac{1}{2r_0}\int d\mu_n^{(j)}(\zeta) = \frac{1}{2r_0}.\end{aligned}$$

☐

Lemma 3.2. *Let $T = \sum_{j=1}^{k} Q_j \frac{d^j}{dz^j} = \sum_{j=1}^{k}\big(\sum_{i=0}^{\deg Q_j} q_{j,i} z^i\big)\frac{d^j}{dz^j}$ be a degenerate ES-operator of order k. Wlog we assume that Q_k is monic, i. e., $q_{k,\deg Q_k} = 1$. Let z_n be a root of p_n with the largest modulus r_n. Then the following inequality holds:*

$$1 \le \sum_{j=1}^{k-1} \sum_{i=0}^{\deg Q_j} |q_{j,i}| 2^{k-j} \frac{r_n^{k-j-\deg Q_k+i}}{(n-k+1)^{k-j}} + \sum_{0 \le i < \deg Q_k} \frac{|q_{k,i}|}{r_n^{\deg Q_k - i}}. \tag{3.4}$$

Proof. From $C_{n,j}(z) = \frac{p_n^{(j+1)}(z)}{(n-j)p^{(j)}(z)}$ we get

$$p^{(j)}(z) = \frac{p_n^{(k)}(z)}{(n-k+1)(n-k+2)\cdots(n-j)\prod_{m=j}^{k-1} C_{n,m}(z)} \tag{3.5}$$

for all $j < k$. Inserting z_n in the eigenvalue equation $Tp_n(z) = \lambda_n p_n(z)$, we obtain

$$\sum_{j=1}^{k-1} \left(\sum_{i=0}^{\deg Q_j} q_{j,i} z_n^i \right) p_n^{(j)}(z_n) + \left(\sum_{i=0}^{\deg Q_k} q_{k,i} z_n^i \right) p_n^{(k)}(z_n) = \lambda_n p_n(z_n) = 0,$$

and after division by $z_n^{\deg Q_k} p_n^{(k)}(z_n)$ we obtain

$$\sum_{j=1}^{k-1} \left(\sum_{i=0}^{\deg Q_j} q_{j,i} \frac{1}{z_n^{\deg Q_k - i}} \right) \frac{p_n^{(j)}(z_n)}{p_n^{(k)}(z_n)} + \sum_{0 \le i < \deg Q_k} q_{k,i} \frac{1}{z_n^{\deg Q_k - i}} + 1 = 0.$$

Thus, applying (3.5) and Lemma 3.1, we obtain

$$\begin{aligned}
1 &= \left| \sum_{j=1}^{k-1} \left(\sum_{i=0}^{\deg Q_j} q_{j,i} \frac{1}{z_n^{\deg Q_k - i}} \right) \frac{p_n^{(j)}(z_n)}{p_n^{(k)}(z_n)} + \sum_{0 \le i < \deg Q_k} q_{k,i} \frac{1}{z_n^{\deg Q_k - i}} \right| \\
&\le \sum_{j=1}^{k-1} \left| \sum_{i=0}^{\deg Q_j} q_{j,i} \frac{1}{z_n^{\deg Q_k - i}} \right| \frac{|p_n^{(j)}(z_n)|}{|p_n^{(k)}(z_n)|} + \sum_{0 \le i < \deg Q_k} \frac{|q_{k,i}|}{r_n^{\deg Q_k - i}} \\
&\le \sum_{j=1}^{k-1} \sum_{i=0}^{\deg Q_j} \frac{|q_{j,i}|}{r_n^{\deg Q_k - i}} \frac{1}{(n-k+1)\cdots(n-j)\prod_{m=j}^{k-1} |C_{n,m}(z_n)|} + \sum_{0 \le i < \deg Q_k} \frac{|q_{k,i}|}{r_n^{\deg Q_k - i}} \\
&\le \sum_{j=1}^{k-1} \sum_{i=0}^{\deg Q_j} \frac{|q_{j,i}|}{r_n^{\deg Q_k - i}} \frac{(2r_n)^{k-j}}{(n-k+1)^{k-j}} + \sum_{0 \le i < \deg Q_k} \frac{|q_{k,i}|}{r_n^{\deg Q_k - i}} \\
&= \sum_{j=1}^{k-1} \sum_{i=0}^{\deg Q_j} |q_{j,i}| 2^{k-j} \frac{r_n^{k-j-\deg Q_k + i}}{(n-k+1)^{k-j}} + \sum_{0 \le i < \deg Q_k} \frac{|q_{k,i}|}{r_n^{\deg Q_k - i}}.
\end{aligned}$$ □

The proof of Theorem 3.2 follows from Theorem 3.1 and Lemma 3.2.

Proof of Theorem 3.2. Applying Theorem 3.1, we see that the last sum on the right-hand side of inequality (3.4) in Lemma 3.2 tends to zero when $n \to \infty$. Now consider the double sum on the right-hand side of (3.4). If, for given i and j, the exponent $(k-j-\deg Q_k+i)$ of r_n is negative or zero, the corresponding term tends to zero when $n \to \infty$ by Theorem 3.1. We now consider the remaining terms in the double sum, namely those for which the exponent $(k-j-\deg Q_k+i)$ of r_n is positive. If $r_n \le c_0(n-k+1)^\gamma$ where $c_0>0$ and $\gamma<\frac{k-j}{k-j+i-\deg Q_k}$ for given $j\in[1,k-1]$ and given $i\in[0,\deg Q_j]$, then the corresponding term

$$\frac{r_n^{k-j+i-\deg Q_k}}{(n-k+1)^{k-j}} = \left(\frac{r_n}{(n-k+1)^{\frac{k-j}{k-j+i-\deg Q_k}}}\right)^{k-j+i-\deg Q_k}$$

in the double sum tends to zero when $n\to\infty$. Thus assume that $r_n \le c_0(n-k+1)^\gamma$ where c_0 is a positive constant and $\gamma<b$. Here

$$b = \min^{+}_{\substack{j\in[1,k-1]\\ i\in[0,j]}} \frac{k-j}{k-j+i-\deg Q_k} = \min^{+}_{j\in[1,k-1]} \frac{k-j}{k-j+\deg Q_j-\deg Q_k}.$$

The notation $\min^+$ means that we only take the minimum over positive terms $(k-j+i-\deg Q_k)$ and $(k-j+\deg Q_j-\deg Q_k)$.[2]

Then every term in the double sum tends to zero when $n\to\infty$, and we obtain a contradiction with (3.4) when $n\to\infty$. Thus, for all sufficiently large integers n, we must have $r_n > c_0(n-k+1)^\gamma$ for all $\gamma<b$, and hence $\liminf_{n\to\infty}\frac{r_n}{n^\gamma} > c_0$ for any $\gamma<b$. But for any such γ, we can form $\gamma'=\frac{\gamma+b}{2}$ for which $\gamma'<b$ and $\gamma<\gamma'$, and thus $\lim_{n\to\infty}\frac{r_n}{n^\gamma}=\infty$ for all $\gamma<b$. □

Proof of Corollary 3.1. For this class of operators, we have

$$\begin{aligned} b &:= \min^{+}_{j\in[1,k-1]} \frac{k-j}{k-j+\deg Q_j-\deg Q_k} \\ &= \min^{+}_{j\in[1,k-1]} \frac{k-j}{k-j+\deg Q_j-j_0} = \frac{k-j_0}{k-j_0} = 1, \end{aligned}$$

and the proof is complete after applying Theorem 3.2. □

Proof of Corollary 3.2. For this class of operators, we have

$$\begin{aligned} b &:= \min^{+}_{j\in[1,k-1]}\left(\frac{k-j}{k-j+\deg Q_j-\deg Q_k}\right) \\ &= \min^{+}_{j\in[1,k-1]}\left(\frac{k-j}{k-j+\deg Q_j}\right) = \min_{j\in[1,j_0]}\frac{k-j}{k} = \frac{k-j_0}{k} \end{aligned}$$

2 On the left-hand side in the expression for b above, we take the minimum over $i\in[0,\deg Q_j]$, so we can put $i=\deg Q_j$ in this expression. Thus with $b=\min^{+}_{j\in[1,k-1]}\frac{k-j}{k-j+\deg Q_j-\deg Q_k}$ we get that $\gamma<\frac{k-j}{k-j+i-\deg Q_k}$ for *every* $j\in[1,k-1]$ and *every* $i\in[0,\deg Q_j]$. Then if $r_n\le c_0(n-k+1)^\gamma$ and $\gamma<b$, every term with positive exponent $(k-j+i-\deg Q_k)$ will tend to zero when $n\to\infty$.

where the third equality follows from choosing any j for which $\deg Q_j = j$, and the minimum is then attained for $j = j_0$ (note that for $j > j_0$ we get $(k-j)/(k-j+\deg Q_j) = 1 > (k-j_0)/k$), and the proof is complete after applying Theorem 3.2. □

Remark 3.2. Note that for the class of operators considered in Corollary 3.1, the growth rate conjecture claims that $\lim_{n\to\infty} \frac{r_n}{n} = c_T$ for some $c_T > 0$ since $d := \max_{j\in[j_0+1,k]}(\frac{j-j_0}{j-\deg Q_j}) = \frac{k-j_0}{k-j_0} = 1$. (The maximum is attained by choosing any $j > j_0$ such that $\deg Q_j = j_0$, e. g., $j = k$.) For the class of operators considered in Corollary 3.2, we claim that $\lim_{n\to\infty} \frac{r_n}{n^{(k-j_0)/k}} = c_T$ for some $c_T > 0$ since

$$d := \max_{j\in[j_0+1,k]}\left(\frac{j-j_0}{j-\deg Q_j}\right) = \max_{j\in[j_0+1,k]}\left(\frac{j-j_0}{j}\right) = \frac{k-j_0}{k}.$$

Remark 3.3. Notice that for a class of operators that contains the operators considered in Corollaries 3.1 and 3.2, we can actually prove that the conjectured upper bound $\lim_{n\to\infty} \sup \frac{r_n}{n^d} \le c_1$ implies the conjectured lower bound $\lim_{n\to\infty} \inf \frac{r_n}{n^d} \ge c_0$ where $c_1 \ge c_0 > 0$, see Theorem 3.5 in § 3.6.2.

Proof of Theorem 3.3. Clearly $\deg Q_{j_0} = j_0$ since there exists at least one such $j < k$. Set

$$T = Q_{j_0}\frac{d^{j_0}}{dz^{j_0}} + Q_k\frac{d^k}{dz^k} = \sum_{i=0}^{j_0} q_{j_0,i}z^i\frac{d^{j_0}}{dz^{j_0}} + \sum_{i=0}^{\deg Q_k} q_{k,i}z^i\frac{d^k}{dz^k},$$

where $q_{j_0,j_0} \neq 0$, and where we wlog assume that Q_k is monic. From inequality (3.4) in Lemma 3.2 we have

$$\begin{aligned}1 &\le \sum_{i=0}^{j_0} |q_{j_0,i}|2^{k-j_0}\frac{r_n^{i-\deg Q_k+k-j_0}}{(n-k+1)^{k-j_0}} + \sum_{0\le i<\deg Q_k} |q_{k,i}|\frac{1}{r_n^{\deg Q_k - i}}\\ &\le \sum_{i=0}^{j_0} |q_{j_0,i}|2^{k-j_0}\frac{r_n^{i-\deg Q_k+k-j_0}}{(n-k+1)^{k-j_0}} + \epsilon,\end{aligned}$$

where we choose n so large that $\epsilon < 1$. (This is possible since $\epsilon \to 0$ when $n \to \infty$ by Theorem 3.1.) Thus, for sufficiently large n, we get

$$\begin{aligned}c_0 &\le \sum_{i=0}^{j_0} |q_{j_0,i}|2^{k-j_0}\frac{r_n^{i-\deg Q_k+k-j_0}}{(n-k+1)^{k-j_0}}\\ &\le \sum_{i=0}^{j_0} |q_{j_0,i}|2^{k-j_0}\frac{r_n^{k-\deg Q_k}}{(n-k+1)^{k-j_0}}\\ &= K\frac{r_n^{k-\deg Q_k}}{(n-k+1)^{k-j_0}},\end{aligned}$$

where $1-\epsilon = c_0 \to 1$ when $n \to \infty$, and $K > 0$ since $q_{j_0,j_0} \neq 0$ (the second inequality follows since $i \leq j_0$). Thus

$$r_n \geq \left(\frac{c_0}{K}\right)^{1/(k-\deg Q_k)} (n-k+1)^{\frac{k-j_0}{k-\deg Q_k}}$$

for sufficiently large integers n, and hence there exists a positive constant $c = (1/K)^{1/(k-\deg Q_k)}$ such that

$$\liminf_{n\to\infty} \frac{r_n}{n^{(\frac{k-j_0}{k-\deg Q_k})}} \geq c.$$

Finally, it is clear that for this two-term operator,

$$d := \max_{j\in[j_0+1,k]} \left(\frac{j-j_0}{j-\deg Q_j}\right) = \frac{k-j_0}{k-\deg Q_k},$$

and we are done. □

Remark 3.4. If Q_k is a monomial (i. e., $Q_k = z^{\deg Q_k}$) in Theorem 3.3, then there exists a positive constant c such that $r_n \geq c(n-k+1)^d$ for *every* n, where $d := \max_{j\in[j_0+1,k]}(\frac{j-j_0}{j-\deg Q_j}) = \frac{k-j_0}{k-\deg Q_k}$. This is easily seen from the calculations in the proof of Theorem 3.3 since $\sum_{0\leq i<\deg Q_k} |\alpha_{k,i}| \frac{1}{r_n^{\deg Q_k - i}}$ on the right-hand side of (3.4) vanishes, and therefore $1 \leq K\frac{r_n^{k-\deg Q_k}}{(n-k+1)^{k-j_0}}$ for *every* n.

As we know, for this class of operators, there exists a unique eigenpolynomial p_n for every sufficiently large n, and the conclusion follows.

Proof of Theorem 3.4. For this class of operators $(j-\deg Q_j) \geq (k-\deg Q_k)$ for every $j > j_0$, and thus

$$d := \max_{j\in[j_0+1,k]} \left(\frac{j-j_0}{j-\deg Q_j}\right) = \frac{k-j_0}{k-\deg Q_k}.$$

Assuming that Q_k is monic, we have the inequality

$$1 \leq \sum_{j=1}^{k-1} \sum_{i=0}^{\deg Q_j} |q_{j,i}| 2^{k-j} \frac{r_n^{k-j+i-\deg Q_k}}{(n-k+1)^{k-j}} + \sum_{0\leq i<\deg Q_k} \frac{|q_{k,i}|}{r_n^{\deg Q_k - i}} \tag{3.6}$$

by Lemma 3.2. The last sum here tends to zero when $n \to \infty$ by Theorem 3.1. Considering the double sum on the right-hand side of (3.6), we see that since $i \leq \deg Q_j$ for every j, we get

$$\begin{aligned}
&\sum_{i=0}^{\deg Q_j} |q_{j,i}| 2^{k-j} \frac{r_n^{k-j+i-\deg Q_k}}{(n-k+1)^{k-j}} \\
&= \sum_{i=0}^{\deg Q_j} |q_{j,i}| 2^{k-j} \frac{r_n^{k-j+\deg Q_j-\deg Q_k}}{(n-k+1)^{k-j}} r_n^{i-\deg Q_j}
\end{aligned}$$

$$= \frac{r_n^{k-j+\deg Q_j-\deg Q_k}}{(n-k+1)^{k-j}}\left(2^{k-j}|q_{j,\deg Q_j}| + \sum_{i<\deg Q_j} 2^{k-j}|q_{j,i}|r_n^{i-\deg Q_j}\right)$$

$$= K_{j,n}\frac{r_n^{k-j+\deg Q_j-\deg Q_k}}{(n-k+1)^{k-j}}, \tag{3.7}$$

where

$$K_{j,n} = 2^{k-j}|q_{j,\deg Q_j}| + \sum_{i<\deg Q_j} 2^{k-j}|q_{j,i}|r_n^{i-\deg Q_j}.$$

In addition, $K_{j,n} > 0$ since $q_{j,\deg Q_j} \neq 0$. Also, $K_{j,n} < \infty$ when $n \to \infty$ since $(i - \deg Q_j) < 0$ for the exponent of r_n, and $K_{j,n} \to 2^{k-j}|q_{j,\deg Q_j}|$ when $n \to \infty$. With the decomposition

$$\begin{aligned} A &= \{j : \deg Q_j = j\},\\ B &= \{j : \deg Q_j < j \text{ and } (k-j+\deg Q_j - \deg Q_k) > 0\},\\ C &= \{j : \deg Q_j < j \text{ and } (k-j+\deg Q_j - \deg Q_k) \le 0\}, \end{aligned}$$

and using (3.7) we see that inequality (3.6) is equivalent to

$$\begin{aligned} 1 &\le \sum_{j=1}^{k-1}\sum_{i=0}^{j} |q_{j,i}|2^{k-j}\frac{r_n^{k-j+i-\deg Q_k}}{(n-k+1)^{k-j}} + \sum_{0\le i<\deg Q_k}\frac{|q_{k,i}|}{r_n^{\deg Q_k - i}}\\ &= \sum_{j\in A} K_{j,n}\frac{r_n^{k-\deg Q_k}}{(n-k+1)^{k-j}} + \sum_{j\in B} K_{j,n}\frac{r_n^{k-j+\deg Q_j-\deg Q_k}}{(n-k+1)^{k-j}}\\ &\quad + \sum_{j\in C} K_{j,n}\frac{r_n^{k-j+\deg Q_j-\deg Q_k}}{(n-k+1)^{k-j}} + \sum_{0\le i<\deg Q_k}\frac{|q_{k,i}|}{r_n^{\deg Q_k - i}}. \end{aligned}$$

The last two sums on the right-hand side of this inequality tend to zero when $n \to \infty$. The last one is due to Theorem 3.1, and the sum over C since $(j - \deg Q_j) \ge (k - \deg Q_k) \Leftrightarrow (k-j+\deg Q_j - \deg Q_k) \le 0$ for every $j \in C$ by assumption, and then applying Theorem 3.1. Therefore, when $n \to \infty$, we get the inequality

$$c_0 \le \sum_{j\in A} K_{j,n}\frac{r_n^{k-\deg Q_k}}{(n-k+1)^{k-j}} + \sum_{j\in B} K_{j,n}\frac{r_n^{k-j+\deg Q_j-\deg Q_k}}{(n-k+1)^{k-j}}, \tag{3.8}$$

where

$$c_0 = 1 - \sum_{j\in C} K_{j,n}\frac{r_n^{k-j+\deg Q_j-\deg Q_k}}{(n-k+1)^{k-j}} - \sum_{0\le i<\deg Q_k}\frac{|q_{k,i}|}{r_n^{\deg Q_k - i}}.$$

Note that $c_0 \to 1$ when $n \to \infty$.

Now assume that B is empty. This corresponds to an operator such that $(j - \deg Q_j) \geq (k - \deg Q_k)$ for every j for which $\deg Q_j < j$. Then the inequality (3.8) becomes

$$\begin{aligned} c_0 &\leq \sum_{j\in A} K_{j,n} \frac{r_n^{k-\deg Q_k}}{(n-k+1)^{k-j}} \\ &= \frac{r_n^{k-\deg Q_k}}{(n-k+1)^{k-j_0}} \left(K_{j_0,n} + \sum_{j\in A\setminus\{j_0\}} K_{j,n} \frac{1}{(n-k+1)^{j_0-j}} \right) \\ &\leq K_A \frac{r_n^{k-\deg Q_k}}{(n-k+1)^{k-j_0}}, \end{aligned} \tag{3.9}$$

where K_A is a positive constant that is finite when $n \to \infty$ since $j_0 - j > 0$ for every $j \in A\setminus\{j_0\}$. (Recall that j_0 is the largest element in A by definition.) Thus, for all sufficiently large integers n, we have

$$r_n \geq \left(\frac{c_0}{K_A}\right)^{1/(k-\deg Q_k)} (n-k+1)^{\frac{k-j_0}{k-\deg Q_k}}.$$

Therefore there exists a positive constant $c = (1/K_A)^{1/(k-\deg Q_k)}$ such that

$$\liminf_{n\to\infty} \frac{r_n}{n^{(\frac{k-j_0}{k-\deg Q_k})}} \geq c,$$

and we are done.

Now assume that B is non-empty. Again inequality (3.8) holds, i. e.,

$$c_0 \leq \sum_{j\in A} K_{j,n} \frac{r_n^{k-\deg Q_k}}{(n-k+1)^{k-j}} + \sum_{j\in B} K_{j,n} \frac{r_n^{k-j+\deg Q_j-\deg Q_k}}{(n-k+1)^{k-j}},$$

where $c_0 \to 1$ when $n \to \infty$. From (3.9) we have

$$\sum_{j\in A} K_{j,n} \frac{r_n^{k-\deg Q_k}}{(n-k+1)^{k-j}} \leq K_A \frac{r_n^{k-\deg Q_k}}{(n-k+1)^{k-j_0}}$$

for the sum over A for large n, and thus

$$\begin{aligned} c_0 &\leq \sum_{j\in A} K_{j,n} \frac{r_n^{k-\deg Q_k}}{(n-k+1)^{k-j}} + \sum_{j\in B} K_{j,n} \frac{r_n^{k-j+\deg Q_j-\deg Q_k}}{(n-k+1)^{k-j}} \\ &\leq K_A \frac{r_n^{k-\deg Q_k}}{(n-k+1)^{k-j_0}} + \sum_{j\in B} K_{j,n} \frac{r_n^{k-j+\deg Q_j-\deg Q_k}}{(n-k+1)^{k-j}} \\ &= \frac{r_n^{k-\deg Q_k}}{(n-k+1)^{k-j_0}} \left(K_A + \sum_{j\in B} K_{j,n} \frac{r_n^{\deg Q_j - j}}{(n-k+1)^{j_0-j}} \right) \\ &\leq K_{AB} \frac{r_n^{k-\deg Q_k}}{(n-k+1)^{k-j_0}}, \end{aligned}$$

where K_{AB} is a positive and finite constant when $n \to \infty$. (Note that $K_{AB} \to K_A$ when $n \to \infty$ since $(\deg Q_j - j) < 0$ and $j_0 - j > 0$ for every $j \in B$.) Thus, for all sufficiently large integers n, we have

$$r_n \geq \left(\frac{c_0}{K_{AB}}\right)^{1/(k-\deg Q_k)} (n-k+1)^{\frac{k-j_0}{k-\deg Q_k}}.$$

Therefore there exists a positive constant $c = (1/K_{AB})^{1/(k-\deg Q_k)}$ such that

$$\liminf_{n\to\infty} \frac{r_n}{n^{\left(\frac{k-j_0}{k-\deg Q_k}\right)}} \geq c. \qquad \square$$

3.4 Empirical deduction of the growth rate conjecture

In this section we explain how we arrived at the growth rate conjecture (see § 3.1) and obtain as a corollary of our method the (conjectural) algebraic equation satisfied by the Cauchy transform of the asymptotic root-counting measure of the scaled eigenpolynomials.

Let $T = \sum_{j=1}^{k} Q_j \frac{d^j}{dz^j} = \sum_{j=1}^{k} (\sum_{i=0}^{\deg Q_j} q_{j,i} z^i) \frac{d^j}{dz^j}$ be an arbitrary degenerate ES-operator of order k and denote by j_0 the largest j for which $\deg Q_j = j$. Wlog we assume that Q_{j_0} is monic, i. e., $q_{j_0,j_0} = 1$. Consider the *scaled eigenpolynomial* $s_n(z) = p_n(n^d z)$, where $p_n(z)$ is the unique monic nth degree eigenpolynomial of T, and d is some positive real number. The goal is to obtain a well-defined algebraic equation for the Cauchy transform of the root-counting measure μ_n of the scaled eigenpolynomial s_n when $n \to \infty$, and as we will see in the process of doing this, we are forced to choose d as in the growth rate conjecture.[3]

Basic assumption

When performing our calculations, we assume that the root-counting measures $\mu_n^{(0)}, \mu_n^{(1)}, \mu_n^{(2)} \dots, \mu_n^{(k-1)}$ of the *scaled eigenpolynomial* $s_n(z)$ and its derivatives up to the kth order exist when $n \to \infty$ and that they are all weakly convergent to the *same* asymptotic root-counting measure μ.[4]

Thus the corresponding Cauchy transforms are all asymptotically identical, and we define $\mathcal{C}(z) := \lim_{n\to\infty} \mathcal{C}_{n,j}(z)$ for all $j \in [0, k-1]$, where $\mathcal{C}(z)$ is the Cauchy transform of μ considered in the complement to the support of μ. Computer experiments strongly indicate that this assumption is true, see § 3.5.2.

3 It is already well known that for the Laguerre polynomials, which appear as eigenpolynomials for a second order ES-operator, the largest root grows as n when $n \to \infty$ and thus $d = 1$ in this case, which is consistent with the growth rate conjecture.

4 Conjecturally supp μ is a tree, see § 3.7.

From the definition of the Cauchy transform we obtain

$$\begin{aligned}\prod_{i=0}^{j-1} \mathcal{C}_{n,i}(z) &= \prod_{i=0}^{j-1} \frac{s_n^{(i+1)}(z)}{(n-j)s_n^{(i)}(z)} \\ &= \frac{s_n^{(1)}(z)}{ns_n(z)} \cdot \frac{s_n^{(2)}(z)}{(n-1)s^{(1)}(z)} \cdot \frac{s_n^{(3)}(z)}{(n-2)s_n^{(2)}(z)} \cdots \\ &\quad \cdot \frac{s_n^{(j-1)}(z)}{(n-j+2)s_n^{(j-2)}(z)} \cdot \frac{s_n^{(j)}(z)}{(n-j+1)s_n^{(j-1)}(z)} \\ &= \frac{s_n^{(j)}(z)}{(n)_j s_n(z)}.\end{aligned}$$

(Recall that $(n)_j = n(n-1)\cdots(n-j+1)$ is the Pochhammer symbol.) Thus our basic assumption implies

$$\mathcal{C}^j(z) = \lim_{n\to\infty} \prod_{i=0}^{j-1} \mathcal{C}_{n,i}(z) = \lim_{n\to\infty} \frac{s_n^{(j)}(z)}{(n)_j s_n(z)}. \tag{3.10}$$

In the above notation consider the eigenvalue equation $Tp_n(z) = \lambda_n p_n(z)$, where the eigenvalue λ_n is given by

$$\lambda_n = \sum_{j=1}^{k} q_{j,j} \frac{n!}{(n-j)!} = \sum_{j=1}^{j_0} q_{j,j} \frac{n!}{(n-j)!} = \sum_{j=1}^{j_0} q_{j,j}(n)_j.$$

(Clearly this sum ends at j_0 since $q_{j,j} = 0$ for all $j > j_0$ by definition of j_0 as the largest j for which $\deg Q_j = j$.) We then have

$$\begin{gathered} Tp_n(z) = \lambda_n p_n(z) \\ \Leftrightarrow \\ \sum_{j=1}^{k} \left(\sum_{i=0}^{\deg Q_j} q_{j,i} z^i \right) p_n^{(j)}(z) = \sum_{j=1}^{j_0} q_{j,j}(n)_j p_n(z). \end{gathered}$$

Substituting $z = n^d z$ in this equation, we obtain

$$\sum_{j=1}^{k} \left(\sum_{i=0}^{\deg Q_j} q_{j,i} n^{di} z^i \right) p_n^{(j)}(n^d z) = \sum_{j=1}^{j_0} q_{j,j}(n)_j p_n(n^d z),$$

and with $s_n(z) = p_n(n^d z)$ we get

$$\sum_{j=1}^{k} \left(\sum_{i=0}^{\deg Q_j} q_{j,i} \frac{z^i}{n^{d(j-i)}} \right) s_n^{(j)}(z) = \sum_{j=1}^{j_0} q_{j,j}(n)_j s_n(z).$$

Dividing this equation by $\frac{n!}{(n-j_0)!}s_n(z) = (n)_{j_0}s_n(z)$, we get

$$\sum_{j=1}^{k}\left(\sum_{i=0}^{\deg Q_j} q_{j,i}\frac{z^i}{n^{d(j-i)}}\right)\frac{s_n^{(j)}(z)}{(n)_{j_0}s_n(z)} = \sum_{j=1}^{j_0} q_{j,j}\frac{(n)_j}{(n)_{j_0}}. \tag{3.11}$$

Consider the right-hand side of (3.11). Since $j \le j_0$ all terms for which $j < j_0$ (if not already zero, which is the case if $q_{j,j} = 0$, i. e., if $\deg Q_j < j$) tend to zero when $n \to \infty$, and therefore the limit of the right-hand side of (3.11) equals

$$\lim_{n\to\infty}\sum_{j=1}^{j_0} q_{j,j}\frac{(n)_j}{(n)_{j_0}} = q_{j_0,j_0} = 1$$

since we assumed that Q_{j_0} is monic. Now consider the jth term in the sum on the left-hand side of (3.11). It equals

$$\begin{aligned}
&\sum_{i=0}^{\deg Q_j} q_{j,i}\frac{z^i}{n^{d(j-i)}}\cdot\frac{s_n^{(j)}(z)}{(n)_{j_0}s_n(z)}\\
&\quad= \sum_{i=0}^{\deg Q_j} q_{j,i}\frac{z^i}{n^{d(j-i)}}\cdot\frac{s_n^{(j)}(z)}{(n)_j s_n(z)}\cdot\frac{(n)_j}{(n)_{j_0}}\\
&\quad= \sum_{i=0}^{\deg Q_j} q_{j,i}\frac{z^i}{n^{d(j-i)}}\cdot\prod_{i=0}^{j-1} C_{n,i}(z)\cdot\frac{(n)_j}{(n)_{j_0}}\\
&\quad= \sum_{i=0}^{\deg Q_j} q_{j,i}\frac{z^i}{n^{d(j-i)+j_0-j}}\prod_{i=0}^{j-1} C_{n,i}(z)\frac{(n)_j}{n^j}\frac{n^{j_0}}{(n)_{j_0}}.
\end{aligned}$$

Taking the limit and using the basic assumption (3.10), we obtain

$$\lim_{n\to\infty}\sum_{i=0}^{\deg Q_j} q_{j,i}\frac{z^i}{n^{d(j-i)}}\cdot\frac{s_n^{(j)}(z)}{(n)_{j_0}s_n(z)} = \lim_{n\to\infty}\sum_{i=0}^{\deg Q_j} q_{j,i}\frac{z^i}{n^{d(j-i)+j_0-j}}\mathcal{C}^j(z)$$

for the jth term and thus, taking the limit of the left-hand side of (3.11), we get

$$\lim_{n\to\infty}\sum_{j=1}^{k}\left(\sum_{i=0}^{\deg Q_j} q_{j,i}\frac{z^i}{n^{d(j-i)}}\right)\frac{s_n^{(j)}(z)}{(n)_{j_0}s_n(z)} = \lim_{n\to\infty}\sum_{j=1}^{k}\left(\sum_{i=0}^{\deg Q_j} q_{j,i}\frac{z^i}{n^{d(j-i)+j_0-j}}\right)\mathcal{C}^j(z).$$

Adding up, the following equation is satisfied by $\mathcal{C}(z)$ for z in the complement to the support of μ:

$$\lim_{n\to\infty}\sum_{j=1}^{k}\left(\sum_{i=0}^{\deg Q_j} q_{j,i}\frac{z^i}{n^{d(j-i)+j_0-j}}\right)\mathcal{C}^j(z) = 1. \tag{3.12}$$

In order to make (3.12) a well-defined algebraic equation, i. e., to avoid infinities in the denominator when $n \to \infty$, we must impose the following condition on the real number d in the exponent of n:

$$d(j-i)+j_0-j \geq 0 \quad \Leftrightarrow \quad d \geq \frac{j-j_0}{j-i}$$

for all $j \in [1,k]$ and all $i \in [0, \deg Q_j]$. Therefore we take $d = \max_{\substack{j\in[1,k]\\ i\in[0,\deg Q_j]}} (\frac{j-j_0}{j-i})$, but this maximum is clearly obtained for the maximal value of i for any given j, so we can take $i = \deg Q_j$. Our condition then becomes $d = \max_{j\in[1,k]}(\frac{j-j_0}{j-\deg Q_j})$, and clearly the maximum is taken only over j for which $Q_j(z)$ is not identically zero. Finally, we observe that since T is degenerate, we have $j_0 < k$, and thus we only need to take this maximum over $j \in [j_0+1,k]$ since there always exists a positive value on d for any operator of the type we consider. Thus our condition becomes

$$d = \max_{j\in[j_0+1,k]}\left(\frac{j-j_0}{j-\deg Q_j}\right).$$

3.4.1 Implications of the growth rate conjecture

In the above notation (recall that $q_{j_0,j_0} = 1$ by monicity of Q_{j_0}), by substituting d defined in the growth rate conjecture in equation (3.12) and letting $n \to \infty$, we get the following algebraic equation satisfied by the Cauchy transform.

Corollary 3.3. *The Cauchy transform $\mathcal{C}(z)$ of the asymptotic root-counting measure μ of the scaled eigenpolynomial $s_n(z) = p_n(n^d z)$ of an arbitrary degenerate ES-operator T satisfies the following algebraic equation for almost all complex z in the usual Lebesgue measure on $\mathbb{C}$:*

$$z^{j_0}\mathcal{C}^{j_0}(z) + \sum_{j\in A} q_{j,\deg Q_j} z^{\deg Q_j}\mathcal{C}^j(z) = 1,$$

where A is the set consisting of all j for which the maximum $d := \max_{j\in[j_0+1,k]}(\frac{j-j_0}{j-\deg Q_j})$ is attained, i. e., $A = \{j : (j-j_0)/(j-\deg Q_j) = d\}$.

For detailed calculations see § 3.6.1.

3.5 Numerical evidence

3.5.1 Evidence for the growth rate conjecture

In Figure 3.3 and in Table 3.1 we present numerical evidence for the growth of $r_n = \max\{|z| : p_n(z) = 0\}$, which supports the choice of d in the growth rate conjecture. We

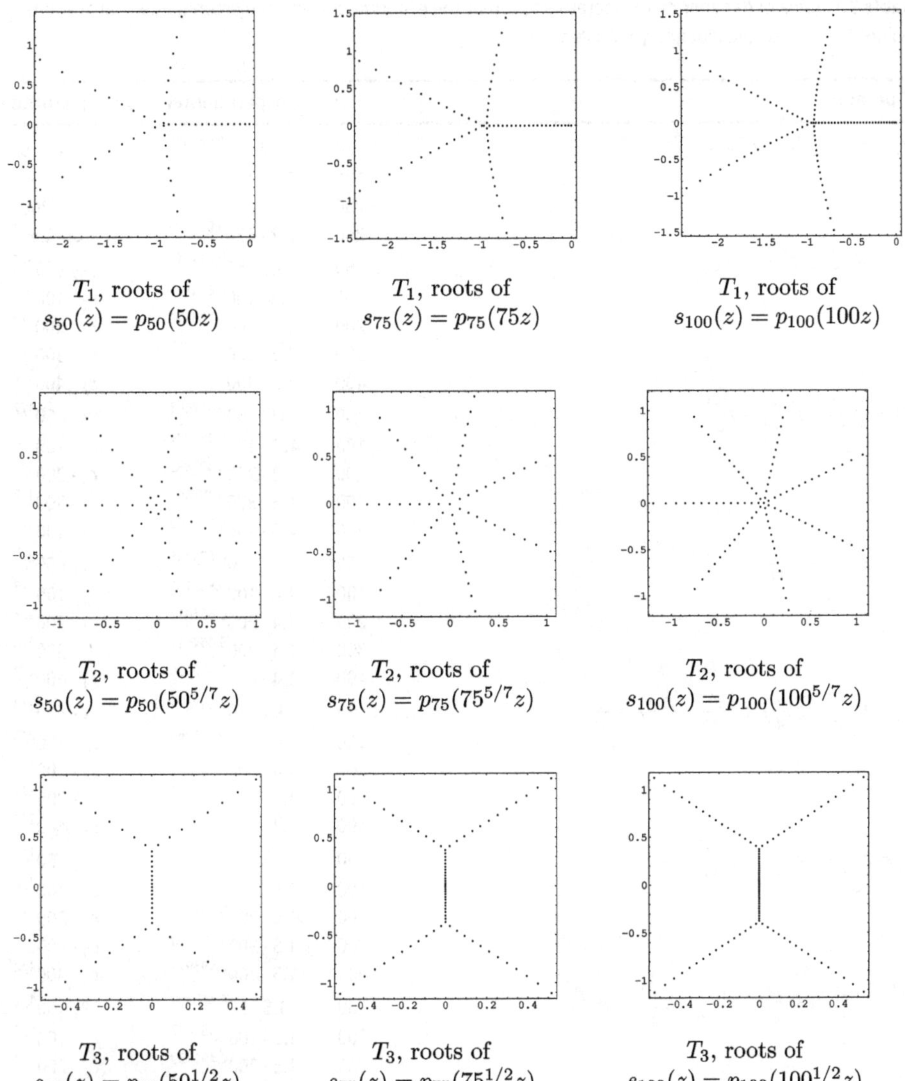

Figure 3.3: Roots of $s_n(z)$ for the operators $T_1 = z\frac{d}{dz} + z\frac{d^2}{dz^2} + z\frac{d^3}{dz^3} + z\frac{d^4}{dz^4} + z\frac{d^5}{dz^5}$, $T_2 = z^2\frac{d^2}{dz^2} + \frac{d^7}{dz^7}$ and $T_3 = z^3\frac{d^3}{dz^3} + z^2\frac{d^4}{dz^4} + z\frac{d^5}{dz^5}$.

have performed similar computer experiments for a large number of other degenerate ES-operators, and in all cases the results are consistent with the growth rate conjecture. Next we present some typical pictures of the zero distribution of the scaled eigenpolynomial $s_n(z) = p_n(n^d z)$ for some degenerate ES-operators in the notation of Table 3.1, where p_n denotes the unique monic eigenpolynomial of degree n for the given operator T. Conjecturally the zeros of all $s_n(z)$ are contained in a compact set.

Table 3.1: Table of degenerate ES-operators together the experimental and conjectured maximal absolute values for their scaled eigenpolynomials $s_n(z)$.

Operator	n	r_n experimental	r_n conjectured
$T_1 = z\frac{d}{dz} + z\frac{d^2}{dz^2} + z\frac{d^3}{dz^3} + z\frac{d^4}{dz^4} + z\frac{d^5}{dz^5}$	50	$2.7 \cdot 50^{0.967595}$	$c_1 \cdot 50^1$
	100	$2.7 \cdot 100^{0.984180}$	$c_1 \cdot 100^1$
	200	$2.7 \cdot 200^{0.992557}$	$c_1 \cdot 200^1$
	250	$2.7 \cdot 250^{0.994272}$	$c_1 \cdot 250^1$
$T_2 = z^2\frac{d^2}{dz^2} + \frac{d^7}{dz^7}$	50	$1.3 \cdot 50^{0.671977}$	$c_2 \cdot 50^{5/7}$
	100	$1.3 \cdot 100^{0.694847}$	$c_2 \cdot 100^{5/7}$
	200	$1.3 \cdot 200^{0.706226}$	$c_2 \cdot 200^{5/7}$
	300	$1.3 \cdot 300^{0.710085}$	$c_2 \cdot 300^{5/7}$
	400	$1.3 \cdot 400^{0.712043}$	$c_2 \cdot 400^{5/7}$
$T_3 = z^3\frac{d^3}{dz^3} + z^2\frac{d^4}{dz^4} + z\frac{d^5}{dz^5}$	50	$4/3 \cdot 50^{0.469007}$	$c_3 \cdot 50^{1/2}$
	100	$4/3 \cdot 100^{0.484824}$	$c_3 \cdot 100^{1/2}$
	200	$4/3 \cdot 200^{0.492832}$	$c_3 \cdot 200^{1/2}$
	300	$4/3 \cdot 300^{0.495592}$	$c_3 \cdot 300^{1/2}$
	400	$4/3 \cdot 400^{0.497009}$	$c_3 \cdot 400^{1/2}$
$T_4 = z^3\frac{d^3}{dz^3} + z^2\frac{d^5}{dz^5}$	50	$1.4 \cdot 50^{0.633226}$	$c_4 \cdot 50^{2/3}$
	100	$1.4 \cdot 100^{0.652141}$	$c_4 \cdot 100^{2/3}$
	200	$1.4 \cdot 200^{0.661412}$	$c_4 \cdot 200^{2/3}$
	300	$1.4 \cdot 300^{0.664511}$	$c_4 \cdot 300^{2/3}$
	400	$1.4 \cdot 400^{0.666066}$	$c_4 \cdot 400^{2/3}$
$\tilde{T}_4 = z^2\frac{d^2}{dz^2} + z^3\frac{d^3}{dz^3} + z\frac{d^4}{dz^4} + z^2\frac{d^5}{dz^5} + \frac{d^6}{dz^6}$	50	$1.4 \cdot 50^{0.632811}$	$\tilde{c}_4 \cdot 50^{2/3}$
	100	$1.4 \cdot 100^{0.651960}$	$\tilde{c}_4 \cdot 100^{2/3}$
	200	$1.4 \cdot 200^{0.661332}$	$\tilde{c}_4 \cdot 200^{2/3}$
	300	$1.4 \cdot 300^{0.664461}$	$\tilde{c}_4 \cdot 300^{2/3}$
	400	$1.4 \cdot 400^{0.666030}$	$\tilde{c}_4 \cdot 400^{2/3}$
$T_5 = z^5\frac{d^5}{dz^5} + z^4\frac{d^6}{dz^6} + z^2\frac{d^8}{dz^8}$	50	$1.5 \cdot 50^{0.462995}$	$c_5 \cdot 50^{1/2}$
	100	$1.5 \cdot 100^{0.481684}$	$c_5 \cdot 100^{1/2}$
	200	$1.5 \cdot 200^{0.491066}$	$c_5 \cdot 200^{1/2}$
	300	$1.5 \cdot 300^{0.494304}$	$c_5 \cdot 300^{1/2}$
	400	$1.5 \cdot 400^{0.495971}$	$c_5 \cdot 400^{1/2}$
$\tilde{T}_5 = z^2\frac{d^2}{dz^2} + z^5\frac{d^5}{dz^5} + z^4\frac{d^6}{dz^6} + z\frac{d^7}{dz^7} + z^2\frac{d^8}{dz^8}$	50	$1.5 \cdot 50^{0.463391}$	$\tilde{c}_5 \cdot 50^{1/2}$
	100	$1.5 \cdot 100^{0.481837}$	$\tilde{c}_5 \cdot 100^{1/2}$
	200	$1.5 \cdot 200^{0.491129}$	$\tilde{c}_5 \cdot 200^{1/2}$
	300	$1.5 \cdot 300^{0.494342}$	$\tilde{c}_5 \cdot 300^{1/2}$
	400	$1.5 \cdot 400^{0.495998}$	$\tilde{c}_5 \cdot 400^{1/2}$
$T_6 = z^3\frac{d^3}{dz^3} + z^2\frac{d^6}{dz^6}$	50	$1.4 \cdot 50^{0.702117}$	$c_6 \cdot 50^{3/4}$
	100	$1.4 \cdot 100^{0.725715}$	$c_6 \cdot 100^{3/4}$
	200	$1.4 \cdot 200^{0.737541}$	$c_6 \cdot 200^{3/4}$
	300	$1.4 \cdot 300^{0.741614}$	$c_6 \cdot 300^{3/4}$
	400	$1.4 \cdot 400^{0.743713}$	$c_6 \cdot 400^{3/4}$
$\tilde{T}_6 = [(1+13i) + (24i-3)z + 11iz^2 + z^3]\frac{d^3}{dz^3} + [(22i-13) -$	50	$1.4 \cdot 50^{0.769260}$	$\tilde{c}_6 \cdot 50^{3/4}$
$(9+14i)z + z^2]\frac{d^6}{dz^6}$	100	$1.4 \cdot 100^{0.760399}$	$\tilde{c}_6 \cdot 100^{3/4}$
	200	$1.4 \cdot 200^{0.756161}$	$\tilde{c}_6 \cdot 200^{3/4}$
	300	$1.4 \cdot 300^{0.754590}$	$\tilde{c}_6 \cdot 300^{3/4}$
	400	$1.4 \cdot 400^{0.753765}$	$\tilde{c}_6 \cdot 400^{3/4}$

3.5.2 Numerical evidence for the basic assumption

In Figure 3.4 we show some examples supporting the basic assumption, namely that when $n \to \infty$ the root-counting measures of $s_n(z)$ and its derivatives up to order k exist and are all weakly convergent to the same measure μ.

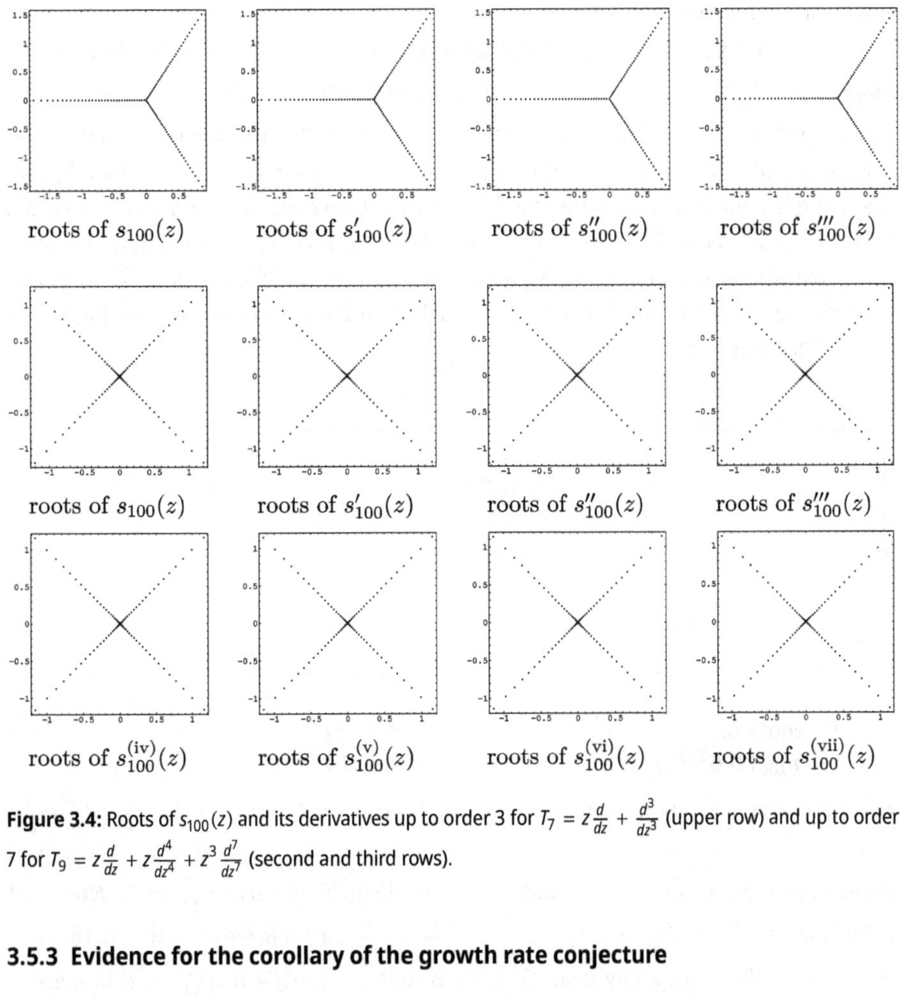

Figure 3.4: Roots of $s_{100}(z)$ and its derivatives up to order 3 for $T_7 = z\frac{d}{dz} + \frac{d^3}{dz^3}$ (upper row) and up to order 7 for $T_9 = z\frac{d}{dz} + z\frac{d^4}{dz^4} + z^3\frac{d^7}{dz^7}$ (second and third rows).

3.5.3 Evidence for the corollary of the growth rate conjecture

The algebraic equation in the corollary of growth rate conjecture satisfied by the Cauchy transform of the asymptotic root-counting measure of the scaled eigenpolynomial indicates that the asymptotic zero distribution depends only on the term $z^{j_0}\frac{d^{j_0}}{dz^{j_0}}$ and the term(s) $q_{j,\deg Q_j} z^{\deg Q_j}\frac{d^j}{dz^j}$ for which

$$d = \max_{j\in[j_0+1,k]}\left(\frac{j-j_0}{j-\deg Q_j}\right)$$

is attained.[5] Thus any term $Q_j \frac{d^j}{dz^j}$ in T for which $j < j_0$ or such that $(j - j_0)/(j - \deg Q_j) < d$ is (conjecturally) irrelevant for the asymptotic zero distribution when $n \to \infty$.

To illustrate this conjectural corollary, we present in Figures 3.4–3.8 pictures of the zero distributions of the scaled eigenpolynomials for several pairwise distinct operators for which the Cauchy transforms $\mathcal{C}$ of the corresponding scaled eigenpolynomials s_n satisfy the *same* equation when $n \to \infty$, namely the equation in the Corollary of the growth rate conjecture in § 3.4.

As a first example consider the operator $T_4 = z^3 \frac{d^3}{dz^3} + z^2 \frac{d^5}{dz^5}$. Here $d = \max_{j\in[j_0+1,k]}(\frac{j-j_0}{j-\deg Q_j}) = (5-3)/(5-2) = 2/3$, the corresponding scaled eigenpolynomial is $s_n(z) = p_n(n^{2/3}z)$, and we have $z^3\mathcal{C}^3 + z^2\mathcal{C}^5 = 1$ for the Cauchy transform of s_n when $n \to \infty$. Now consider the slightly modified operator $\widetilde{T}_4 = z^2 \frac{d^2}{dz^2} + z^3 \frac{d^3}{dz^3} + z \frac{d^4}{dz^4} + z^2 \frac{d^5}{dz^5} + \frac{d^6}{dz^6}$ and note that d is obtained again (only) for $j = 5$ (for $j = 4$ we have $(4-3)/(4-1) = 1/3 < 2/3$ and for $j = 6$ we have $(6-3)/(6-0) = 3/6 = 1/2 < 2/3$). We therefore obtain the same Cauchy transform equation as for T_4, and hence the terms $z^2 \frac{d^2}{dz^2}$, $z \frac{d^4}{dz^4}$ and $\frac{d^6}{dz^6}$ in $\widetilde{T}_4$ can be considered as irrelevant for the zero distribution for sufficiently large n. Figure 3.5 clearly illustrates this.

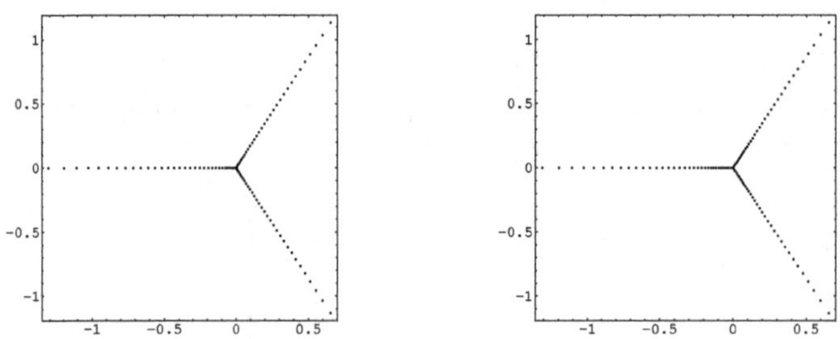

T_4, roots of $s_{100}(z) = p_{100}(100^{2/3}z)$

$\widetilde{T}_4$, roots of $s_{100}(z) = p_{100}(100^{2/3}z)$

Figure 3.5: Comparison of the roots of $s_{100}(z)$ for $T_4 = z^3 \frac{d^3}{dz^3} + z^2 \frac{d^5}{dz^5}$ and $\widetilde{T}_4 = z^2 \frac{d^2}{dz^2} + z^3 \frac{d^3}{dz^3} + z \frac{d^4}{dz^4} + z^2 \frac{d^5}{dz^5} + \frac{d^6}{dz^6}$.

However, instead of $\frac{d^6}{dz^6}$, we may add the more "disturbing" term $z\frac{d^6}{dz^6}$ to T_4. Note that for the operator $\overline{T}_4 = z^2 \frac{d^2}{dz^2} + z^3 \frac{d^3}{dz^3} + z \frac{d^4}{dz^4} + z^2 \frac{d^5}{dz^5} + z \frac{d^6}{dz^6}$ for $j = 6$, we have $(6-3)/(6-1) = 3/5 = 0.6 < 2/3$. Adding any term $Q_j \frac{d^j}{dz^j}$ such that $(j - j_0)/(j - \deg Q_j) < d$ to a given operator, it is clear that the closer the value of $(j - j_0)/(j - \deg Q_j)$ is to d (in this case 2/3), the more disturbing it is in the sense that it requires larger n for the corresponding zero distributions to coincide. See Figure 3.6 where $s_n(z) = p_n(n^{2/3}z)$.

5 Recall that we normalized T by letting Q_{j_0} be monic, i. e., $q_{j_0,j_0} = 1$.

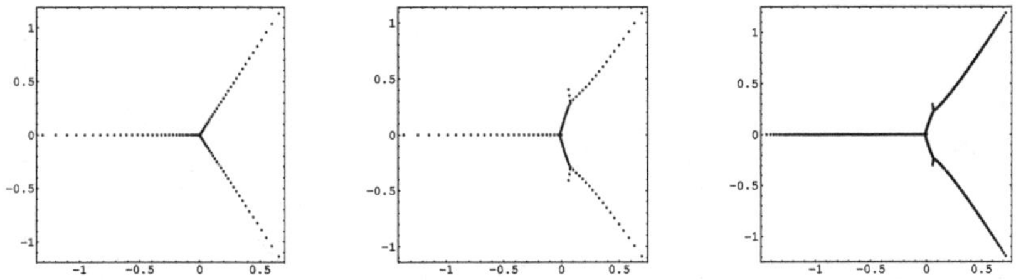

T_4, roots of $s_{100}(z)$ $\qquad$ $\overline{T}_4$, roots of $s_{100}(z)$ $\qquad$ $\overline{T}_4$, roots of $s_{600}(z)$

Figure 3.6: Comparison of the roots of $s_{100}(z)$ and $s_{600}(z)$ for $T_4 = z^3\frac{d^3}{dz^3} + z^2\frac{d^5}{dz^5}$ and $\overline{T}_4 = z^2\frac{d^2}{dz^2} + z^3\frac{d^3}{dz^3} + z\frac{d^4}{dz^4} + z^2\frac{d^5}{dz^5} + z\frac{d^6}{dz^6}$.

However, our experiments indicate that when n increases the zero distributions of the scaled eigenpolynomials of T_4 and $\overline{T}_4$ seem to coincide, as they (conjecturally) should, see the rightmost picture on Figure 3.6.

As a second example shown on Figure 3.7, consider $T_5 = z^5\frac{d^5}{dz^5} + z^4\frac{d^6}{dz^6} + z^2\frac{d^8}{dz^8}$ and $\widetilde{T}_5 = z^2\frac{d^2}{dz^2} + z^5\frac{d^5}{dz^5} + z^4\frac{d^6}{dz^6} + z\frac{d^7}{dz^7} + z^2\frac{d^8}{dz^8}$, whose scaled eigenpolynomials $s_n(z) = p_n(n^{1/2}z)$ both satisfy the Cauchy transform equation $z^5\mathcal{C}^5 + z^4\mathcal{C}^6 + z^2\mathcal{C}^8 = 1$ when $n \to \infty$. In the pictures below we see that the terms $z^2\frac{d^2}{dz^2}$ and $z\frac{d^7}{dz^7}$ of $\widetilde{T}_5$ seem to have no effect on the zero distribution for large n.

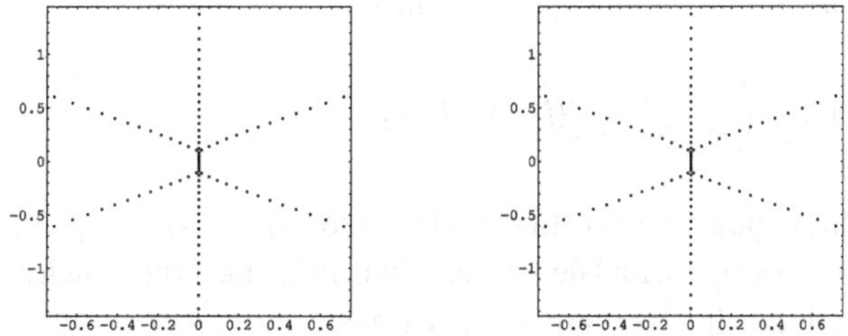

T_5, roots of $s_{100}(z)$ $\qquad$ $\widetilde{T}_5$, roots of $s_{100}(z)$

Figure 3.7: Comparison of the roots of $s_{100}(z)$ for T_5 and $\widetilde{T}_5$.

Finally note that for j_0 and for any j for which d is attained, it is only the highest degree term $q_{j,\deg Q_j}z^{\deg Q_j}$ of Q_j that is involved in the Cauchy transform equation. For example, consider the following case, where adding lower degree terms to $q_{j,\deg Q_j}z^{\deg Q_j}$ in the (relevant) Q_j seems to have no effect on the zero distribution for large n, see Figure 3.8 where $T_6 = z^3\frac{d^3}{dz^3} + z^2\frac{d^6}{dz^6}$, and $\widetilde{T}_6 = [(1+13i)+(24i-3)z+11iz^2+z^3]\frac{d^3}{dz^3} + [(22i-13) + (-9-14i)z + z^2]\frac{d^6}{dz^6}$, and thus $s_n(z) = p_n(n^{3/4}z)$. (Note the difference in scaling between the pictures.)

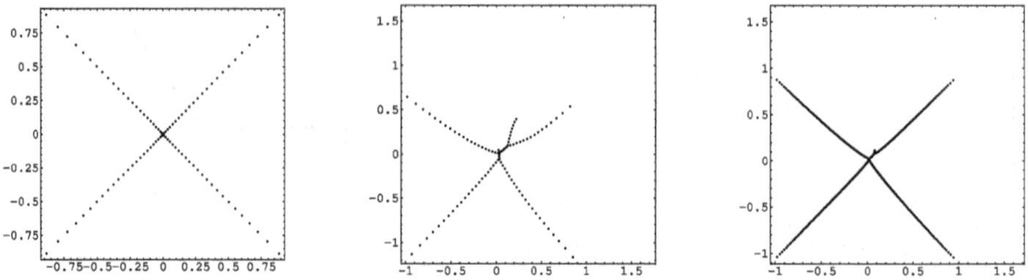

T_6, roots of $s_{100}(z)$ | $\widetilde{T}_6$, roots of $s_{100}(z)$ | $\widetilde{T}_6$, roots of $s_{500}(z)$

Figure 3.8: Comparison of roots for $s_n(z)$ for T_6 and $\widetilde{T}_6$.

3.6 More on the growth rate conjecture

3.6.1 Deducing the corollary of the growth rate conjecture

The algebraic equation in the corollary of the growth rate conjecture follows after inserting d as defined in this conjecture into equation (3.12) in § 3.4 and letting $n \to \infty$. Indeed, if we insert d into (3.12), we get

$$\sum_{j=1}^{k}\left(\sum_{i=0}^{\deg Q_j} q_{j,i}\frac{z^i}{n^{\max_{j\in[j_0+1,k]}\left(\frac{j-j_0}{j-\deg Q_j}\right)(j-i)+j_0-j}}\right)\mathcal{C}^j(z)=1. \tag{3.13}$$

Denote by $N_{j,i}$ the exponent of n in (3.13) for given j and i. Thus

$$N_{j,i}=\max_{j\in[j_0+1,k]}\left(\frac{j-j_0}{j-\deg Q_j}\right)(j-i)+j_0-j.$$

The terms in (3.13) for which this exponent is positive tend to zero as $n \to \infty$.

First we consider j for which $\deg Q_j = j$, and denote, as usual, by j_0 the largest such j. If $j = j_0$, then $i \le \deg Q_{j_0} = j_0$, and thus for $j = j_0$ and $i = j_0$, we get

$$\begin{aligned}N_{j_0,j_0}&=\max_{j\in[j_0+1,k]}\left(\frac{j-j_0}{j-\deg Q_j}\right)(j-i)+j_0-j\\&=\max_{j\in[j_0+1,k]}\left(\frac{j-j_0}{j-\deg Q_j}\right)(j_0-j_0)+j_0-j_0=0,\end{aligned}$$

and for $j = j_0$ and $i < j_0$, we obtain

$$\begin{aligned}N_{j_0,i}&=\max_{j\in[j_0+1,k]}\left(\frac{j-j_0}{j-\deg Q_j}\right)(j-i)+j_0-j\\&>\max_{j\in[j_0+1,k]}\left(\frac{j-j_0}{j-\deg Q_j}\right)(j_0-j_0)+j_0-j_0=0.\end{aligned}$$

Thus $N_{j_0,j_0} = 0$ and $N_{j_0,i} > 0$ for $i < j_0$, and for the term corresponding to $j = j_0$ in (3.13), we get

$$\sum_{i=0}^{j_0} q_{j_0,i} \frac{z^i}{n^{\max_{j\in[j_0+1,k]}(\frac{j-j_0}{j-\deg Q_j})(j_0-i)+j_0-j_0}} \mathcal{C}^{j_0}(z) \to z^{j_0}\mathcal{C}^{j_0}(z)$$

when $n \to \infty$ (recall that $q_{j_0,j_0} = 1$).

Now let j be such that $\deg Q_j = j$ and $j < j_0$. Then $i \le \deg Q_j = j$ and

$$\begin{aligned} N_{j,j} &= \max_{j\in[j_0+1,k]}\left(\frac{j-j_0}{j-\deg Q_j}\right)(j-i)+j_0-j \\ &= \max_{j\in[j_0+1,k]}\left(\frac{j-j_0}{j-\deg Q_j}\right)(j-j)+j_0-j = j_0-j > 0, \end{aligned}$$

and for $i < j$, we get

$$\begin{aligned} N_{j,i} &= \max_{j\in[j_0+1,k]}\left(\frac{j-j_0}{j-\deg Q_j}\right)(j-i)+j_0-j \\ &> \max_{j\in[j_0+1,k]}\left(\frac{j-j_0}{j-\deg Q_j}\right)(j-j)+j_0-j = j_0-j > 0, \end{aligned}$$

that is, $N_{j,i} > 0$ for all $j < j_0$ such that $\deg Q_j = j$ and for all $i \le j$. Thus, for the corresponding terms in (3.13), we get

$$\sum_{j\in\{j<j_0:\deg Q_j=j\}} \sum_{i=0}^{\deg Q_j} a_{j,i} \frac{z^i}{n^{\max_{j\in[j_0+1,j]}(\frac{j-j_0}{j-\deg Q_j})(j-i)+j_0-j}} \mathcal{C}^j(z) \to 0$$

when $n \to \infty$ for every $j < j_0$ for which $\deg Q_j = j$.

Now denote by j_m the value of j for which $d = \max_{j\in[j_0+1,k]}(\frac{j-j_0}{j-\deg Q_j})$ is attained. Note that there may be several distinct j for which this maximum is attained.[6] Then

$$\begin{aligned} N_{j_m,\deg Q_{j_m}} &= \max_{j\in[j_0+1,k]}\left(\frac{j-j_0}{j-\deg Q_j}\right)(j-i)+j_0-j \\ &= \left(\frac{j_m-j_0}{j_m-\deg Q_{j_m}}\right)(j_m-\deg Q_{j_m})+j_0-j_m \\ &= j_m-j_0+j_0-j_m = 0, \end{aligned}$$

6 Consider for example the Laplace-type operator (that is with all linear coefficients Q_j) $T = z\frac{d}{dz} + z\frac{d^2}{dz^2} + \cdots + z\frac{d^k}{dz^k}$. Here $j_0 = 1$ and the equation satisfied by the Cauchy transform of the asymptotic root-counting measure of the scaled eigenpolynomial $s_n(z) = p_n(nz)$ is given by $z\mathcal{C}(z) + z\mathcal{C}^2(z) + \cdots + z\mathcal{C}^k(z) = 1$ since $d = \max_{j\in[2,k]}(\frac{j-j_0}{j-\deg Q_j}) = 1$ is attained for *every* $j = 2, 3, \ldots, k$.

and for $i < \deg Q_{j_m}$, we get

$$\begin{aligned} N_{j_m,i} &= \max_{j\in[j_0+1,k]}\left(\frac{j-j_0}{j-\deg Q_j}\right)(j-i)+j_0-j \\ &> \left(\frac{j_m-j_0}{j_m-\deg Q_{j_m}}\right)(j_m-\deg Q_{j_m})+j_0-j_m \\ &= j_m-j_0+j_0-j_m=0, \end{aligned}$$

i. e., $N_{j_m,\deg Q_{j_m}} = 0$ and $N_{j_m,i} > 0$ for $i < \deg Q_{j_m}$, and for the term corresponding to $j = j_m$ in (3.13) we get

$$\sum_{i=0}^{\deg Q_{jm}} q_{j_m,i}\frac{z^i}{n^{\max_{j\in[j_0+1,k]}(\frac{j-j_0}{j-\deg Q_j})(j_m-i)+j_0-j_m}}\mathcal{C}^{j_m}(z) \to q_{j_m,\deg Q_{j_m}} z^{\deg Q_{jm}}\mathcal{C}^{j_m}(z)$$

when $n \to \infty$. In case of several j for which d is attained, we put $A = \{j : (j-j_0)/(j-\deg Q_j) = d\}$, and for the corresponding terms in (3.13) we get

$$\sum_{j\in A}\sum_{i=0}^{\deg Q_j} q_{j,i}\frac{z^i}{n^{\max_{j\in[j_0+1,k]}(\frac{j-j_0}{j-\deg Q_j})(j-i)+j_0-j}}\mathcal{C}^j(z) \to \sum_{j\in A} q_{j,\deg Q_j} z^{\deg Q_j}\mathcal{C}^j(z)$$

when $n \to \infty$. Now consider the remaining terms in (3.13), namely terms for which $j < j_0$ such that $\deg Q_j < j$, terms for which $j_0 < j < j_m$, and terms for which $j_m < j \le k$ (clearly this last case does not exist if $j_m = k$).

We start with $j < j_0$ such that $\deg Q_j < j$. Then $i \le \deg Q_j < j$ and

$$\begin{aligned} N_{j,i} &= \max_{j\in[j_0+1,k]}\left(\frac{j-j_0}{j-\deg Q_j}\right)(j-i)+j_0-j \\ &> \max_{j\in[j_0+1,k]}\left(\frac{j-j_0}{j-\deg Q_j}\right)(j-j)+j_0-j=j_0-j>0, \end{aligned}$$

and for the corresponding terms in (3.13), we have

$$\sum_{j\in\{j<j_0:\deg Q_j<j\}}\sum_{i=0}^{\deg Q_j} q_{j,i}\frac{z^i}{n^{\max_{j\in[j_0+1,k]}(\frac{j-j_0}{j-\deg Q_j})(j-i)+j_0-j}}\mathcal{C}^j(z) \to 0$$

when $n \to \infty$.

Now assume that $j_m < k$ and consider $j_m < j \le k$. Clearly, $j_m > j_0$ since the maximum is taken over $j \in [j_0+1,k]$, and therefore $i \le \deg Q_j < j$ for $j_m < j \le k$. Also,

$$\max_{j\in[j_0+1,k]}\left(\frac{j-j_0}{j-\deg Q_j}\right) = \left(\frac{j_m-j_0}{j_m-\deg Q_{j_m}}\right) > \left(\frac{j-j_0}{j-\deg Q_j}\right),$$

since the maximum is attained for j_m by assumption. Thus we get

$$\begin{aligned} N_{j,i} &= \max_{j \in [j_0+1,k]} \left(\frac{j-j_0}{j-\deg Q_j} \right)(j-i) + j_0 - j \\ &= \left(\frac{j_m - j_0}{j_m - \deg Q_{j_m}} \right)(j-i) + j_0 - j \\ &> \left(\frac{j-j_0}{j-\deg Q_j} \right)(j-i) + j_0 - j \\ &\geq \left(\frac{j-j_0}{j-\deg Q_j} \right)(j-\deg Q_j) + j_0 - j \\ &= j - j_0 + j_0 - j = 0, \end{aligned}$$

i. e., $N_{j,i} > 0$ for every $j_m < j \leq k$ and every $i \leq \deg Q_j$. For the corresponding terms in (3.13), we therefore get

$$\sum_{j_m < j \leq k} \sum_{i=0}^{\deg Q_j} q_{j,i} \frac{z^i}{n^{\max_{j \in [j_0+1,k]} (\frac{j-j_0}{j-\deg Q_j})(j-i)+j_0-j}} \mathcal{C}^j(z) \to 0$$

as $n \to \infty$.

Finally, we consider $j_0 < j < j_m$. Note that this also covers the case $j_{m_1} < j < j_{m_2}$ where the maximum d is attained for both j_{m_1} and j_{m_2}. Since $i \leq \deg Q_j < j$, we get

$$\begin{aligned} N_{j,i} &= \max_{j \in [j_0+1,k]} \left(\frac{j-j_0}{j-\deg Q_j} \right)(j-i) + j_0 - j \\ &= \left(\frac{j_m - j_0}{j_m - \deg Q_{j_m}} \right)(j-i) + j_0 - j \\ &> \left(\frac{j-j_0}{j-\deg Q_j} \right)(j-i) + j_0 - j \\ &\geq \left(\frac{j-j_0}{j-\deg Q_j} \right)(j-\deg Q_j) + j_0 - j \\ &= j - j_0 + j_0 - j = 0, \end{aligned}$$

i. e., $N_{j,i} > 0$ for every $j_0 < j < j_m$ and every $i \leq \deg Q_j$. Thus for the corresponding terms in (3.13) we get

$$\sum_{j_0 < j < j_m} \sum_{i=0}^{\deg Q_j} q_{j,i} \frac{z^i}{n^{\max_{j \in [j_= +1,k]0} (\frac{j-j_0}{j-\deg Q_j})(j-i)+j_0-j}} \mathcal{C}^j(z) \to 0$$

when $n \to \infty$.

Adding up these results, we finally get the following algebraic equation:

$$z^{j_0} \mathcal{C}^{j_0}(z) + \sum_{j \in A} q_{j,\deg Q_j} z^{\deg Q_j} \mathcal{C}^j(z) = 1,$$

by letting $n \to \infty$ in equation (3.13). Here j_0 is the largest value of j for which $\deg Q_j = j$, and A is the set consisting of all j for which the maximum $d = \max_{j\in[j_0+1,k]}(\frac{j-j_0}{j-\deg Q_j})$ is attained, i. e., $A = \{j : (j-j_0)/(j-\deg Q_j) = d\}$.

3.6.2 Upper bound implies lower bound

Here we prove that for a class of operators containing the operators considered in Corollaries 3.1 and 3.2, the conjectured upper bound $\lim_{n\to\infty}\sup(r_n/n^d) \le c_1$ implies the conjectured lower bound $\lim_{n\to\infty}\inf(r_n/n^d) \ge c_0$ for some constants $c_1 \ge c_0 > 0$ and where d is given by the growth rate conjecture. This follows automatically from inequality (3.4) in Lemma 3.2.

Theorem 3.5. *Let T be a degenerate ES-operator of order k that satisfies the condition*

$$b = \min^{+}_{j\in[1,k-1]}\left(\frac{k-j}{k-j+\deg Q_j - \deg Q_k}\right) = \max_{j\in[j_0+1,k]}\left(\frac{j-j_0}{j-\deg Q_j}\right) = d$$

where the notation $\min^+$ means that the minimum is taken only over positive values of $(k-j+\deg Q_j - \deg Q_k)$. Assume that for all sufficiently large n, the inequality $r_n \le c_1(n-k+1)^d$ holds for some positive constant c_1. Then there exists a positive constant $c_0 \le c_1$ such that $r_n \ge c_0(n-k+1)^d$ holds for all sufficiently large n. Thus

$$\lim_{n\to\infty}\sup\frac{r_n}{n^d} \le c_1 \quad\Rightarrow\quad \lim_{n\to\infty}\inf\frac{r_n}{n^d} \ge c_0.$$

Proof. From inequality (3.4) in Lemma 3.2 we have

$$\begin{aligned}1 &\le \sum_{j=1}^{k-1}\sum_{i=0}^{\deg Q_j}|q_{j,i}|2^{k-j}\frac{r_n^{k-j+i-\deg Q_k}}{(n-k+1)^{k-j}} + \sum_{0\le i<\deg Q_k}\frac{|q_{k,i}|}{r_n^{\deg Q_k - i}}\\ &\le \sum_{j=1}^{k-1}K_j\frac{r_n^{k-j+\deg Q_j-\deg Q_k}}{(n-k+1)^{k-j}} + \sum_{0\le i<i_k}\frac{|q_{k,i}|}{r_n^{i_k-i}}\end{aligned} \tag{3.14}$$

where the K_j are positive constants. The second sum on the right-hand side of (3.14) tends to zero when $n \to \infty$ due to Theorem 3.1. We now decompose the first sum on the right-hand side of (3.14) into three parts. Namely, set

$$A = \left\{j : \frac{k-j}{k-j+\deg Q_j - \deg Q_k} = d\right\},$$

and note that $(k-j+\deg Q_j - \deg Q_k) > 0$ here since $d > 0$.

$$B = \left\{j : \frac{k-j}{k-j+\deg Q_j - \deg Q_k} > d\right\},$$

and note that $(k-j+\deg Q_j - \deg Q_k) > 0$ here since $d > 0$.

$C = \{j : (k-j+\deg Q_j - \deg Q_k) \le 0\}$, and note that $j < k$ in (3.14).

Clearly, due to the condition $b = d$, there are no terms for which both $\frac{k-j}{k-j+\deg Q_j-\deg Q_k} < d$ and $(k - j + \deg Q_j - \deg Q_k) > 0$ hold.

If $j \in A$ then

$$\frac{r_n^{k-j+\deg Q_k-\deg Q_k}}{(n-k+1)^{k-j}} = \left(\frac{r_n}{(n-k+1)^d}\right)^{k-j+\deg Q_j-\deg Q_k}$$

for the corresponding terms in the sum on the right-hand side of (3.14).

If $j \in B$ then $d(k - j + \deg Q_j - \deg Q_k) < (k - j)$, and this inequality together with the upper bound $r_n \le c_1(n - k + 1)^d$, which we assume holds for all sufficiently large n, gives us

$$\frac{r_n^{k-j+\deg Q_j-\deg Q_k}}{(n-k+1)^{k-j}} \le \frac{c_1(n-k+1)^{d(k-j+\deg Q_j-\deg Q_k)}}{(n-k+1)^{k-j}} \to 0$$

when $n \to \infty$ for the corresponding terms in (3.14).

If $j \in C$ then $(k - j + \deg Q_j - \deg Q_k) \le 0$, and we get

$$\frac{r_n^{k-j+\deg Q_j-\deg Q_k}}{(n-k+1)^{k-j}} \to 0$$

when $n \to \infty$ for the corresponding terms in (3.14) due to Theorem 3.1. Note that if $(k - j + \deg Q_j - \deg Q_k) = 0$ the corresponding term tends to zero when $n \to \infty$ since $j < k$ in (3.14).

With this decomposition of the first sum in the right-hand side of the last inequality in (3.14), we can write

$$\begin{aligned}
1 &\le \sum_{j=1}^{k-1} K_j \frac{r_n^{k-j+\deg Q_j-\deg Q_k}}{(n-k+1)^{k-j}} + \sum_{0\le i<i_k} \frac{|q_{k,i}|}{r_n^{i_k-i}} \\
&\le \sum_{j\in A} K_j \left(\frac{r_n}{(n-k+1)^d}\right)^{k-j+\deg Q_j-\deg Q_k} \\
&\quad + \sum_{j\in B} K_j \frac{c_1(n-k+1)^{d(k-j+\deg Q_j-\deg Q_k)}}{(n-k+1)^{k-j}} \\
&\quad + \sum_{j\in C} K_j \frac{r_n^{k-j+\deg Q_j-\deg Q_k}}{(n-k+1)^{k-j}} + \sum_{0\le i<i_k} \frac{|q_{k,i}|}{r_n^{i_k-i}}
\end{aligned}$$

where the last three sums tend to zero when $n \to \infty$ by the above arguments (and the last one due to Theorem 3.1).

Thus, for all sufficiently large n, there exists a positive constant c' such that

$$c' \le \sum_{j\in A} K_j \left(\frac{r_n}{(n-k+1)^d}\right)^{k-j+\deg Q_j-\deg Q_k} \tag{3.15}$$

where $c' \to 1$ when $n \to \infty$. If the set A contains precisely one element, then the sum in (3.15) consists of one single term, and we are done. In other words, there exists a positive constant $c_0 = (c'/K_j)^{1/(k-j+\deg Q_j-\deg Q_k)}$ such that $r_n \geq c_0(n-k+1)^d$ for all sufficiently large n, and thus $\lim_{n\to\infty}\inf(r_n/n^d) \geq c_0$.

But clearly, for some operators T, the set A will contain more than one element. If this is the case, define $m := \min_{j\in A}(k-j+\deg Q_j - \deg Q_k)$ and denote by j_m the corresponding j for which this minimum is attained. Using the upper bound $r_n \leq c_1(n-k+1)^d$, we then get the following inequality[7] from (3.15):

$$\begin{aligned}
c' &\leq \sum_{j\in A} K_j\left(\frac{r_n}{(n-k+1)^d}\right)^{k-j+\deg Q_j-\deg Q_k} = K_{j_m}\left(\frac{r_n}{(n-k+1)^d}\right)^m \\
&\quad + \sum_{j\in A\setminus\{j_m\}} K_j\left(\frac{r_n}{(n-k+1)^d}\right)^m\left(\frac{r_n}{(n-k+1)^d}\right)^{k-j+\deg Q_j-\deg Q_k-m} \\
&\leq K_{j_m}\left(\frac{r_n}{(n-k+1)^d}\right)^m + \sum_{j\in A\setminus\{j_m\}} K_j\left(\frac{r_n}{(n-k+1)^d}\right)^m c_1^{k-j+\deg Q_j-\deg Q_k-m} \\
&= \left(\frac{r_n}{(n-k+1)^d}\right)^m\left(K_{j_m} + \sum_{j\in A\setminus\{j_m\}} K_j\cdot c_1^{k-j+\deg Q_j-\deg Q_k-m}\right) = \left(\frac{r_n}{(n-k+1)^d}\right)^m\cdot K
\end{aligned}$$

where $K > 0$. Thus $r_n \geq (\frac{c'}{K})^{1/m}(n-k+1)^d$ for all sufficiently large n. Therefore there exists a positive constant $c_0 = (1/K)^{1/m}$ such that $\lim_{n\to\infty}\inf\frac{r_n}{n^d} \geq c_0$. (Recall that $c' \to 1$ when $n \to \infty$.) □

---

7 Consider for example the operator $T = z\frac{d}{dz} + \frac{d^2}{dz^2} + z\frac{d^3}{dz^3} + z\frac{d^4}{dz^4}$. Here, by Lemma 3.2,

$$1 \leq \sum_{j=1}^{3}\frac{2^{4-j}r_n^{3-j+\deg Q_j}}{(n-3)^{4-j}} = 8\frac{r_n^3}{(n-3)^3} + 4\frac{r_n}{(n-3)^2} + 2\frac{r_n}{(n-3)},$$

where r_n is the largest modulus of all roots of the unique and monic eigenpolynomial of T. For this operator, $d = 1$, and we see that $(4-j)/(3-j+\deg Q_j) = d$ for $j = 1$ and for $j = 3$. Now, assuming that $r_n \leq c_1(n-3)$ holds for some positive constant c_1 for large n, our inequality becomes

$$\begin{aligned}
1 &\leq 8\frac{r_n^3}{(n-3)^3} + 4\frac{r_n}{(n-3)^2} + 2\frac{r_n}{(n-3)} \\
&\leq 8\frac{r_n}{(n-3)}\cdot\frac{c_1^2(n-3)^2}{(n-3)^2} + 4\frac{c_1(n-3)}{(n-3)^2} + 2\frac{r_n}{(n-3)} \\
&= \left(8c_1^2+2\right)\frac{r_n}{(n-3)} + \frac{4c_1}{(n-3)},
\end{aligned}$$

where the last term tends to zero as $n \to \infty$. Thus $r_n \geq c_0(n-3)$ for sufficiently large choices on n, where $c_0 = 1/(8c_1^2+2)$, and hence $\lim_{n\to\infty}\inf(r_n/n) \geq c_0$.

3.7 Open problems in Chapter 3

3.1. The main challenge is to obtain a complete proof of the growth rate conjecture, see § 3.1. This proof requires both the sharp upper and lower bounds of the largest root. The upper bound can apparently be obtained by a detailed study of the corresponding Riccati equation at ∞. If the growth rate conjecture is settled then to achieve its corollary (the Cauchy transform equation), one can use a technique similar to that of [27] to prove the basic assumption, see § 3.4.

3.2. Estimates similar to that of the growth rate conjecture can be formulated for the sequence of roots $z_{n,i}$ of p_n such that $\lim_{n\to\infty} \frac{r_{n,i}}{r_{n,n}} = \alpha$ where $0 < \alpha < 1$ and $|z_{n,i}| = r_{n,i}$.

3.3. In the case of orthogonal polynomials there is a number of results describing the growth of the largest modulus r_n of the roots as an expansion in powers of n, see, e. g., [108] and [109] and the references therein. In this chapter we conjectured the form of the leading term of r_n in our more general setup. However, the question about the lower terms in the expansion of r_n is natural in our context as well.

3.4. Operators of the type we consider occur in the theory of Bochner–Krall orthogonal systems, i. e., families of polynomials which are both eigenfunctions of some finite order differential operator and orthogonal with respect to some suitable inner product. A lot is known about the asymptotic zero distribution of orthogonal polynomials, and by comparing such known results with results on the asymptotic zero distribution of eigenpolynomials of degenerate exactly solvable operators, we believe it will be possible to gain new insight into the nature of the Bochner–Krall problem. We have previously used our results from Chapter 2 and [27] to prove a special case of a general conjecture describing the leading terms of all Bochner–Krall operators, see Chapter 5 and [28]. Another problem relevant for the Bochner–Krall problem is to describe all ES-operators whose eigenpolynomials have only real roots.

3.5. Numerical evidence indicates that the roots of the scaled eigenpolynomials fill certain curves in the complex plane. The support of the limiting root measure μ seems to be a tree. This is the case for the non-degenerate exactly solvable operators, which was treated in Chapter 2 and [27], but then without such a scaling of the eigenpolynomials. By a tree we mean a connected compact subset Γ of $\mathbb{C}$ that consists of a finite union of analytic curves and where $\hat{\mathbb{C}} \setminus \Gamma$ is simply connected. The (conjectural) algebraic equation satisfied by the Cauchy transform contains a lot of information about μ, and it remains to describe its support explicitly.

3.6. Conjecturally the support of the asymptotic zero distribution of the scaled eigenpolynomial s_n is the union of a finite number of analytic curves in the complex plane, which we denote by $\Xi_T = supp(\mu)$, where μ is the limiting root-counting measure of s_n. Then the following conjecture seems to be quite plausible.

Conjecture 3.2 (Interlacing property). *For any family $\{s_n\}$ of appropriately scaled eigenpolynomials of a degenerate ES-operator, the zeros of any two consecutive polynomials s_{n+1} and s_n interlace along supp(μ) for all sufficiently large integers n.*

When defining the interlacing property some caution is required since the zeros of s_n do not lie exactly on Ξ_T. Thus identify some sufficiently small neighborhood $N(\Xi_T)$ of Ξ_T with the normal bundle to Ξ_T by equipping $N(\Xi_T)$ with the projection onto Ξ_T along the fibers, which are small curvilinear segments orthogonal to Ξ_T. We then say that two sets of points in $N(\Xi_T)$ interlace if their orthogonal projections on Ξ_T interlace in the usual sense. If Ξ_T has singularities, one should first remove some sufficiently small neighbourhoods of these singularities and then proceed as above on the remaining part of Ξ_T. Conjecture 3.2 thus states that for any sufficiently small neighborhood $N(\Xi_T)$ of Ξ_T there exists a number n_0 such that the interlacing property holds for the zeros of s_n and s_{n+1} for all $n \geq n_0$. We conclude this section by showing some pictures illustrating the interlacing property, see Figures 3.9 and 3.10. Small dots represent the roots of s_{n+1}, and large dots represent the roots of s_n for some fixed n.

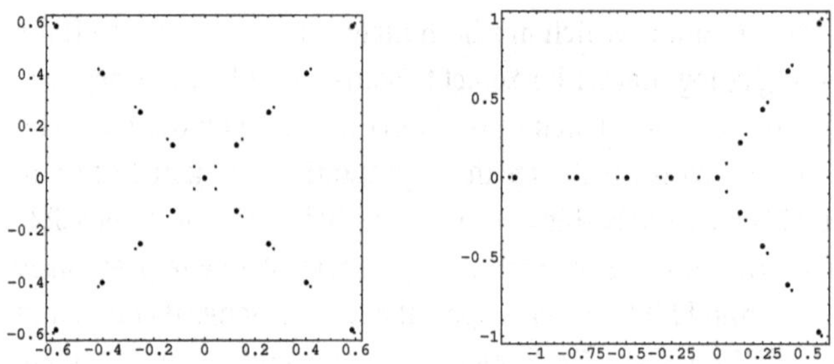

$T = z^2\frac{d^2}{dz^2} + z^3\frac{d^3}{dz^3} + z\frac{d^5}{dz^5}$, roots of s_{25} and s_{24}.

$T = z\frac{d}{dz} + z^2\frac{d^2}{dz^2} + \frac{d^3}{dz^3}$, roots of s_{20} and s_{19}.

Figure 3.9: Examples of root interlacing for $s_n(z)$ and $s_{n+1}(z)$.

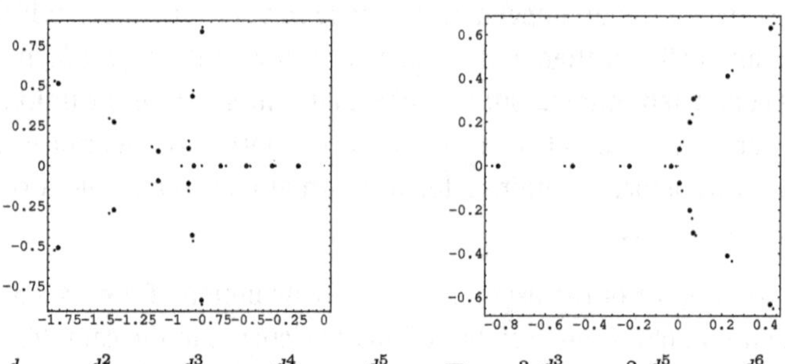

$T = z\frac{d}{dz} + z\frac{d^2}{dz^2} + z\frac{d^3}{dz^3} + z\frac{d^4}{dz^4} + z\frac{d^5}{dz^5}$, roots of s_{23} and s_{22}.

$T = z^3\frac{d^3}{dz^3} + z^2\frac{d^5}{dz^5} + z\frac{d^6}{dz^6}$, roots of s_{20} and s_{19}.

Figure 3.10: More examples of root interlacing for $s_n(z)$ and $s_{n+1}(z)$.

4 WKB-method for eigenpolynomials of ES-operators

(The material of this chapter is based on [40].)

4.1 Preliminaries

In this chapter we prove Property 2.8 of non-degenerate ES-operators formulated in Chapter 2. As above, for a given integer $k \geq 2$, consider an ES-differential operator $T = \sum_{j=1}^{k} Q_j(z) \frac{d^j}{dz^j}$ of order k where $Q_j(z) = \sum_{i=0}^{j} q_{j,i} z^i$ are polynomials in z with $\deg Q_j \leq j$ for $j = 1, 2, \ldots, k$ and at least one of them, say Q_{j^0}, is exactly of degree j^0, Q_k being a monic polynomial. Exactly solvable operators split in two families: non-degenerate, if the leading term of the operator satisfies $\deg Q_k = k$, and degenerate, if $\deg Q_k < k$.

As above, for a polynomial P of degree n, denote by $\mu_P = \frac{1}{n} \sum_{P(z_{k,n})=0} \delta_{z_{k,n}}$ the root-counting measure of the zeros of P_n. For a given sequence $(P_n)_{n=0}^{\infty}$ of polynomials, if the limit $\mu = \lim_{n\to\infty} \mu_{P_n}$ exists (in the sense of the weak convergence), then μ is called the asymptotic zero-counting measure of the sequence.

It is known that for the above operator T, see (1.2) and n large enough, there exists a unique eigenpolynomial p_n of degree n, see Chap. 1, [171, Th. 1], [27, Th. 1], and its eigenvalue is given by

$$\lambda_n = \sum_{1 \leq j \leq k} q_{j,j} (n)_j, \tag{4.1}$$

where $(n)_0 = 1$, $(n)_j = n(n-1) \cdots (n-j+1)$, $k \in \mathbb{N}$. Moreover, the zero counting measures of the eigenpolynomials converge weakly to a measure μ whose support, denoted through this chapter as $supp(\mu)$, is contained in the convex hull of the zeros of Q_k. This support is connected, and its complement is also connected, cf. Chapter 2 and [27, Ths. 3, 4].

The study of asymptotic properties of sequences of polynomials is an important topic in the theory of orthogonal polynomials and approximation theory. The objective of this chapter is the proof of an asymptotic formula for the monic polynomial eigenfunctions p_n of T in compact subsets of $\mathbb{C} \setminus supp(\mu)$ in the sense of Poincaré. The main tool is the WKB-approximation, named after the physicists Wentzel [196], Kramers [107], and Brillouin [45], who efficiently used this technique in their study of asymptotic solutions of linear second order differential equations in quantum physics. This method was known even earlier to Liouville [119], Green [75], and other authors, see [60, Chapter XIII], [195, Chapter 1], and [145, p. 228] for some historical remarks.

In general, the infinite series that define WKB-solutions do not converge in the usual sense, see [145, p. 365] for an example. The use of Borel resummed WKB-solutions to overcome this situation was first suggested for second order equations in [24], developed later in [194], and further in a series of publications [9, 10, 34, 55, 99, 140]. For higher order linear differential equations this method has also been studied considerably, see

https://doi.org/10.1515/9783112207741-005

[7, 68, 100, 136, 183] and the references within. The basic idea behind is the analysis of the Stokes geometry, which describes regions in the complex plane (sometimes called Stokes regions) where WKB-solutions are Borel summable and the connection formulas which describe relations between the Borel sums of WKB-solutions in different Stokes regions. In [136] the authors study the structure of the Stokes geometry of the WKB-solutions of a linear kth order differential equation of the form

$$T\phi(z,\eta) = \sum_{j=0}^{k} a_j(z)\left(\eta^{-1}\frac{\partial}{\partial z}\right)^j \phi(z,\eta) = 0,$$

where a_j are polynomials, a_k is a non-zero complex constant, and $\eta > 0$ is a large real parameter. By reducing higher order linear ordinary differential equations to order 2 via the middle convolution operation introduced in [97] and developed in [58, 146], the authors obtain some information about the Stokes geometry of the WKB-solutions of T. For higher order equations (i. e., $k \geq 3$), the complete structure of the Stokes geometry is not yet fully understood. In this chapter we will not use the approach of the Borel resummed solutions. To obtain our asymptotic expansion, we instead apply a result due to Sibuya on the existence of asymptotic solutions of a perturbed system of first order linear differential equations in the sense of Poincaré, see [90, Th. XII-1-2, p. 374] and [172].

4.1.1 Notation and statement of the results

Denote by $\Delta(z', r)$ the open disk $|z - z'| < r$, and define also the open sector $\mathcal{G}(r, \alpha) = \{\epsilon \in \mathbb{C} : |\arg[\epsilon]| < \alpha, 0 < |\epsilon| < r\}$. Given an open set $U \subset \mathbb{C}$, we denote by $\mathcal{H}(U)$ the space of analytic functions in U. Denote also the topological operators $\mathring{U}$, ∂U, $\overline{U}$, and U^c as the interior, boundary, closure, and complement of U in $\mathbb{C}$ respectively. If γ is a closed Jordan curve, we denote by $int(\gamma)$ and $ext(\gamma)$ the bounded and unbounded components respectively. Let τ be any oriented Jordan arc, then τ^+ will denote the side on the left of τ.

Consider a non-degenerate ES-operator $T = \sum_{j=1}^{k} Q_j(z)\frac{d^j}{dz^j}$ and let (as above) μ denote the asymptotic root-counting measure of the sequence $\{p_n(z)\}$ of its eigenpolynomials. Define $\Omega = \mathbb{C} \setminus supp(\mu)$. For a given Jordan arc τ connecting ∞ and $\omega \in supp(\mu)$, we will denote by Ω_τ the open connected set $\Omega \setminus \tau$. Denote by w_1 the branch of $(\sqrt[k]{Q_k(z)})^{-1}$ in Ω which tends to $\frac{1}{z}$ near ∞. As we stated in Chapter 2, the set $supp(\mu)$ is an embedded tree in $\mathbb{C}$, see also [27, Th. 3]. Thus w_1 is an analytic function outside the $supp(\mu)$. Define also

$$w_\ell(z) = e^{\frac{2(j-1)\pi\imath}{k}} w_1(z), \quad z \in \Omega,\ \ell = 2, \ldots, k. \tag{4.2}$$

Let τ be any Jordan arc from $-\infty$ to $\omega \in supp(\mu)$ such that $]-\infty, p] \subset \tau$ for some $p \in \mathbb{R}$. Let Φ_0 be the primitive of w_1 in Ω_τ such that $\lim_{z\to\infty} \Phi_0(z) - \ln z = 0$, and let Φ_1 be the primitive in Ω_τ of the function

$$\frac{(k-1)Q_k'(z)}{2kQ_k(z)} - \frac{Q_{k-1}(z)}{kQ_k(z)},$$

such that $\lim_{z\to\infty} \Phi_1(z) - (\frac{k-1}{2} - \frac{Q_{k-1,k-1}}{k}) \ln z = 0$. The main result of this chapter is as follows.

Theorem 4.1. *For $k \geq 2$, let T be a non-degenerate ES-operator of order k and p_n be the nth monic eigenpolynomial of T, then as $n \to \infty$*

$$p_n(z) \sim e^{((n-\frac{k-1}{2}+\frac{Q_{k-1,k-1}}{k})\Phi_0(z)+\Phi_1(z))}\left(1+\frac{C_1(z)}{n}+\frac{C_2(z)}{n^2}+\cdots\right),$$

uniformly in compacts subsets $K \subset \Omega$. Here $C_j \in \mathcal{H}(\Omega)$, $j \geq 1$ and the symbol "~" has the same meaning as in [145, (7.03), p. 16] for an asymptotic expansion in the sense of Poincaré.

4.1.2 Overview of the proof

Consider the differential equation

$$T[v](z,\epsilon) - \frac{v(z,\epsilon)}{\epsilon^k} = 0, \quad \epsilon \in \mathbb{C} \setminus \{0\},$$

depending on the parameter ϵ where T is given by (1.2). Writing

$$v^{(k)}(z,\epsilon) + \sum_{j=1}^{k-1} \frac{Q_j(z)}{Q_k(z)} v^{(j)}(z,\epsilon) - \frac{v(z,\epsilon)}{\epsilon^k Q_k(z)} = 0, \tag{4.3}$$

where $Q_j(z) = \sum_{i=0}^{j} q_{j,i} z^i$, we see that p_n is a polynomial solution of (4.3) with $\epsilon_n = \frac{1}{\sqrt[k]{\lambda_n - q_{0,0}}}$.

Reciprocally, if for some quantity ϵ_n, v_n is a polynomial solution of (4.3) of degree n, then v_n is a polynomial eigenfunction of the ES-differential operator (1.2) with eigenvalue $\lambda_n = q_{0,0} + \epsilon_n^{-k}$.

Remark 4.1. As in the previous chapters, we will use the superscript (j) for the jth-order partial derivative of v with respect to z.

The main result of § 4.2 is Proposition 4.1, which guarantees the existence of the exponential series that define the WKB-solution when z varies in a disk and ϵ varies in a sectorial region. More precisely, there exist $a, r, \eta' > 0$, and k linearly independent solutions of (4.3) of the form

$$e^{y_\ell(z,\epsilon;z_1)}, \quad \ell = 1, \ldots, k, \tag{4.4}$$

for $(z,\epsilon) \in \Delta(z',\eta') \times \mathcal{G}(r,\alpha)$, $\Delta(z',\eta') \subset \Omega$, $\ell = 1,\dots,k$ of (4.3) such that

$$y_\ell(z,\epsilon;z_1) = \frac{1}{\epsilon}\int_{z_1}^{z} \mathfrak{w}_\ell(t,\epsilon)dt,$$

with $\mathfrak{w}_\ell(t,\epsilon) \in \mathcal{H}(\Delta(z',\eta') \times \mathcal{G}(r,\alpha))$ and $\mathfrak{w}_\ell(t,\epsilon) \sim \sum_{i=0}^{\infty} \mathfrak{b}_{\ell,i}(z)\epsilon^i$ as $\epsilon \to 0$, $\epsilon \in \mathcal{G}(r,\alpha)$, uniformly in $z \in \Delta(z',\eta')$. Here $z_1 \in \Delta(z',\eta')$ is an arbitrary point, and the path of integration in the above integral is contained in $\Delta(z',\eta')$.

Notation 4.1. Let $\mathbf{B}_{m,j}$ denote the (exponential) partial Bell polynomials [54, (3a) (3d), p. 134]

$$\mathbf{B}_{m,j}(x_1,\dots,x_{m-j+1}) = \sum \frac{m!}{c_1!c_2!\cdots(1!)^{c_1}(2!)^{c_2}\dots} x_1^{c_1}\dots x_{m-j+1}^{c_{m-j+1}},$$
$$\mathbf{B}_{0,0} = 1,$$

where the summation is taken over all integers $c_1, c_2, \dots \geq 0$ such that

$$c_1 + 2c_2 + 3c_3 \cdots = m,$$
$$c_1 + c_2 + c_3 \cdots = j.$$

Let $\mathbf{Y}_m = \mathbf{Y}_m(x_1,\dots,x_m)$, $\mathbf{Y}_0 = 1$ be the (exponential) complete Bell polynomials [54, (3d), p. 134] defined by the generating function in an infinite number of indeterminates $x_1, x_2, \dots$ given by

$$\exp\Bigl(\sum_{i\geq 1} x_i \frac{t^i}{i!}\Bigr) = 1 + \sum_{m\geq 1} \mathbf{Y}_m(x_1,\dots,x_m)\frac{t^m}{m!}.$$

The change of variable $v = e^{\frac{1}{\epsilon}\int^z (w(t,\epsilon)+w_\ell(t))dt}$ transforms (4.3) into a generalized Riccati equation of the form

$$\epsilon^{k-1}w^{(k-1)} = -\epsilon^{k-1}w_\ell^{(k-1)} - \epsilon^{k-1}\sum_{j=1}^{k-1}\frac{Q_j}{Q_k}(w+w_\ell)^{(j-1)}$$
$$-\sum_{j=0}^{k-2}\epsilon^j \sum_{l=k-j}^{k}\frac{Q_l}{Q_k}\mathbf{B}_{l,k-j}(w+w_\ell,\dots,(w+w_\ell)^{(l-k+j)}) + \frac{1}{Q_k}. \qquad (4.5)$$

Next, we transform (4.5) into a non-linear first order system

$$\epsilon\frac{d\vec{g}}{dz} = \vec{G}_\ell(z,\vec{g},\epsilon). \qquad (4.6)$$

Further, we apply to (4.6) a theorem about the existence of asymptotic solutions [90, Th. XII-1-2, p. 374] of Sibuya, see below and also [172]. Using these solutions, we obtain the WKB-expansion.

Theorem 4.2 ([90, Th. XII-1-2, p. 374], [172]). *Consider the system*

$$\epsilon v_i' = f_i(z, v_1, \dots, v_n, \epsilon), \quad i = 1, \dots, n, \tag{4.7}$$

where $f_i(z, v_1, \dots, v_n, \epsilon)$ *are holomorphic functions of the complex variables* $(z, v_1, \dots, v_n, \epsilon)$ *in the domain*

$$|z| < \delta_0, |\epsilon| < \rho_0, |\arg \epsilon| < \alpha_0, |v_i| < \gamma_0, \quad i = 1, \dots, n.$$

Set

$$f_i(z, \vec{v}, \epsilon) = f_{i_0}(z, \epsilon) + \sum_{h=1}^{n} a_{i_h}(z, \epsilon) v_h + \sum_{|p| \geq 2} f_{i_p}(z, \epsilon) \vec{v}^p,$$

where $\vec{v} = (v_1, \dots, v_n) \in \mathbb{C}^n$, $p = (p_1, \dots, p_n)$ *is a multi-index of non-negative integers, and* $\vec{v}^p = v_1^{p_1} \cdots v_n^{p_n}$. *Suppose that the following three conditions hold:*

I) *Each function* $f_i(z, \vec{v}, \epsilon)$ $(i = 1, \dots, n)$ *has an asymptotic expansion*

$$f_i(z, \vec{v}, \epsilon) \sim \sum_{\nu=0}^{\infty} \hat{f}_{i_\nu}(z, \vec{v}) \epsilon^\nu$$

in the sense of Poincaré as $\epsilon \to 0$ *in the sector*

$$|\epsilon| < \rho_0, \quad |\arg \epsilon| < \alpha_0, \tag{4.8}$$

where the coefficients $\hat{f}_{i_\nu}(z, \vec{v})$ *are holomorphic in the domain* $|z| < \delta_0$, $|\vec{v}| < \gamma_0$. *Furthermore, we assume that* $\hat{f}_{i_0}(z, \vec{0}) = 0$, $i = 1, \dots, n$, *for* $|z| < \delta_0$.

Let $A(z, \epsilon)$ *be the* $(n \times n)$*-matrix whose* (j, k)*-entry is* $a_{jk}(z, \epsilon)$. *Then* $A(z, \epsilon)$ *admits an asymptotic expansion*

$$A(z, \epsilon) \sim \sum_{\nu=0}^{\infty} \epsilon^\nu A_\nu(z),$$

as $\epsilon \to 0$ *in sector (4.8), where the entries of the coefficient matrix* $A_\nu(z) = (a_{jk\nu}(z))$ *are holomorphic in the domain* $|z| < \delta_0$, $|\vec{v}| < \gamma_0$.

II) *The eigenvalues of the matrix* $A_0(0)$ *are such that* $\lambda_j \neq 0$, $j = 1, \dots, n$.

Then system (4.7) has a solution $v_j = U_j(z)$, $j = 1, \dots, n$, *such that:*

i) $U_j(z, \epsilon)$ *are holomorphic in a domain*

$$|z| < \delta, \quad |\epsilon| < \rho, \quad |\arg \epsilon| < \alpha,$$

where δ, ρ, *and* α *are suitable positive constants such that* $0 < \delta \leq \delta_0$, $0 < \rho \leq \rho_0$, *and* $0 < \alpha \leq \alpha_0$.

ii) $U_j(z,\epsilon)$ *admit asymptotic expansions*

$$U_j(z,\epsilon) \sim \sum_{\nu=1}^{\infty} U_{j\nu}(z)\epsilon^{\nu}, \quad j = 1,\ldots,n$$

as $\epsilon \to 0$ *in the sector* $0 < |\epsilon| < \rho$, $|\arg \epsilon| < \alpha$, *where the coefficients* $p_{\nu j}(z)$ *are holomorphic in the domain* $|z| < \delta$.

Remark 4.2. Notice that by considering the transformation $z \mapsto z + z'$, the preceding theorem also holds in any disk $|z - z'| < \delta_0$, provided that the hypotheses are true for z varying in a small neighborhood of the origin.

The application of Theorem 4.2 to system (4.6) requires an intermediate transformation. In this case, the matrix A_0 in the hypothesis II of Sibuya's theorem is invertible in Ω, which is studied in Lemmas 4.3 and 4.4.

In § 4.3 we apply the previous results to obtain the asymptotic behavior of the sequence $\{p_n(z)\}$ in compact subsets of Ω, which will be done in several steps. We will denote by z_1, c_n, and n_0 some (non-essential) constants whose values may change from one lemma to another. Using Proposition 4.2 we first prove in Lemmas 4.5 and 4.6 that for every fixed point $z' \in \Omega$, we can find $\Delta(z', \eta) \subset \Omega$ such that the exponential solution e^{y_1} of system (4.4) is dominant and there exist $c_n \in \mathbb{C}$ and $\tilde{y}_1(z, \epsilon_n; z_1), z_1 \in \Delta(z', \eta)$ such that

$$p_n(z) = c_n e^{\tilde{y}_1(z,\epsilon_n;z_1)}, \quad \forall n > n_0, \tag{4.9}$$

where $\tilde{y}_1(z, \epsilon_n; z_1) = \frac{1}{\epsilon_n} \int_{z_1}^{z} \mathfrak{w}_1(t, \epsilon_n)dt$ and $\mathfrak{w}_1(t, \epsilon_n) \sim \sum_{i=0}^{\infty} \mathfrak{b}_{1,i}(z)\epsilon_n^i$ as $n \to \infty$ uniformly in $z \in \Delta(z', \eta)$. Next, in Lemma 4.8 we prove that for any closed Jordan curve γ enclosing $supp(\mu)$, there exist $z_1 \in ext(\gamma), n_0, y$, and a constant c_n such that

$$p_n(z) = c_n e^{y(z,\epsilon_n;z_1)}, \quad \forall n > n_0, \forall z \in ext(\gamma), \tag{4.10}$$

where $y(z, \epsilon_n; z_1) = \frac{1}{\epsilon_n} \int_{z_1}^{z} \mathfrak{w}(t, \epsilon_n)dt$, $\mathfrak{w}(t, \epsilon_n) \in \mathcal{H}(ext(\gamma))$, $\mathfrak{w}(z, \epsilon_n) \sim \sum_{i=0}^{\infty} \mathfrak{b}_i(z)\epsilon_n^i$ as $n \to \infty$ when z varies in a compact subset $K \subset ext(\gamma)$. Here the integration path Γ is chosen so that $\Gamma \subset ext(\gamma) \setminus \tau$, $y(z_0, \epsilon_n; z_1) = \lim_{z \to z_0, z \in \tau^+} y(z, \epsilon_n; z_1)$, when $z_0 \in \tau$, where τ is a Jordan arc connecting ∞ and any point $\omega \in supp(\mu)$ oriented so that ω is the endpoint. To prove this, we consider a covering of K of the form $\{\Delta(z', \eta(z'))\}_{z' \in K}$, where $\Delta(z', \eta(z'))$ is such that the representation (4.9) holds.

Using the Heine–Borel theorem, we obtain a finite subcover of K, and by analytic continuation of a function element $(\tilde{y}_1(z, \epsilon_n; z_1), \Delta(z', \eta(z')))$ we obtain the function y, which also can be continued analytically to $ext(\gamma)$.

Using representation (4.9), in Lemma 4.11 we prove that we can find a primitive of the function $\mathfrak{w}$ such that for n sufficiently large,

$$p_n(z) = e^{\frac{1}{\epsilon_n} y(z,\epsilon_n)}, \quad \forall z \in ext(\gamma), \tag{4.11}$$

where $y(z, \epsilon_n) \sim \sum_{i=0}^{\infty} \Phi_i(z)\epsilon_n^i$, as $n \to \infty$ when z varies in compact subsets $K \subset ext(\gamma)$. Here $\Phi_i(z)$ is the primitive of the function $\mathfrak{b}_i$ in $ext(\gamma)\setminus\tau$ such that $\lim_{z\to\infty} \Phi_i(z) - h_i \ln z = 0$, $\Phi_i(z) = \lim_{\substack{z\to z_0 \\ z\in\tau^+}} \Phi_i(z)$, when $z_0 \in \tau \cap ext(\gamma)$, being τ any Jordan arc from $-\infty$ to $\omega \in supp(\mu)$ such that $]-\infty, p] \subset \tau$ for some $p \in \mathbb{R}$, and h_i is a properly chosen constant. The final step is the proof of Theorem 4.1, which follows by an appropriate transformation of (4.11).

In § 4.4 we discuss the Stokes phenomena, and in § 4.5 we present two applications to ES-operators. In particular, we prove Property 2.8.

4.2 Existence of linearly independent exponential solutions

Lemma 4.1. *For $k \geq 2$, let $U \subset \Omega$ be an open subset, $S \subset \mathbb{C}$, and $w(t,\epsilon) : U \times S \to \mathbb{C}$ be a holomorphic function in the variable t. Then $v = e^{\frac{1}{\epsilon}\int^z (w(t,\epsilon)+w_\ell(t))dt}$ is a solution of*

$$v^{(k)} + \sum_{j=1}^{k-1} \frac{Q_j}{Q_k} v^{(j)} - \frac{v}{\epsilon^k Q_k} = 0, \tag{4.12}$$

if and only if w is a solution of

$$\epsilon^{k-1} w^{(k-1)} = F_\ell(\epsilon, w, \ldots, w^{(k-2)}).$$

Here

$$\begin{aligned} F_\ell(\epsilon, w, \ldots, w^{(k-2)}) = {} & -\epsilon^{k-1} w_\ell^{(k-1)} - \epsilon^{k-1} \sum_{j=1}^{k-1} \frac{Q_j}{Q_k} (w + w_\ell)^{(j-1)} \\ & - \sum_{j=0}^{k-2} \epsilon^j \sum_{l=k-j}^{k} \frac{Q_l}{Q_k} \mathbf{B}_{l,k-j}(w + w_\ell, \ldots, (w + w_\ell)^{(l-k+j)}) + \frac{1}{Q_k}, \end{aligned}$$

and w_ℓ is as in (4.2).

Proof. Let us denote by $P_j(y', \ldots, y^{(j)})$ the polynomial in the variables $(y', \ldots, y^{(j)})$ defined for all integers $1 \leq j \leq k$ by the relation

$$P_j(y', \ldots, y^{(j)}) = e^{-y} (e^y)^{(j)}.$$

According to the Faà di Bruno formula [54, Th. A, p. 137], $P_k(y', \ldots, y^{(k)})$ can be expressed as

$$P_j(y', \ldots, y^{(j)}) = \sum_{l=1}^{j} \mathbf{B}_{j,l}(y', \ldots, y^{(j-l+1)}). \tag{4.13}$$

Relation (4.13) and the variable change $v = e^y$ provide that (4.12) can be expressed as

$$\sum_{j=1}^{k} \frac{Q_j}{Q_k} P_j(y', \ldots, y^{(j)}) - \frac{1}{Q_k \epsilon^k} = 0. \tag{4.14}$$

By multiplying this relation by ϵ^k and using the expression for P_j, one has that $v = e^y$, $y' = (w + w_\ell)\epsilon^{-1}$ is a solution to (4.12) if and only if w solves the generalized Riccati equation

$$\sum_{j=1}^{k} \epsilon^k \frac{Q_j}{Q_k} P_j(\epsilon^{-1}(w + w_\ell), \ldots, \epsilon^{-1}(w + w_\ell)^{(j-1)}) - \frac{1}{Q_k} = 0, \tag{4.15}$$

where

$$P_j(\epsilon^{-1}(w + w_\ell), \ldots, \epsilon^{-1}(w + w_\ell)^{(j-1)}) = \sum_{l=1}^{j} \epsilon^{-l} \mathbf{B}_{j,l}(w + w_\ell, \ldots, (w + w_\ell)^{(j-l)}).$$

After a convenient rearrangement of the terms, we have that relation (4.15) can be equivalently expressed as

$$\sum_{j=0}^{k-1} \epsilon^j \sum_{l=k-j}^{k} \frac{Q_l}{Q_k} \mathbf{B}_{l,k-j}(w + w_\ell, \ldots, (w + w_\ell)^{(l-k+j)}) - \frac{1}{Q_k} = 0.$$

In other words,

$$\epsilon^{k-1}(w + w_\ell)^{(k-1)} + \epsilon^{k-1} \sum_{j=1}^{k-1} \frac{Q_j}{Q_k} (w + w_\ell)^{(j-1)}$$
$$+ \sum_{j=0}^{k-2} \epsilon^j \sum_{l=k-j}^{k} \frac{Q_l}{Q_k} \mathbf{B}_{l,k-j}(w + w_\ell, \ldots, (w + w_\ell)^{(l-k+j)}) - \frac{1}{Q_k} = 0.$$ □

Lemma 4.2. *For $k \geq 3$ and $\ell = 1, \ldots, k$, let F_ℓ be defined as in Lemma 4.1. Then, the differential equation*

$$\epsilon^{k-1} w^{(k-1)} = F_\ell(\epsilon, w, \ldots, w^{(k-2)}) \tag{4.16}$$

of order $k - 1$ can be expressed in the companion form as

$$\epsilon \frac{d\vec{g}}{dz} = \vec{G}_\ell(z, \vec{g}, \epsilon),$$

where $\vec{g} = (g_0, g_1, \ldots, g_{k-2})^t$, $\vec{G}_\ell(z, \vec{g}, \epsilon) = (g_1 + \epsilon w_\ell', g_2, \ldots, g_{k-2}, h_\ell(z, \vec{g}, \epsilon))$, and

$$h_\ell(z, \vec{g}, \epsilon) = -\sum_{j=2}^{k} \mathbf{B}_{M,j}(g_0 + w_\ell, g_1, \ldots, g_{k-j})$$

$$+\frac{1}{Q_k}-\sum_{j=1}^{k-2}\epsilon^j\left(\frac{Q_{k-j}}{Q_k}g_{k-j-1}+\sum_{l=j}^{k-2}\frac{Q_{k-j}}{Q_k}\mathbf{B}_{k-j,k-l}(g_0+w_\ell,g_1,\dots,g_{l-j})\right)$$
$$-\epsilon^{k-1}\left(w_\ell^{(k-1)}+\frac{Q_1}{Q_k}(g_0+w_\ell)\right). \tag{4.17}$$

Proof. Using the change of variables

$$g_0=w,\quad g_1=\epsilon(g_0'+w_j'),\quad g_2=\epsilon g_1',\dots,$$

equation (4.16) transforms into

$$\epsilon g_{k-2}'=F_\ell\left(\epsilon,g_0+w_\ell,\frac{g_1}{\epsilon},\dots,\frac{g_{k-2}}{\epsilon^{k-2}}\right).$$

Further, using [54, Th. A, p. 134], we have that

$$\begin{aligned}
&F_\ell\left(\epsilon,g_0+w_\ell,\frac{g_1}{\epsilon},\dots,\frac{g_{k-2}}{\epsilon^{k-2}}\right)\\
&=-\epsilon^{k-1}\left(w_\ell^{(k-1)}+\frac{Q_1}{Q_k}(g_0+w_\ell)\right)\\
&\quad-\sum_{j=2}^{k-1}\frac{Q_j}{Q_k}\epsilon^{k-j}g_{j-1}-\sum_{j=0}^{k-2}\epsilon^j\left(\sum_{l=j}^{k-2}\frac{Q_{k-j}}{Q_k}\mathbf{B}_{k-j,k-l}(g_0+w_j,g_1,\dots,g_{l-j})\right)+\frac{1}{Q_k}\\
&=-\sum_{j=2}^{k}\mathbf{B}_{k,j}(g_0+w_\ell,g_1,\dots,g_{k-j})+\frac{1}{Q_k}\\
&\quad-\sum_{j=1}^{k-2}\epsilon^j\left(\frac{Q_{k-j}}{Q_k}g_{k-j-1}+\sum_{l=j}^{k-2}\frac{Q_{k-j}}{Q_k}\mathbf{B}_{k-j,k-l}(g_0+w_\ell,g_1,\dots,g_{l-j})\right)\\
&\quad-\epsilon^{k-1}\left(w_\ell^{(k-1)}+\frac{Q_1}{Q_k}(g_0+w_\ell)\right).
\end{aligned}$$

Thus, the differential equation

$$\epsilon^{k-1}w^{(k-1)}=F_\ell(\epsilon,w,\dots,w^{(k-2)})$$

of order $k-1$ can be expressed as

$$\epsilon\frac{d\vec{g}}{dz}=(g_1+\epsilon w_\ell',g_2,\dots,g_{k-2},h_\ell(z,\vec{g},\epsilon))$$

where $\vec{g}=(g_0,g_1,\dots,g_{k-2})^t$ and h_ℓ is given by (4.17). □

The function $\vec{G}_\ell$ defined in Lemma 4.2 will play an important role. Namely, for $k\geq 2$, we define $\vec{G}_\ell$ as follows. Notice that when $k=2$ the differential equation

$$\epsilon^{k-1}w^{(k-1)} = F_\ell(\epsilon, w, \dots, w^{(k-2)})$$

reduces to the usual Riccati equation

$$\epsilon w' = -(w + w_\ell)^2 + \frac{1}{Q_2} - \epsilon w'_\ell - \frac{Q_1}{Q_2}\epsilon(w + w_\ell) \tag{4.18}$$

by application of Lemma 4.1.

Definition 4.1. For $k \geq 3$ and $1 \leq \ell \leq k$, define $\vec{G}_\ell$ as in Lemma 4.2. For $k = 2$, using (4.18), we define $\vec{G}_\ell = G_\ell$ as

$$G_\ell(z, g_0, \epsilon) = -(g_0 + w_\ell)^2 + \frac{1}{Q_2} - \epsilon w'_\ell - \frac{Q_1}{Q_2}\epsilon(g_0 + w_\ell), \quad \ell = 1, 2.$$

Lemma 4.3. *For $\ell = 1, \dots, k$, define $A_\ell(z, \epsilon) := \frac{\partial \vec{G}_\ell}{\partial \vec{g}}(z, \vec{0}, \epsilon)$. Then $A_{\ell,0}(z) = \lim_{\epsilon \to 0} A_\ell(z, \epsilon)$ is invertible in Ω.*

Proof. For $k = 2$, from Definition 4.1 it is immediate that $\lim_{\epsilon\to 0} \frac{\partial G_\ell}{\partial g_0}(z, 0, \epsilon) \neq 0$, $z \in \Omega$, $\ell = 1, 2$. Suppose now that $k \geq 3$. From the expression for $\vec{G}_\ell$, we get

$$\frac{\partial \vec{G}_\ell}{\partial \vec{g}}(z, \vec{g}, \epsilon) = \begin{pmatrix} 0 & 1 & 0 & \cdots & 0 \\ 0 & 0 & 1 & \cdots & 0 \\ & & \vdots & & \\ \frac{\partial h_\ell(z,\vec{g},\epsilon)}{\partial g_0} & \frac{\partial h_\ell(z,\vec{g},\epsilon)}{\partial g_1} & \cdots & & \frac{\partial h_\ell(z,\vec{g},\epsilon)}{\partial g_{k-2}} \end{pmatrix}. \tag{4.19}$$

Using the definition of the complete Bell polynomials, see [54, 3b), p. 134], we get the identity

$$\frac{\partial \mathbf{Y}_n}{\partial x_j}(x_1, \dots, x_n) = \binom{n}{j}\mathbf{Y}_{n-j}(x_1, \dots, x_{n-j}). \tag{4.20}$$

On the other hand, from the identity [54, 3c), p. 134] we can write h_ℓ as

$$h_\ell(z, \vec{g}, \epsilon) = -\mathbf{Y}_j(g_0 + w_\ell, g_1, \dots, g_{j-2}, 0) + \pi_{j-2}(\vec{g}, \epsilon),$$

where $\pi_{j-2}(\vec{g}, \epsilon)$ is a polynomial such that $\frac{\partial \pi_{j-2}}{\partial g_j}(\vec{g}, 0) = 0$, $0 \leq j \leq k-2$. Therefore, using (4.20) one gets

$$\frac{\partial h_\ell}{\partial g_j}(z, \vec{g}, \epsilon) = -\binom{k}{j+1}\mathbf{Y}_{k-j-1}(g_0 + w_\ell, \dots, g_{k-j-1}) + \frac{\partial \pi_{k-2}(\vec{g}, \epsilon)}{\partial g_j}, \quad 0 \leq j \leq k-2.$$

By substituting the latter expression in (4.19) and taking the limit when $\epsilon \to 0$, we obtain

$$A_{\ell,0}(z) = \begin{pmatrix} 0 & 1 & 0 & \cdots & 0 \\ 0 & 0 & 1 & \cdots & 0 \\ & & \vdots & & \\ -\binom{k}{1}w_\ell^{k-1} & -\binom{k}{2}w_\ell^{k-2} & \cdots & -\binom{k}{k-2}w_\ell^2 & -\binom{k}{k-1}w_\ell \end{pmatrix}.$$

Now, to prove that $A_{\ell,0}$ is invertible in Ω, we need to show that 0 is not an eigenvalue. Indeed, $A_{\ell,0}$ is the companion matrix of the polynomial

$$p(\lambda) = \lambda^{k-1} + \binom{k}{k-1} w_\ell \lambda^{k-2} + \cdots + \binom{k}{1} w_\ell^{k-1},$$

cf. [86, Def. 3.3.13, p. 195]. Hence, χ is the characteristic polynomial of the matrix $A_{\ell,0}$. On the other hand, we have that

$$(\lambda + w_\ell)^k = \lambda\chi(\lambda) + w_\ell^k.$$

A straightforward calculation shows that the roots of $\chi(\lambda)$ are given by the products $(\lambda_j w_\ell)_{j=1}^k$, where $\lambda_j = (\omega_j - 1)$ and $\omega_j = e^{\frac{i2k\pi}{k}}$, $j = 1, \ldots, k-1$. Since $\lambda_j w_\ell(z) \neq 0$, $z \in \Omega$, the result follows. □

Lemma 4.4. *Let $\vec{G}_\ell$, $\ell = 1, \ldots, k$ be as in Definition 4.1 and $z' \in \Omega$, $\eta > 0$ satisfy the condition $\Delta(z', \eta) \subset \Omega$. Then there exist $\eta', r, \alpha > 0$ such that the system of differential equations*

$$\epsilon \frac{d\vec{g}}{dz} = \vec{G}_\ell(z, \vec{g}, \epsilon), \quad z \in \Delta(z', \eta)$$

has a solution $\vec{g}_\ell$ holomorphic in $(z, \epsilon) \in \Delta(z', \eta') \times \mathcal{G}(r, \alpha)$ for every $1 \le \ell \le k$.

Moreover it has a uniform asymptotic expansion

$$\vec{g}_\ell(z, \epsilon) \sim \sum_{i=1}^{\infty} \epsilon^i \vec{g}_{\ell,i}(z), \quad \text{as } \epsilon \to 0 \text{ in } \mathcal{G}(r, \alpha), z \in \Delta(z', \eta').$$

Proof. Consider first the case $k \ge 3$. By Definition 4.1, in the domain $(z, \epsilon) \in \Delta(z', \eta) \times \mathcal{G}(r', \alpha')$ the function $\vec{G}_\ell$ can be expressed as

$$\vec{G}_\ell(z, \vec{g}, \epsilon) = \vec{f}_{\ell,0}(z, \epsilon) + A_{\ell,0}(z)\vec{g} + \vec{f}_\ell(z, \epsilon, g_0, \ldots, g_{k-2}), \tag{4.21}$$

where $A_{\ell,0}(z)$ is defined as in Lemma 4.3, $\vec{f}_{\ell,0}(z, \epsilon)$ is a vector-valued function whose components are polynomials in the variable ϵ with coefficients in $\mathcal{H}(\Delta(z', \eta))$ such that $\vec{f}_{\ell,0}(z, 0) = \vec{0}$. Further, $\vec{f}_\ell$ is a vector-valued function whose components are polynomials in the variables $\epsilon, g_0, \ldots, g_{k-2}$ with coefficients in $\mathcal{H}(\Delta(z', \eta))$ such that $\vec{f}_\ell(z, 0, g_0, \ldots, g_{k-2}) = \vec{0}$.

We have that for each $\ell = 1, \ldots, k$, the right-hand side of (4.21) satisfies the hypothesis of Theorem 4.2 in the open set $\Delta(z', \eta) \times \mathcal{G}(r', \alpha')$. Hence, for each fixed $\ell = 1, \ldots, k$, there exist $\eta_\ell \leq \eta, \alpha_\ell \leq \alpha', r_\ell \leq r'$ such that the differential equation (4.21) has a solution $\vec{g}_\ell = (g_{0,\ell}, \ldots, g_{M-2,\ell})$ holomorphic in $(z,\epsilon) \in \Delta(z',\eta_\ell) \times \mathcal{G}(r_\ell,\alpha_\ell)$, where $g_{l,\ell}$; $0 \leq l \leq k-2$ admits a uniform asymptotic expansion

$$g_{l,\ell}(z,\epsilon) \sim \sum_{i=1}^{\infty} \epsilon^i g_{l,\ell,i}(z), \quad \text{as } \epsilon \to 0 \text{ in } \mathcal{G}(r_\ell,\alpha_\ell),\ z \in \Delta(z',\eta_\ell).$$

Hence, if $r = \min_{\ell=1,\ldots,k}\{r_\ell\}$, $\alpha = \min_{\ell=1,\ldots,k}\{\alpha_\ell\}$ and $\eta' = \min_{\ell=1,\ldots,k}\{\eta_\ell\}$, we have settled the required result for $k \geq 3$.

Finally, when $k = 2$ by Definition 4.1 in the domain $z \in (z,\epsilon) \in \Delta(z',\eta) \times \mathcal{G}(r',\alpha')$, one has that

$$\epsilon\frac{dg_0}{dz} = -\epsilon w_\ell' - \frac{Q_1}{Q_2}w_\ell\epsilon - 2w_\ell g_0 + \frac{Q_1}{Q_2}g_0\epsilon - g_0^2, \quad \ell = 1,2.$$

The conclusions of the lemma again follows from Theorem 4.2. □

Proposition 4.1. *Take $k \geq 2$ and $z' \in \Omega$. Then,*

a) *There exist $\alpha, r, \eta' > 0$, and k solutions of (4.3) of the form $e^{y_\ell(z,\epsilon;z_1)}$ such that in the domain $(z,\epsilon) \in \Delta(z',\eta') \times \mathcal{G}(r,\alpha)$, $\Delta(z',\eta') \subset \Omega$, $\ell = 1,\ldots,k$ one has $y_\ell(z,\epsilon;z_1) = \frac{1}{\epsilon}\int_{z_1}^{z} \mathfrak{w}_\ell(t,\epsilon)dt$ where*

$$\mathfrak{w}_\ell(t,\epsilon) \in \mathcal{H}(\Delta(z',\eta') \times \mathcal{G}(r,\alpha)), \quad \mathfrak{w}_\ell(t,\epsilon) \sim \sum_{i=0}^{\infty} \mathfrak{b}_{\ell,i}(t)\epsilon^i$$

as $\epsilon \to 0$, $\epsilon \in \mathcal{G}(r,\alpha)$, uniformly when z varies in $\Delta(z',\eta')$. Here $z_1 \in \Delta(z',\eta')$ is an arbitrary point and the integration path in the above integral is contained in $\Delta(z',\eta')$.

b)

$$\mathfrak{b}_{\ell,0}(z) = w_\ell(z), \quad z \in \Delta(z',\eta'),$$
$$\mathfrak{b}_{\ell,1}(z) = \frac{k-1}{2k}\frac{Q_k'(z)}{Q_k(z)} - \frac{1}{k}\frac{Q_{k-1}(z)}{Q_k(z)}, \quad z \in \Delta(z',\eta'),$$

where w_ℓ is defined as in (4.2).

c) *The k-tuple of functions $(e^{y_\ell(z,\epsilon;z_1)})_{\ell=1}^k$ is linearly independent in $\Delta(z',\eta') \times \mathcal{G}(r,\alpha)$ for r small enough, where $\{y_\ell\}_{\ell=1}^k$ is as in item a).*

Proof. a) By Lemma 4.2 and relation (4.18) we have that the $(k-1)$th order differential equation

$$\epsilon^{k-1}w^{(k-1)} = F_\ell(\epsilon, w, \ldots, w^{(k-2)})$$

can be expressed in the companion form

$$\epsilon\frac{d\vec{g}}{dz} = \vec{G}_\ell(z,\vec{g},\epsilon),$$

where $\vec{G}_\ell$ is as in Definition 4.1. Hence, Lemma 4.4 implies that there exist $\alpha, r, \eta' > 0$ and a holomorphic function u_ℓ in $\Delta(z', \eta') \times \mathcal{G}(r, \alpha), \Delta(z', \eta') \subset \Omega$ such that

$$\epsilon^{k-1} u_\ell^{(k-1)} = F_\ell(\epsilon, u_\ell, \dots, u_\ell^{(k-2)}), \quad 1 \le \ell \le k, \tag{4.22}$$

where u_ℓ admits a uniform asymptotic expansion

$$u_\ell(z, \epsilon) \sim \sum_{i=1}^{\infty} \epsilon^i \mathfrak{b}_{\ell,i}(z), \quad \text{as } \epsilon \to 0 \text{ in } \mathcal{G}(r, \alpha), z \in \Delta(z', \eta'). \tag{4.23}$$

Hence, by Lemma 4.1 one has that if

$$\begin{aligned} y_\ell(z, \epsilon; z_1) &= \epsilon^{-1} \int_{z_1}^{z} w_\ell(t) dt + \epsilon^{-1} \int_{z_1}^{z} u_\ell(t, \epsilon) dt, \\ &:= \epsilon^{-1} \int_{z_1}^{z} \mathfrak{w}_\ell(t, \epsilon) dt \\ &= \sum_{i=0}^{\infty} \epsilon^{i-1} \int_{z_1}^{z} \mathfrak{b}_{\ell,i}(t) dt, \end{aligned} \tag{4.24}$$

then $e^{y_\ell(z,\epsilon)}$, $(z, \epsilon) \in \Delta(z', \eta') \times \mathcal{G}(r, \alpha)$ is a solution of (4.3), and y_ℓ admits a uniform asymptotic expansion

$$y_\ell(z, \epsilon; z_1) \sim \sum_{i=0}^{\infty} \epsilon^{i-1} \int_{z_1}^{z} \mathfrak{b}_{\ell,i}(t) dt, \quad \text{as } \epsilon \to 0 \text{ in } \mathcal{G}(r, \alpha), z \in \Delta(z', \eta'), \tag{4.25}$$

where $\mathfrak{b}_{\ell,0} = w_\ell$.

b) The first relation follows from (4.24) and (4.25). To prove the second assertion, notice that by (4.22) one has

$$\frac{d}{d\epsilon} H_j(\epsilon, u_j, \dots, u_j^{(k-1)}) = 0, \tag{4.26}$$

where $H_j(\epsilon, w, \dots, w^{(k-1)}) = \epsilon^{k-1} w^{(k-1)} - F_j(\epsilon, w, \dots, w^{(k-2)})$. Further

$$\begin{aligned} &\frac{d}{d\epsilon} H_j(\epsilon, u_j, \dots, u_j^{(k-1)}) \\ &= \frac{\partial H_j}{\partial \epsilon}(\epsilon, u_j, \dots, u_j^{(k-1)}) + \sum_{j=0}^{k-1} \frac{\partial H_j}{\partial w^{(k)}}(\epsilon, u_\ell, \dots, u_\ell^{(k-1)}) \frac{du_\ell^{(j)}}{d\epsilon} \\ &= \frac{Q_{j-1}}{Q_j} \mathbf{B}_{j-1,j-1}(u_\ell + w_\ell) + \mathbf{B}_{j,j-1}(u_\ell + w_\ell, (u_\ell + w_\ell)') + I_j(\epsilon, u_\ell, \dots, u_\ell^{(k-1)}) \end{aligned}$$

$$+\frac{\partial H_\ell}{\partial w}(\epsilon, u_\ell, \dots, u_\ell^{(k-1)})\frac{du_\ell}{d\epsilon} + \sum_{j=1}^{k-1}\frac{\partial H_\ell}{\partial w^{(j)}}(\epsilon, u_\ell, \dots, u_\ell^{(k-1)})\frac{du_\ell^{(j)}}{d\epsilon}, \tag{4.27}$$

where

$$I_\ell(\epsilon, u_\ell, \dots, u_\ell^{(k-1)}) = (k-1)\epsilon^{k-2}w_\ell^{(k-1)} + (k-1)\epsilon^{k-2}\sum_{kj=1}^{k-1}\frac{Q_j}{Q_k}(u_\ell + w_\ell)^{(j-1)}$$
$$+\sum_{j=2}^{k-2}k\epsilon^{j-1}\sum_{l=k-j}^{k}\frac{Q_l}{Q_k}\mathbf{B}_{l,k-j}(u_\ell + w_\ell, \dots, (u_\ell + w_\ell)^{(l-k+j)}),$$

and $\frac{\partial H_\ell}{\partial w^{(j)}}$ denotes the partial derivative of H_ℓ with respect to the variable $w^{(j)}$, considering different derivatives of w as independent functions.

On the other hand, using [90, Th. XI-1-11, p. 347], we obtain the relations

$$\lim_{\substack{\epsilon\to 0\\ \epsilon\in\mathcal{G}(r,\alpha)}} I_\ell(\epsilon, u_\ell, \dots, u_\ell^{(k-1)}) = 0,$$
$$\lim_{\substack{\epsilon\to 0\\ \epsilon\in\mathcal{G}(r,\alpha)}} \frac{\partial H_\ell}{\partial w}(\epsilon, u_\ell, \dots, u_\ell^{(k-1)}) = k(w_\ell)^{k-1}, \tag{4.28}$$
$$\lim_{\substack{\epsilon\to 0\\ \epsilon\in\mathcal{G}(r,\alpha)}} \sum_{j=1}^{k-1}\frac{\partial H_\ell}{\partial w^{(j)}}(\epsilon, u_\ell, \dots, u_\ell^{(k-1)})\frac{du_\ell^{(j)}}{d\epsilon} = 0.$$

By taking limits as $\epsilon \to 0$, $\epsilon \in \mathcal{G}(r, \alpha)$ in (4.27) and using (4.23), (4.26), and (4.28), one obtains

$$0 = \frac{Q_{k-1}}{Q_k}\mathbf{B}_{k-1,k-1}(w_j) + \mathbf{B}_{k,k-1}(w_j, w_j') + k(w_j)^{k-1}\mathfrak{b}_{j,1}(z),$$

which proves b).

c) Using b), we see that the k-tuple of functions $\{e^{y_j(z,\epsilon)}\}_{j=1}^k$ can be expressed as

$$\left(e^{\frac{1}{\epsilon}\int_{z_1}^{z} w_1(t)dt + r_1(z,\epsilon)}, \dots, e^{\frac{1}{\epsilon}\int_{z_1}^{z} w_k(t)dt + r_M(z,\epsilon)}\right),$$

where $r_j(z, \epsilon) = O(1)$ when $\epsilon \to 0$, $(z, \epsilon) \in \Delta(z', \eta') \times \mathcal{G}(r, \alpha)$. A straightforward calculation shows that the Wronskian of the system reads as

$$W\left(e^{\frac{1}{\epsilon}\int_{z_1}^{z} w_1(t)dt + r_1(z,\epsilon)}, \dots, e^{\frac{1}{\epsilon}\int_{z_1}^{z} w_k(t)dt + r_k(z,\epsilon)}\right)$$
$$= \prod_{j=1}^{k} e^{\frac{1}{\epsilon}\int_{z_1}^{z} w_j(t)dt + r_j(z,\epsilon)} \begin{vmatrix} 1 & \cdots & 1 \\ \mathbf{Y}_1(\frac{w_1(z)}{\epsilon} + r_1(z,\epsilon)) & \cdots & \mathbf{Y}_1(\frac{w_k(z)}{\epsilon} + r_k(z,\epsilon)) \\ \vdots & \ddots & \vdots \\ \mathbf{Y}_{k-1}(\frac{w_1(z)}{\epsilon} + r_1(z,\epsilon), \dots) & \cdots & \mathbf{Y}_{k-1}(\frac{w_k(z)}{\epsilon} + r_k(z,\epsilon), \dots) \end{vmatrix},$$

using the identity [54, (3c), p. 134], one has $\mathbf{Y}_n(x_1,\dots) = x_1^n + \sum_{i=1}^{n-1} \mathbf{B}_{n,i}(x_1,\dots)$, hence the determinant D in the above expression can be written as

$$D(z,\epsilon) = \epsilon^{-\frac{k(k-1)}{2}} \begin{vmatrix} 1 & 1 & \cdots & 1 \\ w_1(z)+o(\epsilon) & w_2(z)+o(\epsilon) & \cdots & w_k(z)+o(\epsilon) \\ \vdots & \vdots & \ddots & \vdots \\ (w_1(z))^{k-1}+o(\epsilon) & (w_2(z))^{k-1}+o(\epsilon) & \cdots & (w_k(z))^{k-1}+o(\epsilon) \end{vmatrix}$$
$$= \epsilon^{-\frac{k(k-1)}{2}} (V(z)+o(\epsilon)),$$

as $\epsilon \to 0$, $(z,\epsilon) \in \Delta(z',\eta') \times \mathcal{G}(r,\alpha)$, where

$$V(z) = \begin{vmatrix} 1 & 1 & \cdots & 1 \\ w_1(z) & w_2(z) & \cdots & w_k(z) \\ \vdots & \vdots & \ddots & \vdots \\ (w_1(z))^{k-1} & (w_2(z))^{k-1} & \cdots & (w_k(z))^{k-1} \end{vmatrix}.$$

On the other hand, for the Vandermonde determinant V we have that $V(z) \neq 0$, $z \in \Delta(z',\eta')$. Therefore, for r small, one has that $D \neq 0$ when $(z,\epsilon) \in \Delta(z',\eta') \times \mathcal{G}(r,\alpha)$. Hence,

$$W\left(e^{\frac{1}{\epsilon}\int_{z_1}^{z} w_1(t)dt + r_1(z,\epsilon)}, \dots, e^{\frac{1}{\epsilon}\int_{z_1}^{z} w_k(t)dt + r_k(z,\epsilon)}\right) \neq 0,$$

when $(z,\epsilon) \in \Delta(z',\eta') \times \mathcal{G}(r,\alpha)$ and small r. This proves the assertion c). □

4.3 Asymptotic expansion

In this section we settle Theorem 4.1 after proving a series of preliminary lemmas. We denote the small parameter by

$$\epsilon_n = \frac{1}{\sqrt[k]{\lambda_n - q_{0,0}}}, \tag{4.29}$$

where λ_n is defined as in (4.1); here we take the branch of the root in (4.29) for which ϵ_n asymptotically coincides with $\frac{1}{n}$ when $n \to \infty$. Recall that p_n is a polynomial solution to (4.3) such that $\lambda_n = q_{0,0} + \epsilon_n^{-k}$.

Lemma 4.5. *Let $y_j(z,\epsilon;z_1)$, $(z,\epsilon) \in \Delta(z',\eta') \times \mathcal{G}(r,\alpha)$, $j = 1,\dots,k$, be as in item a) of Proposition 4.1. Then,*

a) *There exist open subsets $S_+(z_1), S_-(z_1) \subset \Delta(z',\eta')$ and $0 < \alpha' < \alpha$ such that for $j = 2,\dots,k$,*

$$\Re\left[\epsilon^{-1}\left(\int_{z_1}^{z} w_1(t)dt - \int_{z_1}^{z} w_j(t)dt\right)\right] > 0, \quad \forall z \in S_+(z_1), \epsilon \in \mathcal{G}(r,\alpha'), \tag{4.30}$$

$$\Re\left[\epsilon^{-1}\left(\int_{z_1}^{z} w_1(t)dt - \int_{z_1}^{z} w_j(t)dt\right)\right] < 0, \quad \forall z \in S_-(z_1), \epsilon \in \mathcal{G}(r,\alpha'). \tag{4.31}$$

b) *There exists $z_1 \in \Delta(z',\eta')$ such that $z' \in S_+(z_1)$, where $S_+(z_1)$ is defined as in item a).*

c) *There exists $z'' \in S_-(z_1) \setminus supp(\mu)$ such that $p_n^{(j)}(z'') \neq 0$ for all $n \geq 0$ and $0 \leq j \leq k-2$.*

d)

$$W(\ldots, e^{y_{j-1}(z,\epsilon_n;z_1)}, p_n(z), e^{y_{j+1}(z,\epsilon_n;z_1)}, \ldots) = p_n(z)\epsilon_n^{-\frac{k(k-1)}{2}}(H_j(z,\epsilon_n) + o(\epsilon_n)) \prod_{\substack{\ell=1\\ \ell\neq j}}^{k} e^{\frac{1}{\epsilon_n}\int_{z_1}^{z} w_\ell(t)dt + o(\epsilon_n)},$$

where

$$H_j(z,\epsilon_n) = \begin{vmatrix} \cdots & 1 & 1 & 1 & \cdots \\ \cdots & w_{j-1}(z) & \frac{\epsilon_n p_n'(z)}{p_n(z)} & w_{j+1}(z) & \cdots \\ \vdots & \vdots & \ddots & \vdots & \vdots \\ \cdots & (w_{j-1}(z))^{k-1} & \frac{\epsilon_n^{k-1} p_n^{(k-1)}(z)}{p_n(z)} & (w_{j+1}(z))^{k-1} & \cdots \end{vmatrix}. \tag{4.32}$$

e) *For all n large enough, $W(p_n(z''), e^{y_2(z'',\epsilon_n;z_1)}, e^{y_3(z'',\epsilon_n;z_1)}, \ldots) \neq 0$, where z'' is as in item c).*

f) $\lim_{n\to\infty} \left|\frac{W(\ldots,e^{y_{j-1}(z'',\epsilon_n;z_1)}, p_n(z''), e^{y_{j+1}(z'',\epsilon_n;z_1)},\ldots)}{W(p_n(z''), e^{y_2(z'',\epsilon_n;z_1)}, e^{y_3(z'',\epsilon_n;z_1)},\ldots)}\right| = 0, j > 1.$

Proof. a) Since $b_1(z) = \int_{z_1}^{z} w_1(t)dt$ is holomorphic in $\Delta(z',\eta')$, we have that $U \equiv b_1(\Delta(z',\eta'))$ is an open set containing the origin. Hence, if $t = b_1(z)$ then $\forall z \in \Delta(z',\eta')$, $j = 2,\ldots,k$,

$$\Re\left[\left(1 - \exp\left(\frac{2(j-1)\pi i}{k}\right)\right)t\right] = \Re\left[\int_{z_1}^{z} w_1(t)dt - \int_{z_1}^{z} w_j(t)dt\right]. \tag{4.33}$$

Notice that the sets

$$\left\{t \in U : \Re\left[\left(1 - \exp\left(\frac{2(j-1)\pi i}{k}\right)\right)t\right] > 0, j = 2,\ldots,k\right\}$$

and

$$\left\{t \in U : \Re\left[\left(1 - \exp\left(\frac{2(j-1)\pi i}{k}\right)\right)t\right] < 0, j = 2,\ldots,k\right\}$$

are non-empty. Therefore, for $0 < \alpha' \leq \alpha$ small, one has that if $\epsilon \in \mathcal{G}(r,\alpha')$, then

$$\mathcal{T}_+(z_1) = \left\{t \in U : \Re\left[\epsilon^{-1}\left(1 - \exp\left(\frac{2(j-1)\pi i}{k}\right)\right)t\right] > 0, j = 2, \dots, k\right\},$$
$$\mathcal{T}_-(z_1) = \left\{t \in U : \Re\left[\epsilon^{-1}\left(1 - \exp\left(\frac{2(j-1)\pi i}{k}\right)\right)t\right] < 0, j = 2, \dots, k\right\}$$

are also non-empty sets. Writing $S_+(z_1) = b_1^{-1}(\mathcal{T}_+(z_1))$, $S_-(z_1) = b_1^{-1}(\mathcal{T}_-(z_1))$ and using (4.33), one obtains a).

b) By (4.30) and (4.31), we see that z_1 is in the common boundary of the open sets $S_+(z_1)$ and $S_-(z_1)$. Hence, by taking z_1 sufficiently close to z', we get $z' \in S_+(z_1)$.

c) Let $\mathcal{A}$ be the set of zeros of $p_n^{(j)}$ for all $n \geq 0$ and $0 \leq j \leq k-2$. Since $\mathcal{A}$ is at most countable, we have that $\mathcal{A}$ is of measure 0. Since $supp(\mu)$ is also a set of two-dimensional Lebesgue measure 0, we can find $z'' \in S_-(z_1)$ such that $p_n^{(j)}(z'') \neq 0$ for all $n \geq 0$ and $0 \leq j \leq k-2$.

d) Notice that for n large, $\epsilon_n \in \mathcal{G}(r, \alpha)$. We have that

$$\begin{aligned} &W(\dots, e^{y_{j-1}(z,\epsilon_n;z_1)}, p_n(z), e^{y_{j+1}(z,\epsilon_n;z_1)}, \dots) \\ &= p_n(z) \prod_{\substack{k=1\\k\neq j}}^{M} e^{\frac{1}{\epsilon_n}\int_{z_1}^{z} w_k(t)dt + r_k(z,\epsilon_n)} \\ &\quad \times \begin{vmatrix} \cdots & 1 & 1 & 1 & \cdots \\ \cdots & \mathbf{Y}_1(\frac{w_{j-1}(z)}{\epsilon_n} + r_{j-1}(z,\epsilon_n)) & \frac{Q_n'(z)}{Q_n(z)} & \mathbf{Y}_1(\frac{w_{j+1}(z)}{\epsilon_n} + r_{j+1}(z,\epsilon_n)) & \cdots \\ \vdots & \vdots & \vdots & \vdots & \vdots \\ \cdots & \mathbf{Y}_{k-1}(\frac{w_{j-1}(z)}{\epsilon_n} + r_{j-1}(z,\epsilon_n), \dots) & \frac{p_n^{(k-1)}(z)}{p_n(z)} & \mathbf{Y}_{k-1}(\frac{w_{j+1}(z)}{\epsilon_n} + r_{j+1}(z,\epsilon_n), \dots) & \cdots \end{vmatrix}. \end{aligned}$$

Using identity [54, (3c), p. 134], one has $\mathbf{Y}_n(x_1, \dots) = x_1^n + \sum_{i=1}^{n-1} \mathbf{B}_{n,i}(x_1, \dots)$. Hence

$$\begin{aligned} &W(\dots, e^{y_{j-1}(z,\epsilon_n;z_1)}, Q_n(z), e^{y_{j+1}(z,\epsilon_n;z_1)}, \dots) \\ &= \epsilon_n^{-\frac{M(M-1)}{2}} Q_n(z) \prod_{\substack{k=1\\k\neq j}}^{M} e^{\frac{1}{\epsilon_n}\int_{z_1}^{z} w_k(t)dt + o(\epsilon_n)} \\ &\quad \times \begin{vmatrix} \cdots & 1 & 1 & 1 & \cdots \\ \cdots & w_{j-1}(z) + o(\epsilon_n) & \frac{\epsilon_n Q_n'(z)}{p_n(z)} & w_{j+1}(z) + o(\epsilon_n) & \cdots \\ \vdots & \vdots & \vdots & \vdots & \vdots \\ \cdots & (w_{j-1}(z))^{k-1} + o(\epsilon_n) & \frac{\epsilon_n^{k-1} p_n^{(k-1)}(z)}{p_n(z)} & (w_{j+1}(z))^{k-1} + o(\epsilon_n) & \cdots \end{vmatrix} \\ &= \epsilon_n^{-\frac{k(k-1)}{2}} p_n(z) \prod_{\substack{k=1\\k\neq j}}^{k} e^{\frac{1}{\epsilon_n}\int_{z_1}^{z} w_k(t)dt + o(\epsilon_n)} (H_j(z,\epsilon_n) + o(\epsilon_n)), \end{aligned}$$

where H_j is given by (4.32).

e) Let z'' be as in item c) and $H_j(z, \epsilon_n)$ as in item d). By [27, Lem. 10 & Cor. 4], the root-counting measure of $p_n^{(j)}$, $0 \le j \le k$ converges weakly to a measure μ whose Cauchy transform equals $\frac{1}{\sqrt[k]{Q_k(z)}}$. Hence, for every fixed $1 \le j \le k-1$,

$$\lim_{n\to\infty} \frac{\epsilon_n^j Q_n^{(j)}(z'')}{Q_n(z'')} = \left(\frac{1}{\sqrt[k]{Q_k(z'')}}\right)^j = (w_1(z''))^j. \tag{4.34}$$

Therefore,

$$\lim_{n\to\infty} H_1(z'', \epsilon_n) = H_1(z''),$$

where

$$H_1(z'') = \begin{vmatrix} 1 & \cdots & 1 & 1 & \cdots \\ w_1(z'') & \cdots & w_j(z'') & w_{j+1}(z'') & \cdots \\ \vdots & \vdots & \vdots & \vdots & \vdots \\ (w_1(z''))^{k-1} & \cdots & (w_j(z''))^{k-1} & (w_{j+1}(z''))^{k-1} & \cdots \end{vmatrix}. \tag{4.35}$$

Since $H_1(z'') \neq 0$, then $W(p_n(z''), e^{y_2(z'', \epsilon_n; z_1)}, e^{y_3(z'', \epsilon_n; z_1)}, \dots) \neq 0$ by item d) for all n large enough.

f) Using item d) again, we get

$$\begin{aligned} &\left| \frac{W(\dots, e^{y_{j-1}(z'', \epsilon_n)}, p_n(z''), e^{y_{j+1}(z'', \epsilon_n)}, \dots)}{W(p_n(z''), \dots, e^{y_j(z'', \epsilon_n)}, e^{y_{j+1}(z'', \epsilon_n)}, \dots)} \right| \\ &= \left| \frac{H_j(z'', \epsilon_n) + o(\epsilon_n)}{H_1(z'', \epsilon_n) + o(\epsilon_n)} \right| e^{\Re[\epsilon_n^{-1}(\int_{z_1}^{z''} w_1(t)dt - \int_{z_1}^{z''} w_j(t)dt)]} |1 + o(\epsilon_n)|. \end{aligned} \tag{4.36}$$

Relations (4.32) and (4.34) imply

$$\lim_{n\to\infty} H_j(z'', \epsilon_n) = 0, \quad j > 1. \tag{4.37}$$

On the other hand, for n large enough, $\epsilon_n \in \mathcal{G}(r, \alpha')$. Since $z'' \in S_-(z_1)$, formula (4.31) implies

$$\lim_{n\to\infty} e^{\Re[\epsilon_n^{-1}(\int_{z_1}^{z''} w_1(t)dt - \int_{z_1}^{z''} w_j(t)dt)]} = 0. \tag{4.38}$$

Finally, by (4.35), (4.36), (4.37), and (4.38), we obtain f). □

Lemma 4.6. *Take $z' \in \Omega$. Then there exist $n_0 \in \mathbb{N}$, $0 < \eta$, $z_1 \in \Delta(z', \eta)$, and a unique constant c_n such that*

$$p_n(z) = c_n e^{\tilde{y}_1(z,\epsilon_n;z_1)}, \quad \forall n > n_0, \ \forall z \in \Delta(z',\eta),$$

where $\tilde{y}_1(z,\epsilon_n;z_1) = \frac{1}{\epsilon_n}\int_{z_1}^{z} \tilde{\mathfrak{w}}_1(t,\epsilon_n)dt$, $\tilde{\mathfrak{w}}_1(t,\epsilon_n) \in \mathcal{H}(\Delta(z',\eta))$, *and* $\tilde{\mathfrak{w}}_1(t,\epsilon_n) \sim \sum_{i=0}^{\infty} \mathfrak{b}_{1,i}(z)\epsilon_n^i$ *as* $n \to \infty$, *uniformly in* $z \in \Delta(z',\eta)$.

Proof. Let $\Delta(z',\eta')$ be as in item a) of Proposition 4.1. We have that for n large, $\epsilon_n \in \mathcal{G}(r,\alpha)$. Hence by items a) and c) of Proposition 4.1, there exist k linearly independent solutions $\{e^{y_1(z,\epsilon_n;z_1)},\dots,e^{y_k(z,\epsilon_n;z_1)}\}$ of (4.3) in $\Delta(z',\eta')$ and k unique constants $c_1(n),\dots,c_k(n)$ such that

$$p_n(z) = c_1(n)e^{y_1(z,\epsilon_n;z_1)} + \cdots + c_k(n)e^{y_k(z,\epsilon_n;z_1)}, \quad z \in \Delta(z',\eta'),\ n > n_0, \tag{4.39}$$

where $y_j(z,\epsilon_n;z_1) = \frac{1}{\epsilon_n}\int_{z_1}^{z} \mathfrak{w}_j(t,\epsilon_n)dt$, $\mathfrak{w}_j(t,\epsilon_n) \in \mathcal{H}(\Delta(z',\eta')\times\mathcal{G}(r,\alpha))$, and $\mathfrak{w}_j(t,\epsilon_n) \sim \sum_{i=0}^{\infty} \mathfrak{b}_{j,i}(z)\epsilon_n^i$ as $n \to \infty$, and $z_1 \in \Delta(z',\eta')$.

By item b) of Lemma 4.5 we can take $z_1 \in \Delta(z',\eta')$ such that $z' \in S_+(z_1)$, where $S_+(z_1)$ is defined as in item a) of Lemma 4.5. Choose $0 < \eta < \eta'$ so that $\Delta(z',\eta) \subset S_+(z_1)$, and let z'' be as in item e) of Lemma 4.5, see Figure 4.1. Using Cramer's rule, we get

$$c_j(n) = \frac{W(\dots,e^{y_{j-1}(z'',\epsilon_n;z_1)},p_n(z),e^{y_{j+1}(z'',\epsilon_n;z_1)},\dots)}{W(e^{y_1(z'',\epsilon_n;z_1)},\dots,e^{y_k(z'',\epsilon_n;z_1)})}.$$

Then, for n large and $z \in \Delta(z',\eta)$, we can rewrite (4.39) as

$$\begin{aligned} p_n(z) &= c_1(n)e^{y_1(z,\epsilon_n;z_1)} \\ &\times\left(1 + \sum_{j=2}^{k} \frac{W(\dots,e^{y_{j-1}(z'',\epsilon_n;z_1)},p_n(z''),e^{y_{j+1}(z'',\epsilon_n;z_1)},\dots)}{W(p_n(z''),e^{y_2(z'',\epsilon_n;z_1)},e^{y_3(z'',\epsilon_n;z_1)},\dots)} e^{y_j(z,\epsilon_n;z_1)-y_1(z,\epsilon_n;z_1)}\right). \end{aligned} \tag{4.40}$$

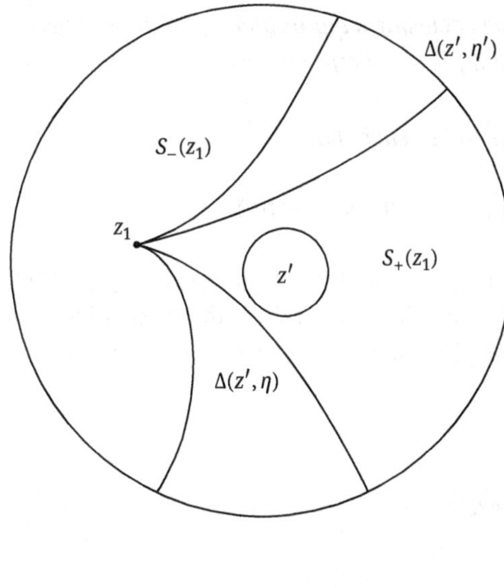

Figure 4.1: Regions $\Delta(z',\eta)$, $S_+(z_1)$ and $S_-(z_1)$.

On the other hand, since $z \in \Delta(z', \eta) \subset S_+(z_1)$, we obtain from relation (4.30) and item f) of Lemma 4.5 that the term in the parenthesis of (4.40) converges uniformly to 1. Hence, for some n_0,

$$p_n(z) = c_n' e^{\tilde{y}_1(z,\epsilon_n;z_1)}, \quad \forall n > n_0, \ \forall z \in \Delta(z', \eta),$$

where $\tilde{y}_1(z, \epsilon_n; z_1) = \frac{1}{\epsilon_n}\int_{z_1}^{z} \tilde{\mathfrak{w}}_1(t, \epsilon_n)dt$, $\tilde{\mathfrak{w}}_1(t, \epsilon_n) \in \mathcal{H}(\Delta(z', \eta))$, and $\tilde{\mathfrak{w}}_1(t, \epsilon_n) \sim \sum_{i=0}^{\infty} \mathfrak{b}_{1,i}(z)\epsilon_n^i$ as $n \to \infty$, and $z_1 \in \Delta(z', \eta')$. Finally, we choose the constant z_1 in the expression for $\tilde{y}_1$ so that it belongs to $\Delta(z', \eta)$, which completes the proof of the lemma. □

Lemma 4.7. *Let $K \subset \Omega$ be a compact set. Then there exist a finite collection $\{\Delta(z_i', \eta_i)\}_{i \in I}$ of open disks such that $K \subset \bigcup_{i \in I} \Delta(z_i', \eta_i)$ and a positive integer n_0 such that*

$$Q_n(z) = c_n(i) e^{\tilde{y}_{1,i}(z,\epsilon_n;z_{1,i})}, \quad \forall n > n_0, \ \forall z \in \Delta(z_i', \eta_i), \ i \in I,$$

where $\tilde{y}_{1,i}(z, \epsilon_n; z_{1,i}) = \frac{1}{\epsilon_n}\int_{z_{1,i}}^{z} \tilde{\mathfrak{w}}_{1,i}(t, \epsilon_n)dt$, $\tilde{\mathfrak{w}}_{1,i}(t, \epsilon_n) \in \mathcal{H}(\Delta(z_i', \eta_i))$, and also $\tilde{\mathfrak{w}}_{1,i}(t, \epsilon_n) \sim \sum_{s=0}^{\infty} \mathfrak{b}_{1,i,s}(z)\epsilon_n^s$ as $n \to \infty$ for z varying in compact subsets of $\Delta(z_i', \eta_i)$, and $z_{1,i} \in \Delta(z_i', \eta_i)$.

Proof. By Lemma 4.6, for each $z' \in \Omega$, there exist $\eta > 0$, $\tilde{y}_1$, and $n_0(z')$ such that

$$p_n(z) = c_n e^{\tilde{y}_1(z,\epsilon_n;z_1')}, \quad \forall n > n_0(z'), \ \forall z \in \Delta(z', \eta(z')),$$

where $\tilde{y}_1(z, \epsilon_n; z_1) = \frac{1}{\epsilon_n}\int_{z_1}^{z} \tilde{\mathfrak{w}}_1(t, \epsilon_n)dt$, $\tilde{\mathfrak{w}}_1(t, \epsilon_n) \in \mathcal{H}(\Delta(z', \eta(z')))$, $\tilde{\mathfrak{w}}_1(t, \epsilon_n) \sim \sum_{i=0}^{\infty} \mathfrak{b}_{1,i}(z)\epsilon_n^i$ as $n \to \infty$ for z varying in compact subsets of $\Delta(z', \eta(z'))$, and $z_1 \in \Delta(z', \eta(z'))$.

Let U be an open set such that $K \subset U$ and consider the open covering $K \subset \bigcup_{z' \in U} \Delta(z', \eta(z'))$. By the Heine–Borel theorem we can find a finite open subcover such that $K \subset \bigcup_{i \in I} \Delta(z_i', \eta_i)$, where I is a finite set of indices. By taking $n_0 = \max_{i \in I} n_0(z_i')$, we complete the proof. □

Lemma 4.8. *Let γ be a closed Jordan curve enclosing supp(μ) and $\Omega_\tau = \Omega \setminus \tau$; where τ is any Jordan arc connecting ∞ and any point $\omega \in$ supp(μ) and oriented so that ω is its endpoint. Then,*

a) *There exist $z_1 \in ext(\gamma)$, n_0, y, and a constant c_n such that*

$$p_n(z) = c_n e^{y(z,\epsilon_n;z_1)}, \quad \forall n > n_0, \ \forall z \in ext(\gamma),$$

where $y(z, \epsilon_n; z_1) = \frac{1}{\epsilon_n}\int_{z_1}^{z} \mathfrak{w}(t, \epsilon_n)dt$, $\mathfrak{w}(t, \epsilon_n) \in \mathcal{H}(ext(\gamma))$, and $\mathfrak{w}(z, \epsilon_n) \sim \sum_{i=0}^{\infty} \mathfrak{b}_i(z)\epsilon_n^i$ as $n \to \infty$ when z varies in compact subsets $K \subset ext(\gamma)$. Here the integration path of the above integral is contained in $ext(\gamma) \setminus \tau$, and $y(z_0, \epsilon_n; z_1) = \lim_{\substack{z \to z_0 \\ z \in \tau^+}} y(z, \epsilon_n; z_1)$, when $z_0 \in \tau$.

b) *Further,*

$$\mathfrak{b}_0(z) = w_1(z), \quad z \in ext(\gamma),$$

$$\mathfrak{b}_1(z) = \left(\frac{k-1}{2k}\frac{Q_k'(z)}{Q_k(z)} - \frac{1}{k}\frac{Q_{k-1}(z)}{Q_k(z)}\right), \quad z \in ext(\gamma).$$

Proof. a) By [27, Lem. 9], we can find a closed Jordan curve γ_1 enclosing γ and such that for $n > n_0$ the zeros of $p_n(z)$ are contained in $int(\gamma_1)$. Define the compact set $K_0 = \overline{int(\gamma_1) \setminus int(\gamma)}$. By Lemma 4.7 there exists a finite covering $\mathfrak{S} = \{\Delta(z_i', \eta_i)\}_{i \in I_0}$ by open disks such that $K_0 \subset \bigcup_{i \in I_0} \Delta(z_i', \eta_i)$ and

$$p_n(z) = c_n(i) e^{y_{1,i}(z,\epsilon_n;z_{1,i})}, \quad \forall n > n_0(i),\ \forall z \in \Delta(z_i', \eta_i),\ i \in I_0,$$

where

$$y_{1,i}(z,\epsilon_n;z_{1,i}) = \frac{1}{\epsilon_n}\int_{z_{1,i}}^{z} \mathfrak{w}_{1,i}(t,\epsilon_n)dt, \quad z, z_{1,i} \in \Delta(z_i',\eta_i),\ \mathfrak{w}_{1,i}(t,\epsilon_n) \in \mathcal{H}(\Delta(z_i',\eta_i)). \tag{4.41}$$

Define the open sets $V_i = \Delta(z_i', \eta_i) \cap \mathring{K}_0$, $i \in I_0$ and set $U_0 \equiv V_{i_0}$, where $i_0 \in I_0$ is such that $V_{i_0} \cap \tau = \emptyset$, and let $z \in (K_0 \setminus \tau)$. Since $(K_0 \setminus \tau)$ is connected, we can find a finite number of sets $\{V_{i_s}\}_{s=0}^m$ so that $V_{i_s} \cap V_{i_{s+1}} \neq \emptyset$ and $z \in V_{i_m}$. If $m > 0$, we write $U_s = V_{i_s} \cap (\mathring{K}_0 \setminus \tau)$, $s = 1, \ldots, m$; since this intersection is not necessarily connected, we choose the component of the intersection from the condition $U_s \cap U_{s-1} \neq \emptyset$. In particular, we have that $c_n(i_s)e^{y_{1,i_s}(z,\epsilon_n;z_{1,i_s})} = c_n(i_{s+1})e^{y_{1,i_{s+1}}(z,\epsilon_n;z_{1,i_{s+1}})}$, $z \in U_s \cap U_{s+1}$ for all $0 \le s \le m-1$. Hence $(c_n(i_m)e^{y_{1,i_m}(z,\epsilon_n;z_{1,i_m})}, U_m)$ is the analytic continuation of $(c_n(i_0)e^{y_{1,i_0}(z,\epsilon_n;z_{1,i_0})}, U_0)$ in the interior of the simply connected set $K_0 \setminus \tau$. If $\Delta(z_j', \eta_j)$ is any disk in the covering $\mathfrak{S}$ such that $\Delta(z_j', \eta_j) \cap U_m \neq \emptyset$, we have

$$y_{1,i_0}(z,\epsilon_n;z_{1,i_0}) = y_{1,j}(z,\epsilon_n;z_{1,j}) + k_{n,j}, \quad z \in U_m \cap \Delta(z_j',\eta_j),\ k_{n,j} \in \mathbb{C}. \tag{4.42}$$

Hence, by (4.41)

$$\mathfrak{w}_{1,i_0}(z,\epsilon_n) = \mathfrak{w}_{1,j}(z,\epsilon_n), \quad z \in U_m \cap \Delta(z_j',\eta_j).$$

Consequently, the analytic continuation of the function element $(\mathfrak{w}_{1,i_0}(z,\epsilon_n), U_0)$ defines a holomorphic function $\mathfrak{w}(z,\epsilon_n)$ when z varies in the interior of the simply connected set $K_0 \setminus \tau$. Hence

$$p_n(z) = c_n e^{y(z,\epsilon_n;z_1)}, \quad \forall n > n_0,\ \forall z \in K_0 \setminus \tau, \tag{4.43}$$

where $y(z,\epsilon_n;z_1) = \frac{1}{\epsilon_n}\int_{z_1}^{z} \mathfrak{w}(t,\epsilon_n)dt$, $n_0 = \max_{i \in I_0}\{n_0(i)\}$, $z_1 = z_{1,i_0}$, and $c_n = c_n(i_0)$.

Given $z_0 \in \tau$, take $\Delta(z_j', \eta_j) \subset \mathfrak{S}$ such that $z_0 \in \Delta(z_j', \eta_j)$. Since $y_{1,j}(z,\epsilon_n;z_{1,j})$ is holomorphic for $z \in \Delta(z_j', \eta_j)$, using (4.42) we get that for $n > n_0$, the following limit exists and is finite:

$$\lim_{\substack{z \to z_0 \\ z \in \tau^+}} y(z,\epsilon_n;z_1) = \lim_{\substack{z \to z_0 \\ z \in \tau^+}} y_{1,j}(z,\epsilon_n;z_{1,j}) + k_{n,j}. \tag{4.44}$$

Hence, the function y is holomorphic in the interior of $K_0\setminus\tau$, and $\lim_{\substack{z\to z_0\\ z\in\tau^+}} y(z,\epsilon_n;z_1)$ exists and is finite. Therefore, by (4.43) and (4.44), if we define $y(z_0,\epsilon;z_1)$ as

$$y(z_0,\epsilon_n;z_1) = \lim_{\substack{z\to z_0\\ z\in\tau^+}} y(z,\epsilon_n;z_1),$$

where $z_0 \in \tau\cap K_0$, then $e^{-y(z,\epsilon_n;z_1)}$ is holomorphic in the interior of K_0 and

$$p_n(z) = c_n e^{y(z,\epsilon_n;z_1)}, \quad \forall n > n_0,\ \forall z \in K_0. \tag{4.45}$$

On the other hand, by (4.45) the function

$$\mathfrak{w}(z,\epsilon_n) = e^{-y(z,\epsilon_n;z_1)} p_n'(z) \tag{4.46}$$

can be analytically continued in the interior of K_0 for every $n > n_0$. Since p_n does not have zeros outside of γ_1 for $n > n_0$, y can be analytically continued to $ext(\gamma)\setminus\tau$. Therefore, by virtue of (4.46), $\mathfrak{w}$ can also be analytically continued to $ext(\gamma)$, and we have that

$$p_n(z) = c_n e^{y(z,\epsilon_n;z_1)}, \quad \forall n > n_0,\ \forall z \in ext(\gamma), \tag{4.47}$$

where $y(z,\epsilon_n;z_1) = \frac{1}{\epsilon_n}\int_{z_1}^{z} \mathfrak{w}(t,\epsilon_n)dt$, $\mathfrak{w}(t,\epsilon_n) \in \mathcal{H}(ext(\gamma))$.

Let $K \subset ext(\gamma)$ be an arbitrary compact subset. Using Lemma 4.7 we can find a finite covering $\mathfrak{S}_1 = \{\Delta(z_i',\eta_i)\}_{i\in I_1}$ by open disks such that $K \subset \bigcup_{i\in I_1}\Delta(z_i',\eta_i)$ and

$$p_n(z) = c_n(i)e^{y_{1,i}(z,\epsilon_n;z_{1,i})}, \quad \forall n > n_1(i),\ \forall z \in \Delta(z_i',\eta_i),\ i \in I_1,$$

where $y_{1,i}(z,\epsilon_n;z_{1,i}) = \frac{1}{\epsilon_n}\int_{z_{1,i}}^{z} \mathfrak{w}_{1,i}(t,\epsilon_n)dt$, $\mathfrak{w}_{1,i}(t,\epsilon_n) \in \mathcal{H}(\Delta(z_i',\eta_i))$ and

$$\mathfrak{w}_{1,i}(z,\epsilon_n) \sim \sum_{s=0}^{\infty} \mathfrak{b}_{1,i,s}(z)\epsilon_n^s, \tag{4.48}$$

as $n \to \infty$ uniformly in $z \in \Delta(z_i',\eta_i)$. By (4.47) we have that for $n > \max\{n_0, n_1(i)\}$,

$$\mathfrak{w}(z,\epsilon_n) = \mathfrak{w}_{1,i}(z,\epsilon_n), \quad z \in \Delta(z_i',\eta_i). \tag{4.49}$$

Hence, by (4.49) and using the fact that the open covering is finite, we get

$$\mathfrak{w}(z,\epsilon_n) \sim \sum_{s=0}^{\infty} \mathfrak{b}_s(z)\epsilon_n^s \quad \text{for } n \to \infty,\ z \in K.$$

Here $\mathfrak{b}_s$ is such that

$$\mathfrak{b}_s(z) = \mathfrak{b}_{1,i,s}(z), \quad z \in \Delta(z_i',\eta_i). \tag{4.50}$$

b) Follows from item b) of Proposition 4.1 and (4.50). □

Lemma 4.9. *Let $\mathfrak{w}$ be defined as in item a) of Lemma 4.8. Then $\mathfrak{w}$ and $(\mathfrak{b}_s)_{s\geq 0}$ are holomorphic when z varies in a neighborhood of ∞, and $\lim_{z\to\infty}\mathfrak{b}_s(z) = 0$, $\lim_{z\to\infty}\mathfrak{w}(z,\epsilon_n) = 0$ for each $n > n_0$ fixed.*

Proof. By the relation

$$\mathfrak{w}(z,\epsilon_n) = \frac{\epsilon_n p_n'(z)}{p_n(z)}, \quad z \in ext(\gamma),\ n > n_0,$$

we obtain that $\mathfrak{w}$ is holomorphic in a neighborhood V_∞ of ∞, and $\lim_{z\to\infty}\mathfrak{w}(z,\epsilon_n) = 0$ for fixed $n > n_0$.

We will prove by induction that $\mathfrak{b}_k$ is holomorphic when $z \in V_\infty$, and $\lim_{z\to\infty}\mathfrak{b}_s(z) = 0$ for all $s \geq 0$. Indeed, by b) of Lemma 4.8 one obtains that the statement is true for $s = 0$.

Assume that $\mathfrak{b}_s$ is holomorphic in V_∞ for $0 \leq s \leq p_0$. Since $\mathfrak{w}(z,\epsilon_n) \sim \sum_{s=0}^{\infty}\mathfrak{b}_s(z)\epsilon_n^s$ as $n\to\infty$ when z varies in compact subsets $K \subset ext(\gamma)$ one has for every integer $p \geq 0$

$$\mathfrak{w}(z,\epsilon_n) = \sum_{s=0}^{p}\epsilon_n^s\mathfrak{b}_s(z) + r_p(z,\epsilon_n), \quad \forall z \in ext(\gamma),$$

where $r_p(z,\epsilon_n) = O(\epsilon_n^{p+1})$ as z varies in K. By the induction hypothesis it follows that $r_{p_0}(z,\epsilon_n) = \epsilon_n^{p_0+1}\mathfrak{b}_{p_0+1}(z) + r_{p_0+1}(z,\epsilon_n)$ is holomorphic in V_∞ and $\lim_{z\to\infty} r_{p_0}(z,\epsilon_n) = 0$. Hence, for all $n > n_0$,

$$\mathfrak{b}_{p_0+1}(z) + \frac{r_{p_0+1}(z,\epsilon_n)}{\epsilon_n^{p_0+1}} = \sum_{k=1}^{\infty}\frac{a_k(\epsilon_n)}{z^k}, \quad z \in V_\infty,\ a_k(\epsilon_n) \in \mathbb{C}. \tag{4.51}$$

Hence,

$$\frac{1}{2\pi\imath}\int_{|t|=R} t^{k-1}\left(\mathfrak{b}_{p_0+1}(t) + \frac{r_{p_0+1}(t,\epsilon_n)}{\epsilon_n^{p_0+1}}\right)dt = a_k(\epsilon_n),$$

by taking limit as $n\to\infty$, we get

$$\frac{1}{2\pi\imath}\int_{|t|=R} t^{k-1}\left(\mathfrak{b}_{p_0+1}(t) + \lim_{n\to\infty}\frac{r_{p_0+1}(t,\epsilon_n)}{\epsilon_n^{p_0+1}}\right)dt = \lim_{n\to\infty} a_k(\epsilon_n).$$

Indeed, since $r_p(z,\epsilon_n) = o(\epsilon_n^p)$, $\forall p \geq 0$, when z varies in compact subsets of Ω, we have that the limit in the left-hand side exists and is finite, whence the limit in the right-hand side exists too. Therefore,

$$\int_{|t|=R} t^{k-1}\mathfrak{b}_{p_0+1}(t)dt = 2\pi\imath a_k, \quad k \geq 1,$$

where $a_k = \lim_{n\to\infty} a_k(\epsilon_n)$. Hence by (4.51), $\mathfrak{b}_{p_0+1}$ is holomorphic in V_∞, which completes the proof of the lemma. □

Lemma 4.10. *For n sufficiently large, we get*

a)

$$\epsilon_n^{-1} = n + \sum_{i=0}^{\infty} \gamma_i n^{-i},$$
$$\epsilon_n^{\ell} = \sum_{i=0}^{\infty} \tau_{i,\ell} n^{-\ell-i}, \quad \ell > 0, \tag{4.52}$$

where $(\gamma_i)_{i=0}^{\infty}$ are defined from the following Laurent expansion about $z = \infty$:

$$f(z) = \sqrt[k]{\sum_{j=1}^{k} Q_{j,j}(z)_j} = z + \sum_{i=0}^{\infty} \gamma_i z^{-i} \tag{4.53}$$

and $\tau_{i,\ell} = \frac{1}{i!}\mathbb{P}_i^{(-\ell)}(\gamma_0, \dots, i!\gamma_{i-1})$, $\gamma_0 = -(\frac{k-1}{2} - \frac{Q_{k-1,k-1}}{k})$, $\tau_{0,\ell} = 1$. Here $\mathbb{P}_n^{(r)}$ denotes the potential polynomials, cf. [54, *Th. B, p. 141*].

b)

$$n = \sum_{i=0}^{\infty} h_i \epsilon_n^{i-1},$$

where $h_0 = 1$, $h_1 = -\gamma_0$, $h_2 = -\gamma_1$, $h_{i+1} = -\gamma_i - \sum_{s=1}^{i-1} h_{s+1}\tau_{i-s,s}$, $i \geq 2$, and $(\gamma_i)_{i=0}^{\infty}$ is defined as in a).

Proof. a) From (4.53) and (4.29), for n sufficiently large we have

$$\epsilon_n^{-1} = n - \left(\frac{k-1}{2} - \frac{q_{k-1,k-1}}{k}\right) + \sum_{i=1}^{\infty} \frac{\gamma_i}{n^i},$$

implying that

$$\epsilon_n^{\ell} = \sum_{i=0}^{\infty} \tau_{i,\ell} n^{-\ell-i}, \quad \ell > 0,$$

where $\tau_{i,\ell} = \frac{1}{i!}\mathbb{P}_i^{(-\ell)}(\gamma_0, \dots, j!\gamma_{i-1})$, $\gamma_0 = -(\frac{k-1}{2} - \frac{q_{k-1,k-1}}{k})$, $\tau_{0,\ell} = 1$; cf. [54, Th. B, p. 141].

b) Assume that n is large enough, then from (4.29) we get

$$\sum_{j=1}^{k} q_{j,j}(n)_j - \frac{1}{\epsilon_n^k} = 0.$$

Changing the variables $n = \omega$ and $u = \epsilon_n$, we obtain the algebraic equation

$$\omega^k + \sum_{j=1}^{k-1} a_j \omega^j - \frac{1}{u^k} = 0, \tag{4.54}$$

where $a_1, \dots, a_{k-1} \in \mathbb{C}$. Multiplying by u^k and changing the variable $w = \omega u$ in (4.54), one obtains

$$F(u, w) = w^k + \sum_{j=1}^{k-1} a_j u^{k-j} w^j - 1 = 0.$$

We have that $F(0,1) = 0$ and $\frac{\partial F}{\partial w}(0,1) \neq 0$. From the implicit function theorem [126, Th. 3.11, Vol. II], there exist a neighborhood V of 0 and a unique analytic function $w(u) = \sum_{i=0}^{\infty} u^i h_i$ such that $w(0) = 1$ and $F(u, w(u)) = 0, \forall u \in V$. Taking into account that $w = \omega u$, $u \neq 0$, we obtain that $\omega(u) = \sum_{i=0}^{\infty} u^{i-1} h_i$ is a solution of (4.54) for u varying in a punctured neighborhood V^* of $u = 0$. Hence,

$$n = \sum_{i=0}^{\infty} h_i \epsilon_n^{i-1}, \tag{4.55}$$

where $h_0 = 1$.

Using (4.52) and (4.55) we get

$$0 = \sum_{j=0}^{\infty} \gamma_j n^{-j} + h_1 + \sum_{i=1}^{\infty} h_{i+1} \sum_{j=0}^{\infty} q_{j,i} n^{-i-j},$$

which after some straightforward transformations results into

$$0 = \gamma_0 + h_1 + \sum_{j=1}^{\infty} \left(\gamma_j + \sum_{i=1}^{j} h_{i+1} \tau_{j-i,i} \right) n^{-j},$$

yielding $h_1 = -\gamma_0$, $h_2 = -\gamma_1$, and $h_{j+1} = -\gamma_j - \sum_{i=1}^{j-1} h_{i+1} \tau_{j-i,i}$, $j \geq 2$. □

Lemma 4.11. *Let τ be any Jordan arc from $-\infty$ to $\omega \in supp(\mu)$ such that $]-\infty, p] \subset \tau$ for some $p \in \mathbb{R}$, and let γ be a closed Jordan curve enclosing $supp(\mu)$. Then, for n sufficiently large, we get*

$$p_n(z) = e^{\frac{1}{\epsilon_n} y(z, \epsilon_n)}, \quad \forall z \in ext(\gamma),$$

where $y(z, \epsilon_n) \sim \sum_{i=0}^{\infty} \Phi_i(z) \epsilon_n^i$, as $n \to \infty$ when z varies in compact subsets $K \subset ext(\gamma)$ and $\Phi_i(z)$ is the primitive of the function $\mathfrak{b}_i$ in $ext(\gamma) \setminus \tau$ such that $\lim_{z\to\infty} \Phi_i(z) - h_i \ln z = 0$, $\Phi_i(z) = \lim_{\substack{z \to z_0 \\ z \in \tau^+}} \Phi_i(z)$, when $z_0 \in \tau \cap ext(\gamma)$, being h_i as in item b) of Lemma 4.10.

Proof. By item a) of Lemma 4.8 there exist $z_1 \in ext(\gamma)$, n_0, y, and a constant c_n such that

$$p_n(z) = c_n e^{y(z, \epsilon_n; z_1)}, \quad \forall n > n_0, \ \forall z \in ext(\gamma),$$

where $y(z, \epsilon_n; z_1) = \frac{1}{\epsilon_n} \int_{z_1}^{z} \mathfrak{w}(t, \epsilon_n) dt$, $\mathfrak{w}(t, \epsilon_n) \in \mathcal{H}(ext(\gamma))$, $\mathfrak{w}(z, \epsilon_n) = \sum_{s=0}^{p} \mathfrak{b}_s(z) \epsilon_n^s + \mathfrak{r}_p(z, \epsilon_n) \epsilon_n^{p+1}$, and as $n \to \infty$,

$$\mathfrak{r}_p(z, \epsilon_n) = O(1), \tag{4.56}$$

for z varying in compact subsets $K \subset ext(\gamma)$, and any fixed $p \geq 0$.

Therefore, for n fixed,

$$\lim_{z\to\infty} \frac{e^{y(z,\epsilon_n;z_1)}}{z^n} = \frac{1}{c_n}.$$

Hence, for every $p \geq 0$ fixed

$$\lim_{z\to\infty} \exp\left(\sum_{s=0}^{p} \epsilon_n^{s-1} \int_{z_1}^{z} \mathfrak{b}_s(t)dt + \epsilon_n^p \int_{z_1}^{z} \mathfrak{r}_p(t, \epsilon_n)dt - n \ln z \right) = \frac{1}{c_n}. \tag{4.57}$$

From item b) of Lemma 4.10, every sufficiently large $n \in \mathbb{N}$ can be uniquely expressed as

$$n = \sum_{s=0}^{p} h_s \epsilon_n^{s-1} + \epsilon_n^p g_p(\epsilon_n), \quad \forall p \in \mathbb{N} \cup \{0\}, \tag{4.58}$$

where h_s is defined as in item b) of Lemma 4.10 and

$$g_p(\epsilon_n) = O(1), \quad \text{as } n \to \infty, \tag{4.59}$$

for all $p \geq 0$ fixed.

From (4.57) and (4.58), for all $p \geq 0$ fixed, we get

$$\lim_{z\to\infty} \exp\left(\epsilon_n^{-1} \left(\int_{z_1}^{z} \mathfrak{b}_0(t)dt - \ln z \right) + \sum_{s=1}^{p} \left(\int_{z_1}^{z} \mathfrak{b}_s(t)dt - h_s \ln z \right) \epsilon_n^{s-1} \right.$$
$$\left. + \left(\int_{z_1}^{z} \mathfrak{r}_p(t, \epsilon_n)dt - g_p(\epsilon_n) \ln z \right) \epsilon_n^p \right) = \frac{1}{c_n}. \tag{4.60}$$

Define h_s, $s \geq 0$ is defined as in item b) of Lemma 4.10. We want to prove that for all $p \geq 0$ there exists a primitive of the function $\mathfrak{b}_p$ in $ext(\gamma) \setminus \tau$ such that $\lim_{z\to\infty} \Phi_p(z) - h_p \ln z = 0$ and

$$p_n(z) = \exp\left(\sum_{s=0}^{p} \Phi_s(z)\epsilon_n^{s-1} + \Psi_p(z, \epsilon_n) \right), \tag{4.61}$$

where $\Psi_p(z, \epsilon_n) = O(\epsilon_n^p)$ as $n \to \infty$ when z varies in compact subsets $K \subset ext(\gamma)$.

The argument follows by induction in p. For $p = 0$, by (4.57) and (4.58), we get

$$\lim_{z\to\infty} \exp\left(\epsilon_n^{-1}\left(\int_{z_1}^{z} \mathfrak{b}_0(t)dt - \ln z \right) + \int_{z_1}^{z} \mathfrak{r}_0(t,\epsilon_n)dt - g_0(\epsilon_n)\ln z \right) = \frac{1}{c_n}. \tag{4.62}$$

By item b) of Lemma 4.8 we have that $\mathfrak{b}_0(z) = w_1(z)$, $z \in ext(\gamma)$. Let Φ_0 be the primitive of $\mathfrak{b}_0$ that asymptotically coincides with $\ln z$ near of ∞. We have that

$$\begin{aligned} \Phi_0(z) &= \int_{z_1}^{z} \mathfrak{b}_0(t)dt + \Phi_0(z_1), \quad z \in ext(\gamma) \setminus \tau, \\ \Phi_0(z_0) &= \lim_{\substack{z\to z_0 \\ z\in\tau^+}} \Phi_0(z), \quad z_0 \in \tau \cap ext(\gamma). \end{aligned} \tag{4.63}$$

On the other hand, relation (4.62) gives

$$\lim_{z\to\infty} \int_{z_1}^{z} \mathfrak{r}_0(t,\epsilon_n)dt - g_0(\epsilon_n)\ln z = d_n \in \mathbb{C},$$

therefore, by taking

$$\begin{aligned} \Psi_0(z,\epsilon_n) &= \int_{z_1}^{z} \mathfrak{r}_0(t,\epsilon_n)dt - d_n, \quad z \in ext(\gamma) \setminus \tau, \\ \Psi_0(z_0,\epsilon_n) &= \lim_{\substack{z\to z_0 \\ z\in\tau^+}} \Psi_0(z,\epsilon_n), \quad z_0 \in \tau \cap ext(\gamma), \end{aligned} \tag{4.64}$$

and by using (4.63) and (4.64), we obtain

$$\lim_{z\to\infty} \exp(\epsilon_n^{-1}(\Phi_0(z) - \ln z) + \Psi_0(z,\epsilon_n) - g_0(\epsilon_n)\ln z) = 1.$$

Notice that by virtue of (4.56) and (4.59), we obtain

$$\Psi_0(z,\epsilon_n) = O(1),$$

as $n \to \infty$ and for z varying in compact subsets of $ext(\gamma)$. Hence, relation (4.61) holds for $p = 0$.

Assume that (4.61) holds for $p = p_0$. From the relation

$$p_n(z) = c_n \exp\left(\sum_{s=0}^{p_0} \Phi_s(z)\epsilon_n^{s-1} + \epsilon_n^{p_0} \int_{z_1}^{z} \mathfrak{b}_{p_0+1}(t)dt + \epsilon_n^{p_0+1} \int_{z_1}^{z} \mathfrak{r}_{p_0+1}(t,\epsilon_n)dt \right),$$

valid $\forall z \in ext(\gamma)$, $n > n_0$ and from the fact that

$$\lim_{z\to\infty} \exp\left(\sum_{s=0}^{p_0} (\Phi_s(z) - h_s \ln z)\epsilon_n^{s-1} + \left(\int_{z_1}^{z} \mathfrak{b}_{p_0+1}(t)dt - h_{p_0+1} \ln z \right)\epsilon_n^{p_0} \right.$$
$$\left. + \left(\int_{z_1}^{z} \mathfrak{r}_{p_0+1}(t, \epsilon_n)dt - g_{p_0+1}(\epsilon_n) \ln z \right)\epsilon_n^{p_0+1} \right) = \frac{1}{c_n}, \tag{4.65}$$

one gets

$$d_n' = \lim_{\substack{z\to\infty \\ z\in ext(\gamma)\setminus\tau}} \left(\int_{z_1}^{z} \mathfrak{b}_{p_0+1}(t)dt - h_{p_0+1} \ln z \right)\epsilon_n^{p_0} + \left(\int_{z_1}^{z} \mathfrak{r}_{p_0+1}(t, \epsilon_n)dt - g_{p_0+1}(\epsilon_n) \ln z \right)\epsilon_n^{p_0+1}. \tag{4.66}$$

By Lemma 4.9, there exists a neighborhood V_∞ of ∞ such that for $z \in V_\infty \setminus \tau$,

$$\int_{z_1}^{z} \mathfrak{b}_{p_0+1}(t)dt = a_{-1,p_0+1} \ln z + a_{0,p_0+1} + \sum_{s=2}^{\infty} \frac{a_{-s,p_0+1}}{z^{s-1}}, \tag{4.67}$$

$$\int_{z_1}^{z} \mathfrak{r}_{p_0+1}(t, \epsilon_n)dt = m_{-1,p_0+1}(\epsilon_n) \ln z + m_{0,p_0+1}(\epsilon_n) + \sum_{s=2}^{\infty} \frac{m_{-s,p_0+1}(\epsilon_n)}{z^{s-1}}. \tag{4.68}$$

Hence, by (4.66), (4.67), and (4.68) we deduce that for all fixed $n > n_0$, the limit

$$\lim_{\substack{z\to\infty \\ z\in V_\infty\setminus\tau}} \sum_{s=2}^{\infty} \frac{a_{-s,p_0+1}\epsilon_n^{p_0} + m_{-s,p_0+1}(\epsilon_n)\epsilon_n^{p_0+1}}{z^{s-1}} + ((a_{-1,p_0+1} - h_{p_0+1})\epsilon_n^{p_0}$$
$$+ (m_{-1,p_0+1}(\epsilon_n) - g_{p_0+1}(\epsilon_n))\epsilon_n^{p_0+1}) \ln z + (a_{0,p_0+1}\epsilon_n^{p_0} + m_{0,p_0+1}(\epsilon_n)\epsilon_n^{p_0+1}) \tag{4.69}$$

is finite.

Since $\mathfrak{r}_{p_0+1}$ is holomorphic for $z \in V_\infty$ and R large enough, one has

$$m_{-1,p_0+1}(\epsilon_n)2\pi\imath = \int_{|t|=R} \mathfrak{r}_{p_0+1}(t, \epsilon_n)dt.$$

Hence by (4.56)

$$\lim_{n\to\infty} m_{-1,p_0+1}(\epsilon_n) \tag{4.70}$$

exists and is finite.

On the other hand, by (4.69)

$$a_{-1,p_0+1} - h_{p_0+1} = -(m_{-1,p_0+1}(\epsilon_n) - g_{p_0+1}(\epsilon_n))\epsilon_n. \tag{4.71}$$

Therefore, by (4.59) and (4.70)

$$\lim_{n\to\infty} -(m_{-1,p_0+1}(\epsilon_n) - g_{p_0+1}(\epsilon_n))\epsilon_n = 0 = \alpha_{-1,p_0+1} - h_{p_0+1},$$

that is,

$$\alpha_{-1,p_0+1} = h_{p_0+1}. \tag{4.72}$$

Hence, from (4.71)

$$m_{-1,p_0+1}(\epsilon_n) = g_{p_0+1}(\epsilon_n). \tag{4.73}$$

By (4.72) and (4.73) we deduce that there exist a primitive Φ_{p_0+1} of the function $\mathfrak{b}_{p_0+1}$ and a primitive Ψ_{p_0+1} of the function $\mathfrak{r}_{p_0+1}$ in $ext(\gamma) \setminus \tau$ such that

$$\lim_{z\to\infty} \Phi_{p_0+1}(z) - h_{p_0+1} \ln z = 0,$$
$$\lim_{z\to\infty} \Psi_{p_0+1}(z,\epsilon_n) - g_{p_0+1}(\epsilon_n) \ln z = 0.$$

Hence, if $\Phi_{p_0+1}(z_0) = \lim_{\substack{z\to z_0\\ z\in\tau^+}} \Phi_{p_0+1}(z)$ and $\Psi_{p_0+1}(z_0,\epsilon_n) = \lim_{\substack{z\to z_0\\ z\in\tau^+}} \Psi_{p_0+1}(z,\epsilon_n)$ when $z_0 \in \tau \cap ext(\gamma)$, we obtain for all fixed $n > n_0$

$$\lim_{z\to\infty} e^{\sum_{s=0}^{p_0}(\Phi_s(z)-h_s \ln z)\epsilon_n^{s-1} + (\Phi_{p_0+1}(z)-h_{p_0+1}\ln z)\epsilon_n^{p_0} + (\Psi_{p_0+1}-g_{p_0+1}(\epsilon_n)\ln z)\epsilon_n^{p_0+1}} = 1.$$

Notice that by (4.56) and (4.59)

$$\Psi_{p_0+1}(z,\epsilon_n) = O(1),$$

as $n \to \infty$ when z varies in compact subsets of $ext(\gamma)$. Hence (4.61) holds for $p = p_0 + 1$. This completes the proof of the lemma. □

Proof of Theorem 4.1. Notice that by [27, Th. 3], the support $supp(\mu)$ is an embedded tree in $\mathbb{C}$, see Chapter 2. It suffices to consider any Jordan curve γ enclosing $supp(\mu)$ and prove the asymptotic relation in compact subsets of $ext(\gamma)$. By Lemma 4.11,

$$p_n(z) = \exp\left(\sum_{s=0}^{p} \Phi_s(z)\epsilon_n^{s-1} + \Psi_p(z,\epsilon_n)\epsilon_n^p\right), \quad \forall z \in ext(\gamma),\ \forall p \in \mathbb{N}\cup\{0\}, \tag{4.74}$$

where Φ_s is the primitive of the function $\mathfrak{b}_s$ in $ext(\gamma)\setminus\tau$ such that $\lim_{z\to\infty}\Phi_s(z) - h_s \ln z = 0$, h_s as in b) of Lemma 4.10, and $\Psi_p(z,\epsilon_n) = O(1)$ as $n \to \infty$ for z varying in compact subsets $K \subset ext(\gamma)$ and fixed $p \geq 0$.

Using item a) of Lemma 4.10, $p > 1$, and $z \in ext(\gamma)$, write (4.74) as

$$p_n(z) = \exp\left(\left(n - \left(\frac{k-1}{2} - \frac{Q_{k-1,k-1}}{k}\right)\right)\Phi_0(z) + \Phi_1(z) + \Phi_0(z)\sum_{j=1}^{p-1}\frac{\gamma_j}{n^j}\right.$$

$$+\sum_{s=1}^{p-1}\sum_{j=s}^{p-1}\frac{q_{j-s,s}}{n^j}\Phi_{s+1}(z)+\Psi_p^*(z,n)\frac{1}{n^p}\Bigg)$$

$$=\exp\left(n\Phi_0(z)-\left(\frac{k-1}{2}-\frac{q_{k-1,k-1}}{k}\right)\Phi_0(z)+\Phi_1(z)+\sum_{j=1}^{p-1}\Phi_j^*(z)\frac{1}{n^j}+\Psi_p^*(z,n)\frac{1}{n^p}\right),$$

where $\Phi_j^*(z)=\gamma_j\Phi_0(z)+\sum_{s=1}^{j}\tau_{j-s,s}\Phi_{s+1}(z)$. Since $\Psi_p^*(z,\epsilon_n)=O(1)$ as $n\to\infty$, then for z varying in compact subsets $K\subset ext(\gamma)$, we can write

$$p_n(z)=e^{\Phi_0(z)(n-\frac{k-1}{2}+\frac{q_{k-1,k-1}}{k})+\Phi_1(z)}\left(1+\sum_{j=1}^{p-1}\frac{C_j(z)}{n^j}+O\left(\frac{1}{n^p}\right)\right). \tag{4.75}$$

Here $C_j(z)=\frac{1}{j!}\mathbf{Y}_j(\Phi_1^*(z),2\Phi_2^*(z),\dots,j!\Phi_j^*(z))$.

To prove that $C_j\in\mathcal{H}(\Omega)$, $\forall j\in\mathbb{N}$, we need that $\exp((\frac{k-1}{2}-\frac{q_{k-1,k-1}}{k})\Phi_0+\Phi_1)$ and $\exp(\Phi_0)$ belong to $\mathcal{H}(\Omega)$. We may assume that $\Phi_0(z)\neq\ln z$, or equivalently that $Q_k(z)\neq z^k$, otherwise there is nothing to prove. By Lemma 4.9, $\mathfrak{b}_0\in\mathcal{H}(V_\infty)$ and $\lim_{z\to\infty}\mathfrak{b}_s(z)=0$, being V_∞ a neighborhood of ∞. Since Φ_0 is the primitive of the function $\mathfrak{b}_0$ in $ext(\gamma)$ such that $\lim_{z\to\infty}\Phi_0(z)-\ln z=0$. The function e^{Φ_0} admits a Laurent expansion

$$e^{\Phi_0(z)}=\beta z+\sum_{s=0}^{\infty}\frac{\beta_s}{z^s},\quad z\in V_\infty.$$

Therefore

$$e^{\Phi_0(z)}-\beta z=\sum_{s=0}^{\infty}\frac{\beta_s}{z^s},\quad z\in V_\infty. \tag{4.76}$$

Take $\widetilde{\Omega}=\mathcal{T}(\Omega)$, where $\mathcal{T}(z)=\frac{1}{z}$ and define the boundary point $\widetilde{\omega}$ of $\widetilde{\Omega}$ by the condition $|\widetilde{\omega}|=\min_{\omega\in\partial\widetilde{\Omega}}|\omega|$. Taking a cut in $\widetilde{\Omega}$ using the line segment $\widetilde{\mathfrak{l}}$ from 0 to $\widetilde{\omega}$, we see that $\mathfrak{l}=\mathcal{T}^{-1}(\widetilde{\mathfrak{l}})$ is a Jordan arc in the region Ω. By (4.76), we have that the left-hand side is holomorphic in V_∞, hence by [126, Th. 8.5, p. 269 & Cor. p. 272, Vol. III] it can be prolonged analytically in the simply connected open set $V_\infty\bigcup(\Omega\setminus\mathfrak{l})$. Now, by considering $z\mapsto\frac{1}{z}$, we see that the right-hand side of (4.76) is a convergent power series in the disk $\Delta(0,|\widetilde{\omega}|)\subset\widetilde{\Omega}$. Therefore $e^{\Phi_0(z)}\in\mathcal{H}(\Omega)$.

On the other hand, by (4.74) one has

$$\lim_{n\to\infty}\frac{p_n(z)}{e^{n\Phi_0(z)}}=\exp\left(\left(\frac{k-1}{2}-\frac{q_{k-1,k-1}}{k}\right)\Phi_0(z)+\Phi_1(z)\right)$$

in compact subsets of Ω. Since $e^{\Phi_0}\in\mathcal{H}(\Omega)$, it follows from the Weierstrass theorem about uniformly convergent sequences of holomorphic functions that

$$\exp\left(\left(\frac{k-1}{2}-\frac{q_{k-1,k-1}}{k}\right)\Phi_0+\Phi_1\right)\in\mathcal{H}(\Omega),$$

cf. [126, Th. 15.8, p. 330, Vol. I].

Now, from (4.75) and since $\exp(\Phi_0)$, $\exp((\frac{k-1}{2} - \frac{q_{k-1,k-1}}{k})\Phi_0 + \Phi_1)$ belong to $\mathcal{H}(\Omega)$, we have that

$$C_1(z) = \lim_{n\to\infty} \frac{nQ_n(z)}{\exp(n\Phi_0(z) - (\frac{k-1}{2} - \frac{q_{k-1,k-1}}{k})\Phi_0(z) + \Phi_1(z))} - n - O\left(\frac{1}{n}\right)$$

in compact subsets of Ω. Hence, $C_1 \in \mathcal{H}(\Omega)$. A straightforward inductive argument shows that $C_j \in \mathcal{H}(\Omega)$, $\forall j \in \mathbb{N}$, which completes the proof of the theorem. □

4.4 Stokes phenomenon

Recall that the Stokes phenomenon or the Stokes discontinuity refers to the discontinuous change in an asymptotic expansion when parameter crosses certain rays. (It was named after its discoverer George Stokes, see [177, 178].) The Stokes phenomenon has been a topic of active studies for over a century, e. g., [1, 60, 147].

From the results about the asymptotic expansion, we see that for $Q_k(z) = z^k$ and in compact subset of $\mathbb{C} \setminus \{0\}$, the asymptotic expansion reads as

$$p_n(z) \sim z^n e^{\frac{1}{k}\sum_{j=1}^{k-1} \frac{q_{k-1,k-1-j}}{jz^j}}.$$

Hence, if $Q_{k-1}(z) = z^{k-1}$ or $Q_{k-1}(z) = 0$, then $Q_n(z) \sim z^n$, and we have that ∞ is the unique singularity. Therefore there is no Stokes phenomenon in $\mathbb{C}$. If $Q_{k-1}(z) \neq z^{k-1}$, we have a discontinuity in the asymptotic expansion at $z = 0$ and $z = \infty$. On the other hand, when $Q_k(z) \neq z^k$, we have that the Stokes phenomenon occurs at the embedded tree given by $supp(\mu)$. The description of the Stokes phenomena of the present asymptotic expansion by means of the Stokes curves, cf. [68, 85, 136], is certainly an interesting and non-trivial problem worth to be studied further.

4.5 Two applications

4.5.1 Proving Property 2.8 for ES-operators

Let Q_k be a monic polynomial of degree k and consider the ES-linear differential operator

$$T[v] = \frac{d^k}{dz^k}(Q_k v). \tag{4.77}$$

This operator has been considered in Chapter 2, where a series of conjectures about its polynomial eigenfunctions $\{p_n\}$ has been formulated and settled. We assume that each p_n is monic without loss of generality, see Chapter 2 and [122]. Most of these conjectures have been settled in [26, 27, 28], see Chapter 3, but some of them remained. In particular, we are interested in Property 2.8 (see Conjecture 11 of loc. cit).

Conjecture 4.1 ([122]). *Let BC be a branch cut in $\mathbb{C}$ consisting of rays from all distinct roots of Q_k to ∞ such that they do not intersect each other or the embedded tree given by supp(μ). Select the branch of $\frac{1}{\sqrt[k]{Q_k(z)}}$ in $\Omega_{Q_k} = \mathbb{C} \setminus BC$, which asymptotically coincides with $\frac{1}{z}$ near ∞, and choose $z_0 \in \Omega_{Q_k}$. Then*

$$\lim_{n\to\infty} \frac{p_n(z)}{e^{n\Psi_{Q_k}(z)}} = \left(\frac{e^{\Psi_{Q_k}(z)}}{\sqrt[k]{Q_k(z)}} \right)^{\frac{k+1}{2}},$$

uniformly in compact subsets of $\mathbb{C} \setminus supp(\mu)$, where $\Psi_{Q_k}(z) = \int_{z_0}^{z} \frac{1}{\sqrt[k]{Q_k(t)}} dt$.

The result in Theorem 4.1 can be applied to the operator defined by (4.77). Indeed, let Φ_0 be as in Theorem 4.1, a straightforward calculation shows that

$$\Phi_1(z) = -\frac{k+1}{2k} \ln Q_k(z),$$

hence,

$$\lim_{n\to\infty} \frac{p_n(z)}{e^{n\Phi_0(z)}} = \left(\frac{e^{\Phi_0(z)}}{\sqrt[k]{Q_k(z)}} \right)^{\frac{k+1}{2}},$$

in compact subsets of $\mathbb{C} \setminus supp(\mu)$. Observe that the results do not coincide in the factor $\sqrt[k]{Q_k(z)}$ since the branch cuts are different.

4.5.2 Asymptotic behavior of eigenpolynomials of a differential operator of order four

As a concrete example, consider the linear differential operator of order four [115]

$$T[v] = (z^2-1)^2 u^{(4)} + 4z(z^2-1)v^{(3)} + 2(z-1)((1+2c)z + 2c + 3)v^{(2)},$$

denote by $\{p_n\}_{n=0}^{\infty}$ the sequence of its monic polynomial eigenfunctions with eigenvalues $\lambda_n = n(n-1)(n^2 - n + 4c)$. It is known that this sequence is orthogonal with respect to the Sobolev inner product given by

$$\langle P, Q\rangle = P(1)\overline{Q}(1) + \frac{1}{c} P'(1)\overline{Q}'(1) + \int_{-1}^{1} P'\overline{Q}' dx, \quad c > 0.$$

The results of the preceding section can be applied to study the strong asymptotic behavior of the sequence $\{p_n\}_{n=0}^{\infty}$. Let γ be any closed Jordan curve enclosing $[-1,1]$. From Theorem 4.1 we obtain immediately

$$\Phi_0(z) = \int^{z} \frac{d\zeta}{\sqrt{\zeta^2-1}} = \ln\left(\frac{\varphi(z)}{2}\right), \quad z \in ext(\gamma),$$

where $\varphi(z) = z + \sqrt{z^2-1}$, here we take the branch of φ for which $|\varphi(z)| > 1$ whenever $z \in \mathbb{C} \setminus [-1,1]$. In a similar way,

$$\begin{aligned}\Phi_1(z) &= \int^{z} \left(\frac{3Q_M'(\zeta)}{8(\zeta^2-1)^2} - \frac{z}{\zeta^2-1}\right) d\zeta \\ &= \frac{1}{4}\ln(z^2-1), \quad z \in ext(\gamma).\end{aligned}$$

Therefore,

a) $p_n(z) \sim \sqrt{2}(\frac{\varphi(z)}{2})^n \frac{\sqrt[4]{z^2-1}}{\sqrt{\phi(z)}}(1 + \frac{C_1(z)}{n} + \frac{C_2(z)}{n^2} + \cdots)$,

b) $\lim_{n\to\infty} \frac{p_{n+1}(z)}{p_n(z)} = \frac{\varphi(z)}{2}$,

c) $\lim_{n\to\infty} \sqrt[n]{|p_n(z)|} = \frac{|\varphi(z)|}{2}$,

uniformly in compact subsets $K \subset ext(\gamma)$.

4.6 Open problems in Chapter 4

4.1. The main result of this chapter has been basically derived from a classical theorem due to Sibuya on the existence of asymptotic solutions for non-linear systems of first order differential equations depending on a small parameter and from the results of [27].

Problem 4.1. Is it possible to extend the results/methods of the chapter to solutions of ES-operators other than eigenpolynomials?

Notice that some key ingredients such as the ones given in Lemma 4.5 probably would not be valid in a general case. The study of the Stokes geometry for the analysis of the Borel summability of WKB-solutions seems a promising path to attack this problem.

5 Two results on Bochner–Krall problem

(The material of this chapter is based on [28] and [89].)

5.1 Preliminaries

In the first part of this chapter we consider a polynomial sequence $\{p_n(z)\}_{n=0}^{\infty}$ in a variable z, where $\deg p_n = n$. This sequence is called *orthogonal with respect to a positive measure* ν supported on $\mathbb{R}$ if $\int_{\mathbb{R}} p_n(z)p_m(z)\,d\nu(z)$ is non-zero precisely when $n = m$. We are here concerned with polynomials orthogonal with respect to a measure of the *Nevai class*, see [141] and below.

Definition 5.1. The Nevai class for the interval $[-1,1]$ consists of all families $\{p_n\}$ of orthogonal polynomials (i. e., all orthogonality measures ν on the real line) with a 3-term recurrence relation

$$xp_n(x) = \alpha_{n+1}p_{n+1}(x) + \beta_n p_n(x) + \alpha_n p_{n-1}(x),$$

for which

$$\lim_{n\to\infty} \alpha_n = \frac{1}{2}, \quad \lim_{n\to\infty} \beta_n = 0.$$

For an arbitrary interval $[a,b]$, we should apply to the above definition the affine translation of $[-1,1]$, which sends -1 to a and 1 to b.

As in Chapter 1, we say that $\{p_n\}$ is a *sequence of eigenpolynomials* if there exists a linear differential operator $T = \sum_{j=1}^{k} Q_j(z)\frac{d}{dz^j}$ where Q_j are polynomials in z such that $Tp_n = \lambda_n p_n$ with $\lambda_n \in \mathbb{C}$. Finally, the polynomial sequence $\{p_n(z)\}_{n=0}^{\infty}$ is called a *Bochner–Krall system* if it is both orthogonal and a sequence of eigenpolynomials. If it is orthogonal and belongs to the Nevai class, we say that it is a *Bochner–Krall system of Nevai type.*

As we already mentioned in Chapter 1 it is an open problem to classify all Bochner–Krall systems. The following conjecture has been formulated in [66].

Conjecture 5.1. *For any Bochner–Krall system, the leading coefficient Q_k of the operator T is a power of a polynomial of degree at most 2.*

The main result of the first part of this chapter is an affirmative answer to this conjecture for Bochner–Krall systems of Nevai type.

Theorem 5.1. *Let $\{p_n\}$ be a compact type BKS, orthogonal with respect to a measure ν consisting of eigenpolynomials for a differential operator T. If ν is of the Nevai class and the convex hull of supp(ν) is the interval $[a,b]$, then k is even and $Q_k(z)$ is a constant multiple of $((z-a)(z-b))^{k/2}$.*

https://doi.org/10.1515/9783112207741-006

This result is obtained by applying the results on the asymptotic distribution of zeros from Chapter 2, see [27] and [141].

In the second part of this chapter we address the following generalized and complexified version of the Bochner–Krall problem.

Problem 5.1. Which ES-operators have sequences of eigenpolynomials which satisfy some finite order recurrence relation?

It is well-known that the usual orthogonal polynomials satisfy linear 3-term recurrence relations, see (5.1). Other interesting families (e. g., multi-orthogonal polynomials) satisfy finite linear recurrences of higher order. Therefore Problem 5.1 generalizes the classical Bochner–Krall set-up and is purely algebraic. Below we suggest a general conjecture characterizing such ES-operators and settle it for operators of order 3, see details in § 5.3.

5.2 On the leading coefficient of a Bochner–Krall system

Recall that by a trivial generalization of Favard's theorem, every sequence of monic orthogonal polynomials satisfies a 3-term recurrence relation of the form:

$$zp_n(z) = p_{n+1}(z) + u(n)p_n(z) + v(n)p_{n-1}(z), \tag{5.1}$$

where $u(n)$ and $v(n)$ are real numbers. Observe that the weight function $w(z)$ is non-negative in $\mathbb{R}$ if and only if $v(n) > 0$ for all $n > 0$.

In this section we prove Theorem 5.1. The following results, which characterize the asymptotic distribution of zeros for Nevai type OPS and for systems of eigenpolynomials respectively, are crucial to our treatment.

Suppose that a polynomial system $\{p_n\}$ is orthogonal with respect to a positive measure ν, and that the convex hull of $supp(\nu)$ is a compact interval $[a, b]$. It is well known (see [50]) that the zeros of every p_n are contained in the interval $[a, b]$. The following more precise result describes the asymptotic distribution of zeros for orthogonal systems of Nevai type.

Theorem A (See [141, Theorem 3, p. 50]). *Let the polynomial system $\{p_n\}$ be orthogonal with respect to a measure ν of Nevai class on $\mathbb{R}$, and let the convex hull of $supp(\nu)$ be $[a, b]$. Then the asymptotic distribution of zeros of $\{p_n\}$ is an absolutely continuous measure μ that depends only on $[a, b]$. The support of μ is precisely $[a, b]$, and its density in this interval is given by*

$$\rho(z) = \frac{1}{\pi\sqrt{(b-z)(z-a)}}.$$

Next we describe the asymptotic root distribution for a sequence of eigenpolynomials.

Theorem B (See Chapter 2, and [27, Theorems 2–4]). *Let $\{p_n\}$ be a sequence of eigenpolynomials for a non-degenerate ES-operator T with $Q_k(z)$ being monic of degree k. Then the asymptotic distribution of zeros of $\{p_n\}$ is a probability measure μ with the following properties:*

a) *μ has compact support.*
b) *Its Cauchy transform $\mathcal{C}(z) = \int \frac{d\mu(\zeta)}{z-\zeta}$ satisfies the equation*

$$\mathcal{C}^k(z) = 1/Q_k(z)$$

for almost all $z \in \mathbb{C}$.

These properties determine μ uniquely.

Note that the limiting measure μ is independent of all terms in (2) except the leading term $Q_k(z)\frac{d^k}{dz^k}$.

To derive from these two results a statement about Nevai type BKS, we will need the following.

Proposition 5.1. *Let $\{p_n\}$ be a sequence of eigenpolynomials for a differential operator T, and assume that all the zeros of p_n are real. Then there exists a compact set containing all the zeros of every p_n if and only if $\deg Q_k(z) = k$. (In other words, T is a non-degenerate ES-operator).*

We first prove Proposition 5.1 and then derive Theorem 5.1 as its corollary. Since the proof of Theorem B, in the situation where we will need it, follows easily along the same lines, we will include such a proof for the convenience of the reader.

Consider a polynomial system $\{p_n\}$ with the associated root-counting measures $\{\mu_n\}$. Assume that the supports of the measures μ_n are all contained in the same compact set, and that $\mu_n \to \mu$ in the weak topology. Let $\mathcal{C}_n(z)$ be the Cauchy transform of μ_n. It is given by

$$\mathcal{C}_n(z) = \int \frac{d\mu_n(\zeta)}{z-\zeta} = \frac{p_n'(z)}{np_n(z)}.$$

If $\mathcal{C}(z)$ denotes the Cauchy transform of μ, it follows that

$$\frac{p_n'(z)}{np_n(z)} \to \mathcal{C}(z)$$

for almost every $z \in \mathbb{C}$.

Suppose now that $\{p_n\}$ is a system of eigenpolynomials for an operator T and that $supp(\mu_n)$ are all contained in the same compact subset of the real line. Then there exists at least a subsequence of $\{\mu_n\}$ converging weakly to some measure μ. Moreover, if we let $\mu_n^{(j)}$ denote the root-counting measure of the jth derivative of p_n, then it follows from

Rolle's theorem that (a subsequence of) $\mu_n^{(j)}$ converges weakly to the same μ for every $j > 0$. In particular,

$$\frac{p_n^{(j+1)}(z)}{(n-j)p_n^{(j)}(z)} \to \mathcal{C}(z)$$

for almost every $z \in \mathbb{C}$, where $\mathcal{C}(z)$ is the Cauchy transform of μ. If we divide both sides of the differential equation $Tp_n = \lambda_n p_n$ by $n(n-1)\cdots(n-k+1)p_n = (n)_k P_n$, we obtain

$$Q_k(z)\prod_{j=0}^{k-1}\frac{p_n^{(j+1)}(z)}{(n-j)p_n^{(j)}(z)} + \frac{Q_{n-1}(z)}{n-k+1}\prod_{j=0}^{k-2}\frac{p_n^{(j+1)}(z)}{(n-j)p_n^{(j)}(z)} + \cdots = \frac{\lambda_n}{(n)_k}.$$

When $n \to \infty$ all the terms on the left-hand side but the first tend to zero. Therefore, for almost all $z \in \mathbb{C}$,

$$Q_k(z)\mathcal{C}^k(z) = \lim_{n\to\infty}\frac{\lambda_n}{(n)_k}.$$

We have already mentioned several times that

$$\lambda_n = \sum_{j=1}^{k} q_{j,j}(n)_j$$

where $q_{j,j}$ is the coefficient of z^j in $Q_j(z)$, see (1.3). In particular, if $\deg Q_k(z) < k$, i. e., $q_{k,k} = 0$, then $\lambda_n/(n)_k \to 0$ when $n \to \infty$. Therefore $\mathcal{C}(z) = 0$ for almost all z implying that $\mu = 0$, a contradiction. This argument proves one of the implications in Proposition 5.1. For the other implication, we refer to [27, Lemma 9]. Moreover, if $Q_k(z)$ is monic of degree k, then $\lambda_n/(n)_k \to 1$, and it follows that $\mathcal{C}^k(z) = 1/Q_k(z)$.

Proof of Theorem 5.1. Suppose that $\{p_n(z)\}$ is a Nevai type BKS as in the assumptions of this theorem. By the remark preceding Theorem A, the zeros of every $p_n(z)$ are contained in the interval $[a, b]$. It follows from Proposition 5.1 that $\deg Q_k(z) = k$, and we might as well assume that $Q_k(z)$ is monic. Hence the Cauchy transform $\mathcal{C}(z)$ of the limit $\mu = \lim_{n\to\infty}\mu_n$ satisfies the equation

$$\mathcal{C}^k(z) = 1/Q_k(z).$$

On the other hand, a direct computation of the Cauchy transform, using the expression for μ in Theorem A, gives $\mathcal{C}^2(z) = 1/(z-a)(z-b)$. Comparing these results yields $Q_k(z) = ((z-a)(z-b))^{k/2}$. ☐

5.3 Generalizing Bochner–Krall problem

In the second part of this chapter we suggest a conjectural answer to a natural generalization/algebraization of the classical Bochner–Krall problem 1.1, see Chapter 1.

5.3.1 Algebraization and generalization

The following natural algebraic version of the classical Bochner–Krall problem was studied in substantial detail in [78, 79].

Problem 5.2 (Algebraic Bochner–Krall problem). Describe all ES-operators such that their sequence of monic eigenpolynomials satisfies recurrence relation (5.1) with some complex-valued $u(n)$ and $v(n)$.

Observe that contrary to the case of the classical Bochner–Krall problem 1.1, Problem 5.2 is purely algebraic and (hopefully) has an easier solution compared to the classical one, see Conjecture 5.2 below. On the other hand, the connection between the algebraic Bochner–Krall problem and the classical one is rather straightforward.

Proposition 5.2. *A differential operator solves the classical Bochner–Krall problem with a positive weight function $w(x)$ if and only if it solves Problem 5.2 and all its eigenpolynomials have only real and interlacing roots, i. e., for any natural n, roots of p_n and p_{n+1} interlace.*

Remark 5.1. Which additional conditions on an ES-operator T can guarantee the real-rootedness of its eigenpolynomials is unclear at the moment, but this could be related to the major results in the Pólya–Schur theory, see, e. g., [36].

Remark 5.2. Observe that if we allow coefficients of the operator to depend on n, then the whole problem trivializes, i. e., more or less any sequence of polynomials can be obtained in such a way.

In what follows, we also discuss the following natural generalization of Problem 5.2.

Problem 5.3 (Generalized algebraic Bochner–Krall problem). Describe the set of ES-operators such that their sequence of eigenpolynomials satisfies a finite recurrence relation of the form

$$xp_n(x) = p_{n+1}(x) + \sum_{j=0}^{d} b_j(n)p_{n-j}(x) = \Lambda(n)p_n, \tag{5.2}$$

where d is independent of n and the coefficients $b_j(n)$ are independent of x. Here

$$\Lambda(n) = S + \sum_{j=0}^{d} b_j(n)S^{-j}$$

is a difference operator, and $Sf(n) := f(n+1)$ is the shift operator.

Remark 5.3. By a theorem of Maroni [127], the latter condition means that the sequence $\{p_n(x)\}$ of polynomials is d-orthogonal, see, e. g., [190]. The study of d-orthogonal and multiple orthogonal polynomials has been a popular area of research during the last three decades, see, e. g., [11].

Remark 5.4. It is clear that if an operator T solves Problem 5.2 or 5.3, then for any univariate polynomial s of positive degree, the operator $s(T)$ also solves the same problem with the same sequence of eigenpolynomials $\{p_n(x)\}$ and the sequence of eigenvalues coinciding with $\{s(\lambda_n)\}$. Therefore in what follows we will always assume that T is the operator of the lowest order with a given sequence of eigenpolynomials.

Remark 5.5. Another way to produce new solutions to Problems 5.2 and 5.3 from the existing ones is to apply a bispectral Darboux transformation to the difference operator Λ. This will produce a new difference operator $\widehat{\Lambda}$ and a new differential operator $\widehat{T}$. However, the order of $\widehat{T}$ could increase and, in fact, this is what happens in all known examples. We will call a differential operator T *irreducible* if it cannot be obtained by applying a bispectral Darboux transformation to a differential operator of a lower order.

Most of the above situations allow application of Darboux transformations on the side of the difference operator. More precisely, let us present the operator Λ as a matrix, acting on the vector $(p_0, p_1, \dots)$. Then in certain situations there exists a factorization of Λ as

$$\Lambda = D_1 \circ D_2,$$

where D_1 and D_2 are rational functions of n, and, additionally, D_1 is an upper triangular while D_2 is a lower triangular matrices. Set

$$\widehat{\Lambda} = D_2 \circ D_1$$

and define a new sequence of polynomials by using

$$\widehat{p}_n(x) = D_2 p_n(x).$$

It is easy to see that $\deg \widehat{p}_n(x) = n$ and also

$$x\widehat{p}_n(x) = \widehat{\Lambda}\widehat{p}_n(x).$$

In such a situation a general procedure described in [18] allows us to obtain a differential operator $\widehat{T}$ such that

$$\widehat{T}\widehat{p}_n(x) = \mu_n \widehat{p}_n(x).$$

To carry out this procedure in concrete cases is a non-trivial problem. However, in [78, 79] it has been successfully performed in the situations when one starts either from the Laguerre or from the Jacobi polynomials. In particular, using a 1-step Darboux transformation, one obtains Krall's polynomials. Analogously, many-step Darboux transformations lead to differential operators of higher order, but they do not increase the order of the difference operator.

In this language, the classical result of S. Bochner [43] can be stated as follows, see § 1.3.

Theorem C. *Every orthogonal polynomial system $\{p_n(x)\}$ obtained as a sequence of eigenpolynomials of a linear differential operator of order 2 coincides with one of the sequences of Hermite, Laguerre, Jacobi, or Bessel orthogonal polynomials, i. e., with one of the classical orthogonal polynomial systems (up to the action of the affine group on the variable x and scaling).*

Similarly, the result of H. Krall [105] from § 1.3 claims the following.

Theorem D. *Every linear differential operator of order 4 solving the classical Bochner–Krall problem and which cannot be obtained as a function of a linear differential operator of order 2 is given either by*

(a) *A 1-step Darboux transformation applied to an operator of order 2 corresponding to a system of Jacobi polynomials with at least one of the parameters α, β being an integer, or by*
(b) *A 1-step Darboux transformation applied to an operator corresponding to a system of Laguerre polynomials with an integer value of its parameter α.*

Let us now present our novel conjectures and results related to Problems 5.2 and 5.3. Concerning Problem 5.2, we propose the following general guess supported by the results of [78, 79].

Conjecture 5.2. *Any sequence $\{p_n\}$ of monic polynomials obtained as a sequence of eigenpolynomials of a linear differential operator solving Problem 5.2 belongs to one of the following two classes:*

1) *A classical sequence of orthogonal polynomials with, in general, complex-valued parameters, i. e.,*
   (a) *Hermite polynomials H_n.*
   (b) *Laguerre polynomials $L_n^{(\alpha)}$.*
   (c) *Jacobi polynomials $P_n^{(\alpha,\beta)}$.*
   (d) *Bessel polynomials Y_n.*
2) *A sequence of polynomials that can be obtained from*
   (e) *Laguerre polynomials $L_n^{(\alpha)}$ with positive integer values of α by applying a finite number of Darboux transformations to the difference operator*

$$\Lambda(n) := S + u(n)Id + v(n)S^{-1}$$

occurring in the right-hand side of (5.1).

(f) *Jacobi polynomials* $P_n^{(\alpha,\beta)}$ *with positive integer values of either* α, β, *or both* α *and* β *by applying a finite number of Darboux transformations to the operator* $\Lambda(n)$.

Recall that in this chapter S denotes the shift operator.

Remark 5.6. The fact that families (a)–(d) solve Problem 5.2 is classical. Analogous statement for families (e)–(f) has been proven in [78] and [79]. The major difficulty in settling Conjecture 5.2 is to show that the above list of cases is exhaustive; while possible in principle, this is a computationally challenging problem.

Concerning Problem 5.3, we have the following two conjectural claims.

Conjecture 5.3. *For any irreducible differential operator* T *of order* k *solving Problem* 5.3, *the order of the corresponding difference operator* Λ *is also* k.

The next claim similar to Conjecture 5.2 gives a description of all irreducible operators solving Problem 5.3.

Conjecture 5.4. *For any positive integer* k, *the irreducible differential operators of order* k *solving Problem* 5.3 *belong to one of the following two types:*

1)
$$T = \sum_{j=1}^{k} a_j x^{j-1} \partial^j + x\partial, \quad a_j \in \mathbb{C},\ a_k \neq 0,$$

generating the $(k-1)$*-orthogonal polynomials, see* [23].

2) *For any divisor* ℓ *of* k, *set* $G := (\sum_{m=0}^{\ell-1} a_m (x\partial)^m)\partial$, $a_m \in \mathbb{C}$, $a_{\ell-1} \neq 0$. *Let* $q(t)$ *be any complex polynomial of degree* k/ℓ *without a constant term. Then the operator*
$$T = q'(G)G + x\partial$$

is an irreducible operator of order k.

Remark 5.7. Both types of operators appearing in Conjecture 5.4 together with their properties were discussed in some detail in recent papers [87, 88].

Remark 5.8. Notice that the operators
$$T = \sum_{j=1}^{k} a_j \partial^j + x\partial$$

generating the Appell polynomials [6] are included in Type 2 with $\ell = 1$, see [88].

Example 5.1. Conjecturally, all the irreducible operators of order 4 solving Problem 5.3 are given by

1)
$$T = \sum_{j=1}^{4} a_j x^{j-1} \partial^j + x\partial, \quad a_j \in \mathbb{C},\ a_4 \neq 0$$

generating the 3-orthogonal polynomials.

2′)
$$T = \sum_{j=1}^{4} a_j \partial^j + x\partial, \quad a_j \in \mathbb{C},\ a_4 \neq 0$$

generating the Appell polynomials, see [6, 74].

2″)
$$T = b_2 G^2 + b_1 G + x\partial, \quad b_2 \neq 0,$$

where $G = (a_1 x\partial + a_0)\partial$, $a_1 \neq 0$.

The difference operator Λ in all these cases is of order 4, i. e., $d = 3$.

The main result of this part of the present chapter is a proof of Conjecture 5.4 in the first non-trivial case of differential operators T of order 3.

Theorem 5.2. *All irreducible differential operators T of order 3 solving Problem 5.3 whose corresponding difference operators Λ also have order 3 are given by:*

1)
$$T = \sum_{j=1}^{3} a_j x^{j-1} \partial^j + x\partial, \quad a_j \in \mathbb{C}, a_3 \neq 0,$$

generating the 2-orthogonal polynomials, see [23]. They satisfy the 4-term recurrence relation

$$\begin{aligned} xp_n(x) = p_{n+1}(x) &- (a_1 + 2(n-1)a_2 + 3(n-1)(n-2)a_3)p_n(x) + (n-1)(a_2 + (3n-6)a_3) \\ &\times (a_1 + (n-2)a_2 + (n-2)(n-3)a_3)p_{n-1}(x) \\ &- (n-1)(n-2)a_3(a_1 + (n-3)a_2 + (n-3)(n-4)a_3) \\ &\times (a_1 + (n-2)a_2 + (n-2)(n-3)a_3)p_{n-2}(x), \end{aligned}$$

with the standard initial conditions: $p_{-2}(x) = p_{-1}(x) = 0$, $p_0(x) = 1$.

2)
$$T = \sum_{j=1}^{3} a_j \partial^j + x\partial, \quad a_j \in \mathbb{C}, a_3 \neq 0$$

generating the Appell polynomials, see [6, 74]. *They satisfy the 4-term recurrence relation*

$$xp_n(x) = p_{n+1}(x) - a_1 p_n(x) - a_2(n-1)p_{n-1}(x) - a_3(n-1)(n-2)p_{n-2}(x),$$

with the standard initial conditions: $p_{-2}(x) = p_{-1}(x) = 0$, $p_0(x) = 1$.

Remark 5.9. Theorem 5.2 will follow from the detailed study of three special cases presented in § 5.5, § 5.6, and § 5.7 respectively. Its proof contains several new ideas, but also a substantial amount of explicit calculations which we were able to carry out by hand in § 5.5 and § 5.7. In § 5.6 we had to use Mathematica package since these calculations were too heavy. It is of course highly desirable to find an alternative proof avoiding such an amount of explicit calculations and which might be applicable for a possible attack on our general Conjecture 5.4, but at the moment we do not see such a possibility.

5.4 Initial facts

Proof of Proposition 5.2. Notice that a solution to the classical Bochner–Krall problem with a positive weight function on the real line produces a sequence $\{p_n\}$ of (orthogonal) polynomials that satisfy a 3-term recurrence relation (5.1) with real coefficients $u(n)$ and $v(n) > 0$. Under this condition, each p_n can be taken as a real and real-rooted polynomial. On the other hand, requiring that all polynomials satisfying (5.1) are real-rooted will imply that they can be taken real, which means that $u(n)$ and $v(n)$ are real. Additionally, the real-rootedness of all p_n implies that $v(n) > 0$ for all n. □

Problem 5.3 asks to describe all ES-linear differential operators L that have a sequence of polynomial eigenfunctions $\{p_n(x)\}$ with eigenvalues λ_n, i. e.,

$$Tp_n(x) = \lambda_n p_n(x) \tag{5.3}$$

and which at the same time satisfy a difference equation of the form

$$\Lambda p_n(x) = xp_n(x). \tag{5.4}$$

In the next lemma we deduce a simple, but very useful necessary condition for the solvability of the latter problem for differential operators of order 3. Let us assume that a polynomial system $\{p_n(x)\}$ satisfying (5.3) and (5.4) exists. Define

$$\mathrm{ad}_x T := [x, T] \quad \text{and} \quad \mathrm{ad}_x^{j+1} T := \mathrm{ad}_x(\mathrm{ad}_x^j T).$$

It is obvious that $\mathrm{ad}_x T$ is a differential operator of order less than the order of T. Below we assume that the order of T equals 3, i. e.,

$$T = Q_3(x)\partial^3 + Q_2(x)\partial^2 + Q_1(x)\partial. \tag{5.5}$$

Then it is obvious that

$$\mathrm{ad}_x^4 T = 0.$$

The following result obtained in [62] will be crucial in all our calculations. We state it in a form suitable for the present paper.

Lemma 5.1. *The eigenvalues λ_n satisfy the condition*

$$\mathrm{ad}_\Lambda^4 \lambda_n = 0. \tag{5.6}$$

Proof. The expression $[x, T]p_n(x)$ can be expanded as

$$[x, T]p_n(x) = xTp_n(x) - Txp_n(x) = x\lambda_n p_n(x) - T\Lambda p_n(x).$$

Using the fact that both λ_n and Λ do not depend on x, we can rewrite the above as

$$x\lambda_n p_n(x) - T\Lambda p_n(x) = \lambda_n x p_n(x) - \Lambda T p_n(x).$$

Applying the operators x and T to the right-hand side of the latter formula, we obtain

$$[x, T]p_n(x) = \lambda_n \Lambda p_n(x) - \Lambda\lambda_n p_n(x) = [\lambda_n, \Lambda]p_n(x),$$

i. e., $\mathrm{ad}_\Lambda \lambda = -\mathrm{ad}_x T$. The fact that $\mathrm{ad}_x^j T = (-1)^j \mathrm{ad}_\Lambda^j \lambda_n$ implies that $\mathrm{ad}_\Lambda^4 \lambda_n = 0$. □

Equation (5.6) will be referred to as the ad-*condition* and will be our main tool.

Notation 5.1. Below we will impose the restriction that Λ is a monic difference operator of order 3, i. e.,

$$\Lambda = S + \sum_{i=0}^{2} b_i(n) S^{-i}, \tag{5.7}$$

see Conjecture 5.3. In what follows we will use the following notation for the coefficients of the polynomials $Q_3(x)$, $Q_2(x)$ and $Q_1(x)$ respectively:

$$\begin{cases} Q_3(x) = q_{33}x^3 + q_{32}x^2 + q_{31}x + q_{30}; \\ Q_2(x) = q_{22}x^2 + q_{21}x + q_{20}; \\ Q_1(x) = q_{11}x + q_{10}. \end{cases}$$

5.4.1 The ad-condition

Notice that λ_n is a polynomial of degree at most 3 in the variable n. Indeed, the entries of λ_n come from the differentiation of the term x^n in $p_n(x)$. More exactly,

$$\lambda_n = (n)_3 q_{33} + (n)_2 q_{22} + n q_{11}.$$

We will now derive a polynomial form of $\mathrm{ad}_\Lambda^3 \lambda_n$.

Lemma 5.2. *The operator* $\mathrm{ad}_\Lambda^3 \lambda_n$ *is of the form*

$$\mathrm{ad}_\Lambda^3 \lambda_n = \alpha\Lambda^3 + \beta\Lambda^2 + \gamma\Lambda + \delta I = 6Q_3(\Lambda), \tag{5.8}$$

where the coefficients $\alpha, \beta, \gamma, \delta \in \mathbb{C}$ *are independent of* n.

Proof. Put $f(x,n) := p_n(x)$. From the condition $\mathrm{ad}_\Lambda^4 \lambda_n = 0$ it follows that $\mathrm{ad}_\Lambda^3 \lambda_n$ is a polynomial in Λ, which we denote by $q(\Lambda)$. The degree of $q(\Lambda)$ is at most 3 since it contains S in at most the third degree. Also notice that

$$\mathrm{ad}_x^3 T = (-1)^3 6Q_3(x).$$

Having in mind that

$$\mathrm{ad}_x^3 T(p_n) = (-1)^3 \mathrm{ad}_\Lambda^3 \lambda(p_n),$$

we obtain

$$(-1)^3 6Q_3(\Lambda)p_n = (-1)^3 \mathrm{ad}_\Lambda^3 \lambda p_n,$$

i. e., $q(x) = 6Q_3(x)$. □

Remark 5.10. Formula (5.8) says that the leading polynomial coefficient $a_3(x)$ coincides with $\frac{1}{6}(\alpha x^3 + \beta x^2 + \gamma x + \delta)$, i. e., in the above notation $a_{33} = \alpha/6$, $a_{32} = \beta/6$, $a_{31} = \gamma/6$, $a_{30} = \delta/6$, which we will use below.

5.5 Case when λ is a linear polynomial

As we mentioned above, λ_n is a polynomial in n of degree at most 3. Here we will discuss the simplest case when λ_n is linear. In terms of the differential operator T given by (5.5), the linearity of λ_n means that $\deg Q_3(x) < 3$ (which is equivalent to $q_{33} = \alpha/6 = 0$), $\deg Q_2(x) < 2$ (which is equivalent to $q_{22} = 0$), and $\deg Q_1(x) = 1$ (which is equivalent to $q_{11} \neq 0$). Additionally, observe that in this case we can without loss of generality assume that $\lambda_n \equiv n$. Indeed, if $\lambda_n = \mu n + \nu$, then we can consider $T - \nu$ instead of T. Finally, we can divide both sides by μ, which results in $\lambda_n \equiv n$.

Assuming that $\lambda_n \equiv n$, we will separately study the following three (sub)cases:

(i) $Q_3(x)$ is a quadratic polynomial, i. e., $q_{33} = 0$; $q_{32} = \beta/6 \neq 0$;
(ii) $Q_3(x)$ is a linear polynomial, i. e., $q_{33} = q_{32} = 0$, $q_{31} = \gamma/6 \neq 0$;
(iii) $Q_3(x)$ is a non-vanishing constant, i. e., $q_{33} = q_{32} = q_{31} = \gamma/6 = 0$, $q_{30} = \delta/6 \neq 0$.

5.5.1 Subcase $\deg Q_3(x) = 2 \Leftrightarrow \beta \neq 0$

Since $\lambda_n \equiv n$, we get

$$T = (q_{32}x^2 + q_{31}x + q_{30})\partial_x^3 + (q_{21}x + q_{20})\partial^2 + (x + q_{10})\partial_x,$$

where $q_{32} \neq 0$. Rescaling T we can achieve $q_{32} = 1$. Shifting x we can additionally assume that $q_{10} = 0$. (Both changes only result in somewhat shorter formulas, but otherwise are not essential.) Thus, without loss of generality, we can assume that the operator T is of the form

$$T = (x^2 + q_{31}x + q_{30})\partial_x^3 + (q_{21}x + q_{20})\partial^2 + x\partial_x. \tag{5.9}$$

To prove the required result, we do the following. Expanding the polynomial $p_n(x)$ as

$$p_n(x) = x^n + p_1(n)x^{n-1} + \cdots + p_k(n)x^{n-k} + \cdots,$$

we are going to compute the coefficients $p_k(n)$ for the first few values of k. Then we compute the coefficients $b_j(n)$ of the recurrence relation Λ. The condition that the polynomials $p_n(x)$ satisfy a 4-term relation implies that $b_3(n) \equiv 0$. We will use the following identities:

$$\begin{cases} p_1(n) = p_1(n+1) + b_0(n); \\ p_2(n) = p_2(n+1) + b_0(n)p_1(n) + b_1(n); \\ p_3(n) = p_3(n+1) + b_0(n)p_2(n) + b_1(n)p_1(n) + b_2(n); \\ p_4(n) = p_4(n+1) + b_0(n)p_3(n) + b_1(n)p_2(n) + b_2(n)p_1(n) + b_3(n). \end{cases} \tag{5.10}$$

These relations imply the following expressions for $b_j(n)$:

$$\begin{cases} b_0(n) = p_1(n) - p_1(n+1); \\ b_1(n) = p_2(n) - p_2(n+1) - b_0(n)p_1(n); \\ b_2(n) = p_3(n) - p_3(n+1) - b_0(n)p_2(n) - b_1(n)p_1(n); \\ b_3(n) = p_4(n) - p_4(n+1) - b_0(n)p_3(n) - b_1(n)p_2(n) - b_2(n)p_1(n). \end{cases} \tag{5.11}$$

We want to conclude that under our assumptions we get $q_{31} = q_{30} = 0$. The following statement, which is the major result of this section, claims exactly this.

Proposition 5.3. *If a differential operator L of the form*

$$T = (x^2 + q_{31}x + q_{30})\partial_x^3 + (q_{21}x + q_{20})\partial^2 + x\partial_x \tag{5.12}$$

has a sequence of polynomial eigenfunctions $\{p_n(x)\}$, $n = 0, 1, \ldots$ that satisfies a 4-term recurrence relation, then $a_{31} = a_{30} = a_{20} = 0$. The operator T will have the form 1) *from Theorem* 5.2, *i. e.,*

$$T = \sum_{j=1}^{3} a_j x^{j-1} \partial^j + x\partial.$$

Proof. We will actually show that if $b_3(n) \equiv 0$ (which is a necessary condition for the sequence $\{p_n(x)\}$ to satisfy a 4-term recurrence), then all three parameters q_{31}, q_{30}, q_{20} vanish. To do this, we will not need the explicit form of $b_3(n)$, which is very cumbersome. It would be enough to obtain polynomial coefficients of q_{31}, q_{30}, and q_{20} in the presentation of $b_3(n)$.

The reason for the latter sufficiency is as follows. It has been shown in [88] that if $q_{31} = q_{30} = q_{20} = 0$, then $b_3(n) \equiv 0$. Below we will refer to q_{31}, q_{30}, q_{20} as *parameters*. In the general case, we have

$$b_3(n) \sim q_{31} n^{m_1} + q_{30} n^{m_2} + q_{20} n^{m_3},$$

where the numbers m_j are some non-negative integers and the relation itself means that $b_3(n)$ is equal to the r. h. s. modulo some terms that contain higher powers of parameters. Here n^{m_j} is the leading power of n of the corresponding parameter.

To compute these leading terms, we first calculate the corresponding terms in $4p_4(n)$. Set

$$4p_4(n) = cn^{q_0} + q_{31} n^{q_1} + q_{30} n^{q_2} + q_{20} n^{q_3},$$

where $c \in \mathbb{C}$ does not depend on the parameters. (Later we are going to use a similar expansion for $b_3(n)$ itself as well.)

We will use formulas (5.11). Applying operator (5.9) to $p_n(x)$, we get

$$\begin{aligned} T(p_n) &= (x^2 + q_{31}x + q_{30})((n)_3 x^{n-3} + p_1(n)(n-1)_3 x^{n-4} + \cdots + p_k(n)(n-k)_3 x^{n-k-3} + \cdots) \\ &\quad + (q_{21}x + q_{20})((n)_2 x^{n-2} + p_1(n)(n-1)_2 x^{n-3} + \cdots + p_k(n)(n-k)_2 x^{n-k-2} + \cdots) \\ &\quad + x(nx^{n-1} + p_1(n)(n-1)x^{n-2} + \cdots + p_k(n)(n-k)x^{n-k-1} + \cdots) \\ &= n(x^n + p_1(n)x^{n-1} + \cdots + p_k(n)x^{n-k} + \cdots). \end{aligned}$$

Comparing the coefficients of x^{n-1} in both sides of the latter equality, we obtain

$$p_1(n) = (n)_3 + (n)_2 q_{21}.$$

From (5.11) we obtain

$$b_0(n) = -\Delta(p_1(n)) = -3(n)_2 - 2nq_{21}.$$

We see that both $p_1(n)$ and $b_0(n)$ do not contain our parameters. We observe that in order to calculate $b_3(n)$, we need first to compute $b_1(n)$ and $b_2(n)$.

Let us start with $p_2(n)$. We have

$$2p_2(n) = (n)_3 q_{31} + (n)_2 q_{20} + p_1(n-1)p_1(n).$$

For later use, notice that

$$p_2(n) \sim \frac{1}{2}[n^3 q_{31} + n^2 q_{20} + n^6].$$

Using (5.11), we obtain

$$\begin{aligned} b_1(n) &= -\Delta(p_2(n)) - b_0(n)p_1(n) \\ &= -\frac{3(n)_2}{2} q_{31} - nq_{20} + \frac{1}{2}p_1(n)\Delta^2(p_1(n-1)), \end{aligned}$$

which gives

$$b_1(n) \sim -\frac{3n^2}{2} q_{31} - nq_{20} + 3n^4.$$

Let us apply a similar procedure to find $p_3(n)$. From the above expression for $T(p_n(x))$ we get

$$\begin{aligned} 3p_3(n) &= (n)_3 q_{30} + p_1(n)(n-1)_3 q_{31} + p_1(n)(n-1)_2 q_{20} + p_1(n-2)p_2(n) \\ &= \frac{1}{2}p_1(n-2)[(n)_3 q_{31} + (n)_2 q_{20} + p_1(n-1)p_1(n)] \\ &\quad + (n)_3 q_{30} + p_1(n)(n-1)_3 q_{31} + p_1(n)(n-1)_2 q_{20}. \end{aligned}$$

The latter relation implies that

$$p_3(n) \sim \frac{n^3}{3} q_{30} + \frac{n^6}{2} q_{31} + \frac{n^5}{2} q_{20} + \frac{1}{3!} n^9.$$

For $b_2(n)$, we get

$$b_2(n) = -\Delta(p_3(n)) - b_0(n)p_2(n) - b_1(n)p_1(n),$$

which implies that

$$b_2(n) \sim n^4 q_{31} - n^2 q_{30} + \frac{2n^3}{3} q_{20}.$$

Further, computing $p_4(n)$, we get

$$4p_4(n) = p_2(n)(n-2)_3 q_{31} + p_1(n)(n-1)_3 q_{30} + p_2(n)(n-2)_2 q_{20} + p_1(n-3)p_3(n),$$

which implies that

$$p_4(n) \sim \frac{n^9}{4} q_{31} + \frac{n^6}{3} q_{30} + \frac{n^8}{4} q_{20} + \frac{1}{4!} n^{12}.$$

Finally, we obtain

$$b_3(n) = -\Delta(p_4(n)) - b_0(n)p_3(n) - b_1(n)p_2(n) - b_2(n)p_1(n)$$

by plugging the expressions for $p_j(n)$, $b_j(n)$ from the above formulas. This gives

$$b_3(n) \sim \frac{9n^7}{2}q_{31} + \frac{9n^4}{2}q_{30} + 3n^6 q_{20}.$$

As the consequence of the latter expansion, we see that if we assume that $b_3(n) \equiv 0$, then, in particular, we get that $q_{31} = q_{30} = q_{20} = 0$. □

5.5.2 Subcase $\deg Q_3(x) \le 1 \Leftrightarrow q_{32} = 0$

Our goal in this subsection is to prove the following statement.

Proposition 5.4. *The case* $\deg Q_3(x) \le 1$ *is possible if and only if* $q_{31} = q_{21} = 0$ *and* $q_{30} \ne 0$. *In this case we obtain the Appell polynomials.*

Proof. We use the same approach as in § 5.5.1. The operator T is of the form

$$T = (q_{31}x + q_{30})\partial_x^3 + (q_{21}x + q_{20})\partial^2 + (x + q_{10})\partial_x.$$

By translation of x we can make $q_{10} = 0$ and work with

$$T = (q_{31}x + q_{30})\partial_x^3 + (q_{21}x + q_{20})\partial^2 + x\partial_x.$$

(Again such change of variables only results in somewhat shorter formulas but otherwise does not restrict the generality.)

Expanding

$$p_n(x) = x^n + p_1(n)x^{n-1} + \cdots + p_k(n)x^{n-k} + \cdots,$$

let us apply T to the polynomial $p_n(x)$. Straightforward calculation gives for $T(p_n)$ the expression

$$\begin{aligned}
T(p_n) &= (q_{31}x + q_{30})((n)_3x^{n-3} + p_1(n)(n-1)_3x^{n-4} + \cdots + p_k(n)(n-k)_3x^{n-k-3} + \cdots) \\
&\quad + (q_{21}x + q_{20})((n)_2x^{n-2} + p_1(n)(n-1)_2x^{n-3} + \cdots + p_k(n)(n-k)_2x^{n-k-2} + \cdots) \\
&\quad + x(nx^{n-1} + p_1(n)(n-1)x^{n-2} + \cdots + p_k(n)(n-k)x^{n-k-1} + \cdots) \\
&= n(x^n + p_1(n)x^{n-1} + \cdots + p_k(n)x^{n-k} + \cdots).
\end{aligned}$$

Comparing the coefficients of x^{n-1} in both sides, we find

$$p_1(n) = (n)_2 q_{21}.$$

From the expansion of Λ we conclude that

$$b_0(n) = -2nq_{21}. \tag{5.13}$$

From the coefficients at x^{n-2} we get

$$p_2(n) = \frac{1}{2}[(n)_3 q_{31} + (n)_2 q_{20} + p_1(n)p_1(n-1)].$$

Again from the expansion of Λ, we obtain

$$b_1(n) = \frac{-3(n)_2 q_{31} - 2nq_{20} + 4(n)_2 q_{21}^2}{2} = \frac{(n)_2(2q_{21}^2 - 3q_{31}) - 2nq_{20}}{2}. \tag{5.14}$$

On the other hand, we can use the formula

$$\mathrm{ad}_\Lambda^3 \lambda = 6Q_3(\Lambda) = 6(q_{31}\Lambda + q_{30})$$

following from Lemma 5.2. (Observe that $\alpha = \beta = 0$.)

We need only the terms containing $S^1, S^0 = Id$ and S^{-1} in the latter equation. They can be computed by hand using the appropriate expressions for $\mathrm{ad}_\Lambda^1\lambda$, $\mathrm{ad}_\Lambda^2\lambda$, and $\mathrm{ad}_\Lambda^3\lambda$, which we provide in the next lemma.

Lemma 5.3. *The expressions for* ad_Λ^j, $j = 1, 2, 3$ *are given by*

$$\mathrm{ad}_\Lambda^1(n) = S - b_1(n)S^{-1} + \cdots \tag{5.15}$$
$$\mathrm{ad}_\Lambda^2(n) = 2q_{21}S - 2\Delta b_1(n)S^0 + \{-3\Delta b_2(n) + b_1(n)2a_{21}\}S^{-1} + \cdots \tag{5.16}$$
$$\mathrm{ad}_\Lambda^3(n) = \{4q_{21}^2 - 2\Delta b_1(n)\}S^1 - 3\Delta^2 b_2(n)S^0 + \{b_1(n)[\Delta^2 b_1(n-1) - 4q_{21}^2] - 6\Delta b_2(n)\}S^{-1} + \cdots \tag{5.17}$$

where by $\cdots$ *we denote all terms containing* S^j *for* $j < -1$.

Proof. Direct computation using the equality $\Delta b_0(n) = -2q_{21}$. □

Lemma 5.3 leads to the following system of equations:

$$\begin{cases} 4q_{21}^2 - 2\Delta^2(b_1(n)) = 6q_{31}; \\ -3\Delta^2(b_2(n)) = 6\gamma b_0(n) + 6q_{30}; \\ -b_1(n)6\gamma + 6\Delta b_2(n) = 6q_{30}b_1(n). \end{cases} \tag{5.18}$$

Observe that the first equation implies the trivial identity

$$4q_{21}^2 - 2(2q_{21}^2 - 3q_{31}) = 6q_{31},$$

which additionally confirms the correctness of our calculations.

From the second equation, using (5.13) and the relation $\gamma = 6q_{31}$, we obtain

$$-3\Delta^2(b_2(n)) = 6\gamma b_0(n) + 6q_{30} = -12nq_{21}q_{31} + 6q_{30}$$

from which we see that $b_2(n)$ is at most cubic in n. Expanding

$$b_2(n) = A_3(n)_3 + A_2(n)_2 + \cdots,$$

we obtain

$$3A_3 = 2q_{21}q_{31}; \quad A_2 = -q_{30}. \tag{5.19}$$

Using the expressions for $b_0(n)$ and $b_1(n)$, we transform the third equation into

$$b_1(n)[\gamma + q_{30}] = \Delta(b_2(n)). \tag{5.20}$$

To move further, we apply the obvious fact that the coefficient at S^{-5} in $\mathrm{ad}_\Lambda^3 = \gamma\Lambda + \delta I$ vanishes. Indeed, observe that terms containing S^{-5} come from the lowest terms in $[\Lambda, \mathrm{ad}_\Lambda^2]$, which implies that the operators $b_2(n)S^{-2}$ and $\{b_2(n)b_1(n-2) - b_1(n)b_2(n-1)\}S^{-3}$ commute.

Assume now that both operators are not identically zero. Then they can be expressed as univariate polynomials of one and the same operator which in the case under consideration can be written as $s(n)S^{-1}$. Thus we obtain

$$\begin{cases} b_2(n)S^{-2} = (s(n)S^{-1})^2; \\ \{b_2(n)b_1(n-2) - b_1(n)b_2(n-1)\}S^{-3} = \ell(s(n)S^{-1})^3, \quad 0 \neq \ell \in \mathbb{C}. \end{cases} \tag{5.21}$$

The first equation gives $b_2(n) = s(n)s(n-1)$. Such a relation is only possible if $b_2(n)$ is a polynomial of degree 2 in n, which implies that the coefficient A_3 vanishes. Due to the relation $3A_3 = 2q_{31}q_{21}$, we get either $q_{21} = 0$ or $q_{31} = 0$.

Let us first consider the case when $q_{21} = 0$. Then we obtain that $b_0 = 0$ and

$$b_1 = \frac{(n)_2(-3q_{31}) - 2nq_{20}}{2}.$$

However if $q_{31} \neq 0$, we conclude that $\gamma + q_{30}$ must vanish since otherwise b_2 will be of degree 3 in n. But if $\gamma + q_{30} = 0$, we get that b_2 is a constant. The second equation from (5.18) together with the fact that $b_0 = 0$ gives that $q_{30} = 0$. Using that $\gamma + q_{30} = 0$, we come to the conclusion that $\gamma = 6q_{31} = 0$. This is a contradiction since under such assumptions the order of T drops from 3 to 2.

We see that only the case $q_{31} = 0$ is possible. The third equation from (5.18) gives that $\Delta b_2 = q_{30}b_1$. In this case the formula for b_1 gives

$$b_1(n) = (n)_2 q_{21}^2 - nq_{20},$$

which is only possible if $q_{21} = 0$ since $q_{30} \neq 0$. □

5.6 Case when λ is a quadratic polynomial

This case corresponds to $\deg Q_3(x) < 3$, $\deg Q_2(x) = 2$, and $\deg Q_1(x) \leq 1$. We want to show that in this situation there are no differential operators whose eigenpolynomials satisfy a finite recurrence relation.

Proposition 5.5. *The case when λ_n is quadratic is impossible, i. e., no linear differential operator T of order* 3 *satisfying this condition can solve Problem* 5.3.

Proof. Our arguments are partially computer-aided due to the complexity of calculations. (The corresponding Mathematica code and its results are available upon request.) The algorithm of what we are doing is presented below.

We are going to compute the coefficients of the 4-term recurrence in terms of the coefficients of the operator T. In the case under consideration, we can write the operator T in the form

$$T = (q_{32}x^2 + q_{31}x + q_{30})\partial^3 + (x^2 + q_{21}x + q_{20})\partial^2 + (q_{11}x + q_{10})\partial.$$

We can also assume that $q_{21} = 0$, which can be achieved by a translation of x. (By a slight abuse of notation we use the same letters for the coefficients of T.) From the above form of T we get that

$$\lambda_n = n(n-1) + \nu n.$$

As above, introduce

$$p_n(x) = x^n + p_1(n)x^{n-1} + \cdots + p_k(n)x^{n-k} + \cdots.$$

We are going to compute the coefficients $p_k(n)$ of $p_n(x)$ by considering $T(p_n(x))$ and using the formulas (5.10) and (5.11). $T(p_n(x))$ satisfies the relation

$$\begin{aligned} T(p_n) &= (q_{32}x^2 + q_{31}x + q_{30})((n)_3x^{n-3} + p_1(n)(n-1)_3x^{n-4} \cdots + p_k(n)(n-k)_3x^{n-k-3} + \cdots) \\ &\quad + (x^2 + q_{20})((n)_2x^{n-2} + p_1(n)(n-1)_2x^{n-3} \cdots + p_k(n)(n-k)_2x^{n-k-2} + \cdots) \\ &\quad + (q_{11}x + q_{10})(nx^{n-1} + p_1(n)(n-1)x^{n-2} \cdots + p_k(n)(n-k)x^{n-k-1} + \cdots) \\ &= (n(n-1) + \nu n)(x^n + p_1(n)x^{n-1} + \cdots + p_k(n)x^{n-k} + \cdots). \end{aligned}$$

We now equalize the coefficients at the same powers of n in the right-hand and the left-hand sides of the latter equation. For our purposes it would be enough to find expressions for $p_1(n)$, $p_2(n)$, $p_3(n)$, and $p_4(n)$ only. Knowing $p_1(n)$, $p_2(n)$, $p_3(n)$, $p_4(n)$ and using the above formulas, we can express $b_0(n)$, $b_1(n)$, $b_2(n)$ and $b_3(n)$. As before, to obtain a recurrence relation of order at most 4 for the eigenpolynomials, we need to ensure that $b_3(n) \equiv 0$.

We will normalize the above expression for T using the action of the affine group on x. First, let us consider the case when $q_{32} \neq 0$. By rescaling we can make $q_{32} = 1$. Then the leading of $b_3(n)$ equals $11184n^{13}$, see (5.24) which implies that $b_3(n)$ cannot vanish identically.

If $q_{3,2} = 0$ but $q_{3,1} \neq 0$, by rescaling we can assume that $q_{3,1} = 1$. In this case the leading coefficient of $b_3(n)$ equals $4224n^{11}$, which again implies that $b_3(n)$ cannot vanish identically.

Next assume that $q_{3,2} = q_{3,1} = 0$. Let us rescale x to obtain $q_{3,0} = 1$. We still need to find if there exist values of q_{20}, q_{11}, q_{10} for which $m_9 = m_8 = m_7 = m_6 = 0$, where m_i is the coefficient at n^i in the expression (5.26) for $b_3(n)$.

First, we get that $m_9 = 8(96q_{20}^2 + 288q_{10})$. Assuming that $m_9 = 0$ we obtain $q_{10} = -\frac{q_{20}^2}{3}$. Inserting the latter expression for q_{10} in $b_3(n)$, and using $m_8 = 0$, we get that either $q_{11} = 2$ or $q_{20} = 0$. In the former case, setting $q_{10} = -\frac{q_{20}^2}{3}$ and $q_{11} = 2$ in the expressions for m_7 and m_6, we get

$$\begin{cases} m_6 = -\frac{64}{3}q_{20}^2(27 + 4q_{20}^3); \\ m_7 = \frac{64}{3}q_{20}^2(9 + 2q_{20}^3). \end{cases}$$

Thus, if $q_{20} \neq 0$, then the equations for $m_6 = 0$ and $m_7 = 0$ are obviously incompatible, i. e., have no common solutions.

Now let us consider the case when $q_{32} = q_{31} = q_{21} = q_{20} = 0$. We will show that $b_3 = 0$, which however is not enough to claim that all $b_j = 0$ for $j > 2$. For this reason we are going to compute them up to $j = 5$. We know that in this case also $q_{10} = 0$. Again applying the operator T to the polynomials $p_n(x)$, we obtain

$$\begin{aligned} T(p_n) &= (n)_3x^{n-3} + p_1(n)(n-1)_3x^{n-4} \cdots + p_k(n)(n-k)_3x^{n-k-3} + \cdots \\ &\quad + x^2((n)_2x^{n-2} + p_1(n)(n-1)_2x^{n-3} \cdots + p_k(n)(n-k)_2x^{n-k-2} + \cdots) \\ &\quad + q_{11}x(nx^{n-1} + p_1(n)(n-1)x^{n-2} \cdots + p_k(n)(n-k)x^{n-k-1} + \cdots) \\ &= (n(n-1) + q_{11}n)(x^n + p_1(n)x^{n-1} + \cdots + p_k(n)x^{n-k} + \cdots). \end{aligned}$$

Then, for $k > 3$, we find the following formulas for $p_j(n)$:

$$\begin{cases} p_1(n)(2 - 2n - q_{11}) & = 0; \\ p_2(n)(6 - 4n - 2q_{11}) & = 0; \\ p_3(n)(12 - 6n - 3q_{11}) & = -(n)_3; \\ \vdots \quad \vdots \quad \vdots \quad \vdots & \quad \vdots \\ p_k(n)(k(k+1) - 2kn - kq_{11}) & = -(n-k+3)_3p_{k-3}(n). \end{cases}$$

By induction, we see that $p_{3j} \neq 0$ while $p_i = 0$ for $i \neq 3j$. In particular,

$$p_6(n)(42-12n-6q_{11}) = -(n-3)_3 p_3(n),$$

i. e.,

$$p_6(n) = \frac{(n)_5}{18(2n+q_{11}-4)(2n+q_{11}-7)}.$$

From this we easily compute the several first b_js getting

$$\begin{cases} b_0 = 0; \\ b_1 = 0; \\ b_2 = -\Delta p_3(n) \neq 0; \\ b_3 = -\Delta p_4(n) - b_0 p_3(n) - b_1 p_2(n) - b_2 p_1(n) = 0; \\ b_4 = -\Delta p_5(n) - b_0 p_4(n) - b_1 p_3(n) - b_2 p_2(n) - b_3 p_1(n) = 0; \\ b_5 = -\Delta p_6(n) - b_0 p_5(n) - b_1 p_4(n) - b_2 p_3(n) - b_3 p_2(n) - b_4 p_1(n). \end{cases}$$

In the expression for b_5 there are two non-vanishing terms: $\Delta p_6(n)$ and $b_2 p_3$. Both terms are of order n^3, so we only need to show that their sum is not identically zero. Straightforward calculations give

$$b_2 p_3(n) = \frac{1}{9} \frac{(n)_3 (n)_2 (4n-q_{11})}{(2n+q_{11}-2)(2n+q_{11}-4)^2}$$

and

$$\Delta p_6(n) = \frac{(n)_4 R(n)}{(2n+q_{11}-2)(2n+q_{11}-4)(2n+q_{11}-5)(2n+q_{11}-7)},$$

where $R(n)$ is a polynomial of degree at most 2, whose explicit expression is irrelevant for our purposes. We see that the sum of $b_2 p_3(n)$ and $\Delta p_6(n)$ cannot vanish since they contain different factors.

In fact, the explicit expression for b_5 can be obtained even without computer. Our symbolic computations give

$$b_5(n) = \frac{n(n-1)(n-2)(n-3)(n-4)(3q_{1,1}+5n-16)}{9(q_{1,1}+2n-2)(q_{1,1}+2n-7)(a_{1,1}+2n-5)(q_{1,1}+2n-4)}. \tag{5.22}$$

□

5.7 Case when λ is a cubic polynomial

In case when λ_n is a cubic polynomial in n using translation and scaling, we can without loss of generality assume that

$$\lambda_n = n(n-1)(n-2) + \nu n(n-1) + \mu n.$$

As before, one can easily observe that $\nu = q_{22}$ and $\mu = q_{11}$.

Using this ansatz, we first formulate some important preliminary statements about the coefficients of Λ and T.

Lemma 5.4. *The coefficient a in (5.8) equals 6.*

Proof. To prove this, we compute the term of the highest degree in T in the two expressions for $\mathrm{ad}^3_\Lambda \lambda_n$, which we presented earlier. On one hand, from (5.8) we obtain that $\mathrm{ad}^3_\Lambda \lambda_n = aS^3 + \cdots$. Let us compute the term containing S^3 in $\mathrm{ad}^3_\Lambda \lambda_n$. Simple computation shows that

$$\begin{cases} \mathrm{ad}^1_\Lambda(\lambda) = \Delta(\lambda_n)S + \cdots; \\ \mathrm{ad}^2_\Lambda(\lambda) = \Delta^2(\lambda_n)S^2 + \cdots; \\ \mathrm{ad}^3_\Lambda(\lambda) = \Delta^3(\lambda_n)S^3 + \cdots. \end{cases}$$

Since $\Delta^3(\lambda_n) = 6$, the result follows. □

Lemma 5.5. *The coefficient $b_2(n)$ in the expansion (5.7) of Λ vanishes identically.*

Proof. Suppose that $b_2(n) \not\equiv 0$. Then again we can use the above approach and compute the term in $\mathrm{ad}^3_\Lambda \lambda_n$ containing S^{-6} in two different ways. On one hand, if taken from $6\Lambda^3$, this term is given by

$$ab_2(n)b_2(n-2)b_2(n-4)S^{-6} = 6b_2(n)b_2(n-2)b_2(n-4)S^{-6}.$$

On the other hand, the computation of the same term from $\mathrm{ad}^3_\Lambda \lambda_n$ shows that it coincides with

$$b_2(n)b_2(n-2)b_2(n-4)(-\lambda(n-6) + 3\lambda(n-4) - 3\lambda(n-2) + \lambda_n)S^{-6}.$$

However, $-\lambda(n-6)+3\lambda(n-4)-3\lambda(n-2)+\lambda_n = 48$, which is only possible if $b_2(n) \equiv 0$. □

Lemma 5.6. *Under our current assumptions, the coefficient $b_1(n)$ in the expansion (5.7) of Λ vanishes identically.*

Proof. The argument is similar to the above computation of $b_2(n)$. Assume that $b_1(n) \not\equiv 0$. Notice that by Lemma 5.5,

$$\Lambda = S + b_0(n)S^0 + b_1(n)S^{-1}.$$

The coefficient of S^{-3} in the expression for $6\Lambda^3$ is given by $6b_1(n)b_1(n-1)b_1(n-2)$.

From the expression for $\mathrm{ad}^3_\Lambda \lambda_n$, we find that the same term coincides with

$$b_1(n)b_1(n-1)b_1(n-2)(\lambda(n-3) - 3\lambda(n-2) + 3\lambda(n-1) - \lambda_n) = -6b_1(n)b_1(n-1)b_1(n-2).$$

Again we get a contradiction. □

Lemma 5.7. *The coefficient $b_0(n)$ in the expansion (5.7) of Λ is an arbitrary constant.*

Proof. By Lemmas 5.5 and 5.6, $\Lambda = S + b_0(n)$, which makes the computation of both Λ^3 and ad^3_Λ very easy. Moreover, we only need the coefficient at $S^0 = Id$. We have

$$\Lambda^j = \cdots + b_0^j(n)S^0, \quad j \in \mathbb{N}.$$

For $\mathrm{ad}^j_\Lambda \lambda_n$, $j \in \mathbb{N}$, it is obvious that it does not contain S^0. Comparing the coefficients at S^0, we obtain

$$\alpha b_0^3(n) + \beta b_0^2(n) + \gamma b_0(n) + \delta = 0.$$

However this is only possible if $b_0(n)$ is a constant. □

Denoting this above constant by p, we get the following claim.

Proposition 5.6. *The only linear differential operators of order* 3 *solving Problem* 5.3, *which have polynomial eigenfunctions with* $\lambda_n = n(n-1)(n-2) + \nu n(n-1) + \mu n$, *are of the form*

$$T = 6(x-p)^3\partial_x^3 + \nu(x-p)^2\partial_x^2 + \mu(x-p)\partial.$$

Hence any such operator T is reducible.

Proof. By the last lemma we have

$$xp_n(x) = p_{n+1}(x) + pp_n(x).$$

Hence $p_{n+1} = (x-p)^n$. The corresponding differential operator of order 3 is as announced. □

Remark 5.11. Our proof of the main Theorem 5.2 is based on consideration of a large number of special subcases and is partially computer-aided. To be able to approach our general conjectures, such method is not very effective, and a more conceptual understanding of our proof is needed.

5.8 Computer-aided formulas relevant for § 5.6

$$T = (q_{32}x^2 + q_{31}x + q_{30})\partial^3 + (x^2 + q_{21}x + q_{20})\partial^2 + (q_{11}x + q_{10})\partial,$$
$$xp_n(x) = p_{n+1}(x) + \sum_{j=0}^{n} b_j(n)p_{n-j}(x). \tag{5.23}$$

Putting $q_{3,2} = 1$, we get

$$b_3(n) = \frac{(11184\mathbf{n}^{13} + \mathcal{O}(\mathbf{n}^{12}))}{24(q_{1,1} + 2\mathbf{n} - 5)(q_{1,1} + 2\mathbf{n} - 4)(q_{1,1} + 2\mathbf{n} - 3)^2(q_{1,1} + 2\mathbf{n} - 2)^4(q_{1,1} + 2\mathbf{n} - 1)}. \tag{5.24}$$

With $q_{3,2} = 0$ and $q_{3,1} = 1$ we get

$$b_3(n) = \frac{(4224\mathbf{n}^{11} + \mathcal{O}(\mathbf{n}^{10}))}{24(q_{1,1} + 2\mathbf{n} - 5)(q_{1,1} + 2\mathbf{n} - 4)(q_{1,1} + 2\mathbf{n} - 3)^2(q_{1,1} + 2\mathbf{n} - 2)^4(q_{1,1} + 2\mathbf{n} - 1)} \tag{5.25}$$

and with $q_{3,2} = 0$, $q_{3,1} = 0$, and $q_{3,0} = 1$ (recall that $q_{21} = 0$) we arrive at

$$\begin{aligned} b_3(n) = {} & \frac{1}{24(q_{1,1} + 2\mathbf{n} - 5)(q_{1,1} + 2\mathbf{n} - 4)(q_{1,1} + 2\mathbf{n} - 3)^2(q_{1,1} + 2\mathbf{n} - 2)^4(q_{1,1} + 2\mathbf{n} - 1)} \\ & \times \Big(8\mathbf{n}^9(96q_{2,0}^2 + 288q_{1,0}) + 8\mathbf{n}^8(144(3q1,1 - 8)q_{2,0}^2 - 3200q_{1,0} + 1168q_{1,0}q_{1,1}) \\ & - 4\mathbf{n}^7\Big(-96q_{2,0}q_{1,0}^2 + 2(-2032q_{1,1}^2 + 11200q_{1,1} - 15064)q_{1,0} \\ & - 96(17q_{1,1}^2 - 92q_{1,1} + 122)q_{2,0}^2\Big) \\ & - 4\mathbf{n}^6\Big(-48(7q_{1,1} - 18)q_{2,0}q_{1,0}^2 + 2(-1960q_{1,1}^3 + 16416q_{1,1}^2 - 44596q_{1,1} + 39280)q_{1,0} \\ & - 48(35q_{1,1}^3 - 290q_{1,1}^2 + 782q_{1,1} - 688)q_{2,0}^2\Big) + \mathcal{O}(\mathbf{n}^5)\Big). \end{aligned} \tag{5.26}$$

(In the above formulas we use $\mathbf{n}$ instead of n to make it easier for a reader to follow which powers of the variable n these formulas contain.)

5.9 Open problems in Chapter 5

5.1. The major problem in the context of this chapter is whether every BKS, orthogonal with respect to a positive measure with compact support, is a Jacobi-type OPS, compare [66, Conjecture 7.3]. For the constant leading coefficient, the analogous fact was proved in [117].

5.2. Is there an analog of Theorem A on the asymptotic zero distribution for a signed measure μ with compact support? What is the situation for a probability measure with a non-compact support as well as for Sobolev orthogonal polynomial systems. (There exists a literature on this topic.)

5.3. Generalize the results of [27] to operators with $\deg Q_k(z) < k$.

5.4. Extend the results of the second part of this chapter to the case of operators of order 4.

5.5. Settle Conjecture 5.3.

6 Homogenized spectral problem for ES-operators

(The material of this chapter is based on [39].)

6.1 Preliminaries

In spectral theory and WKB-analysis it is typical to consider more general types of families of linear operators than those obtained by adding the identity operator to a given one. Already G. D. Birkhoff considered such problems in the beginning of the 20th century, see [31]. They also appeared many times later in the writings of M. Fedoryuk, W. Wasow, F. W. J. Olver, and many other specialists in the asymptotic analysis. In this chapter we study the properties of asymptotic root-counting measures for families of (non-standard) polynomial eigenfunctions of ES-operators. Using this information, we describe a class of algebraic functions possessing a branch near $\infty \in \mathbb{CP}^1$, which is representable (up to a constant factor) as the Cauchy transform of a compactly supported probability measure.

Notation 6.1. Since we consider non-standard eigenvalue problems in this chapter, we need to assume here that ES-operators might have a non-vanishing constant term, i. e., they are of the form $T = \sum_{j=0}^{k} Q_j(x)\frac{d^j}{dz^j}$ with $\deg Q_j \le j$.

An ES-operator T of order k is called *strictly exactly solvable* if the flag $\mathbb{C}_0[x] \subset \mathbb{C}_1[x] \subset \mathbb{C}_2[x] \subset \cdots$ is strictly preserved. One can show that the coefficients $Q_j(z) := \sum_{i=0}^{j} q_{j,i}z^i$ of a strictly exactly solvable T satisfy the condition that the equation $\sum q_{j,j}t(t-1)\cdots(t-j+1) = 0$ has no positive integer solutions.

Let ES_k denote the linear space of all ES-operators of order at most k.

Consider a spectral pencil of ES_k-operators $T_\lambda := \sum_{j=0}^{k} Q_j(z,\lambda)\frac{dj}{dz^j}$, i. e., a parametrized curve $T : \mathbb{C} \to ES_k$. Here $Q_j(z,\lambda)$ is a polynomial of degree at most j in z for every value of the spectral parameter λ. Each value of the parameter λ for which the equation $T_\lambda p(z) = 0$ has a polynomial solution is called a *(generalized) eigenvalue*, and the corresponding polynomial solution is called a *(generalized) eigenpolynomial*.

The main problem that we address in this chapter is as follows.

Problem 6.1. Given a spectral pencil T_λ describe the asymptotics of its eigenvalues and the asymptotics of the root distribution of the corresponding eigenpolynomials.

Motivated by the necessities of the asymptotic theory of linear ordinary differential equations, see, e. g., [68, Chapter 5], we concentrate below on the fundamental special case of *homogenized spectral pencils*, i. e., rational normal curves in ES_k of the form

$$T_\lambda = \sum_{j=0}^{k} Q_j(z)\lambda^{k-j}\frac{d^j}{dz^j}, \tag{6.1}$$

https://doi.org/10.1515/9783112207741-007

where each $Q_j(z)$ is a polynomial of degree at most j. Consider the algebraic curve Γ given by the equation

$$\sum_{j=0}^{k} Q_j(z)w^j = 0, \tag{6.2}$$

where the polynomials $Q_j(z) = \sum_{i=0}^{j} q_{j,i}z^i$ are the same as in (6.1). Given a curve Γ as in (6.2) and the pencil T_λ as in (6.1) with the same coefficients $Q_j(z)$, $0 \le j \le k$, we call Γ the *plane curve associated with* T_λ and we say that T_λ is the *spectral pencil associated with* Γ.

The curve Γ and its associated pencil T_λ are of *general type* if the following two non-degeneracy requirements are satisfied:

(i) $\deg Q_k(z) = k$ (i. e., $q_{k,k} \neq 0$).

(ii) No two roots of the (characteristic) equation

$$q_{k,k} + q_{k-1,k-1}t + \cdots + q_{0,0}t^k = 0 \tag{6.3}$$

lie on a line through the origin (in particular, 0 is not a root of (6.3)).

The first statement of the chapter is as follows.

Proposition 6.1. *If the characteristic equation (6.3) has k distinct solutions $\alpha_1, \alpha_2, \dots, \alpha_k$ and satisfies the above non-degeneracy assumptions (in particular, these imply that $q_{0,0} \neq 0$ and $q_{k,k} \neq 0$), then*

(i) *For all sufficiently large n, there exist exactly k distinct eigenvalues $\lambda_{n,i}$, $i = 1, \dots, k$, such that the associated spectral pencil T_λ has a polynomial eigenfunction $p_{n,i}(z)$ of degree exactly n.*

(ii) *For all large n, the eigenvalues $\lambda_{n,i}$ split into k distinct families labeled by the roots of (6.3). In particular, there exists n_* such that for all $n \ge n_*$ we can relabel the generalized eigenvalues in such a way that $|\frac{\lambda_{n,i}}{n} - \alpha_i|$ is the smallest among all $|\frac{\lambda_{n,j}}{n} - \alpha_i|$. Therefore*

$$\lim_{n\to\infty} \frac{\lambda_{n,i}}{n} = \alpha_i, \quad i = 1, \dots, k.$$

Theorem 6.1. *In the notation of Proposition 6.1 for any pencil T_λ of general type and every $i = 1, \dots, k$, there exists a subsequence $\{n_{\ell,i}\}$, $\ell = 1, 2, \dots$, such that the limits*

$$\Psi_i(z) := \lim_{\ell\to\infty} \frac{p'_{n_{\ell,i}}(z)}{\lambda_{n_{\ell,i}} p_{n_{\ell,i}}(z)}, \quad i = 1, \dots, k,$$

exist almost everywhere in $\mathbb{C}$ and are analytic functions in some neighborhood of ∞. Each $\Psi_i(z)$ satisfies equation (6.2), i. e., $\sum_{j=0}^{k} Q_j(z)\Psi_i^j(z) = 0$ almost everywhere in $\mathbb{C}$, and the functions $\Psi_1(z), \dots, \Psi_k(z)$ are independent sections of the curve Γ considered as a branched covering over $\mathbb{CP}^1$ in a sufficiently small neighborhood of ∞.

As we explain in § 6.5, a key ingredient in the proof of Theorem 6.1 is the following localization result for the roots of the above eigenpolynomials.

Theorem 6.2. *For any general type pencil T_λ, all the roots of all its polynomial eigenfunctions $p_{n,i}(z)$ lie in a certain disk in $\mathbb{C}$ centered at the origin.*

The sketch of the proof of this fundamental result is as follows. We convert the differential equation satisfied by the eigenpolynomials into a non-linear Riccati type equation for the logarithmic derivatives of the eigenpolynomials. This equation is then solved recursively and the formal solution is shown to be analytic in a neighborhood of ∞. This proves that the zeros of the eigenpolynomials (coinciding with the poles of the logarithmic derivatives) must lie in some compact subset of $\mathbb{C}$.

In fact, we conjecture that for each j, the above convergence result holds for the whole sequence of corresponding eigenpolynomials.

Conjecture 6.1. *In the notation of Theorem* 6.1, *for every $i = 1, \dots, k$, the limit*

$$\Psi_i(z) = \lim_{n\to\infty} \frac{p'_{n,i(z)}}{\lambda_{n,i} p_{n,i}(z)}$$

exists and has all the properties stated in Theorem 6.1 *almost everywhere in $\mathbb{CP}^1$.*

We emphasize the fact that the *proof* of Theorem 6.1 actually shows that Conjecture 6.1 is valid (at least) in a neighborhood of infinity.

Let us formulate our further results and reinterpret Theorems 6.1–6.2 and Conjecture 6.1.

Notation 6.2. The function $L_{n,i}(z) = \frac{p'_{n,i}(z)}{\lambda_{n,i} p_{n,i}(z)}$ is called the *normalized logarithmic derivative* of the polynomial $p_{n,i}(z)$. The limit $\Psi_i(z) = \lim_{n\to\infty} L_{n,i}(z)$ (if it exists) will be referred to as the *asymptotic (normalized) logarithmic derivative* of the family $\{p_{n,i}(z)\}$.

As above, given a sequence $\{P_m(z)\}$ of polynomials we call the limit $\mu = \lim_{m\to\infty} \mu_{P_m}$ (if it exists in the sense of the weak convergence of functionals) *the asymptotic root-counting measure* of the sequence $\{P_m(z)\}$.

If μ exists and is compactly supported in $\mathbb{C}$, it is also a probability measure and its support $supp(\mu)$ is the limit (in the set-theoretic sense) of the sequence $\{\mathcal{Z}(P_m)\}$ of the zero loci to $\{P_m(z)\}$.

A reinterpretation of Theorem 6.1 in the above terms is as follows.

Proposition 6.2. *In the notation of Theorem* 6.1, *for each $i = 1, \dots, k$, there exists a subsequence $\{n_{\ell,i}\}$, $\ell = 1, 2, \dots$, such that the asymptotic root-counting measure μ_i of the family $\{p_{n_{\ell,i}}(z)\}_i$ exists and has compact support with vanishing Lebesgue area. The Cauchy transform of μ_i coincides with $a_i\Psi_i$ almost everywhere in $\mathbb{C}$.*

As an illustration we present below some numerical results showing a rather complicated behavior of the zero loci $\mathcal{Z}(p_{n,i})$ for $i = 1, \dots, k$. Recall that $supp(\mu_i) =$

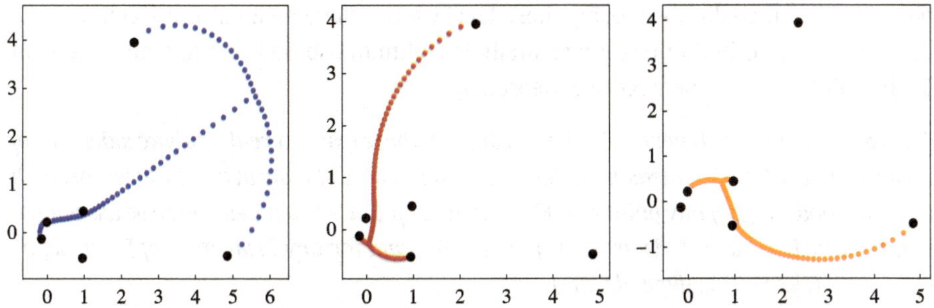

Figure 6.1: Approximations of the three asymptotic root-counting measures for a homogenized spectral pencil of order 3.

$\lim_{n\to\infty} \mathcal{Z}(p_{n,i})$ and note that there are slight differences between the scalings used in the four pictures shown in Figure 6.1.

Explanations to Figure 6.1

The three pictures show the roots of three eigenpolynomials of degree 55 for the (ad hoc chosen) homogenized spectral pencil

$$T_\lambda = (z^3 - (5+2i)z^2 + (4+2i)z)\frac{d^3}{dz^3} + \lambda(z^2 + iz + 2)\frac{d^2}{dz^2} + \lambda^2\frac{1}{5}(z - 2 + i)\frac{d}{dz} + \lambda^3$$

consisting of ES_3-operators. (Here i is the imaginary unit.) The six fat points on all pictures are the branching points of the associated algebraic curve Γ given by

$$(z^3 - (5+2i)z^2 + (4+2i)z)w^3 + (z^2 + iz + 2)w^2 + \frac{1}{5}(z - 2 + i)w + 1 = 0$$

and considered as the branched covering of the z-plane. Numerical comparisons of the eigenfunctions of different degrees show that the above three root distributions already give a very good approximation of the three corresponding limiting probability measures μ_1, μ_2, μ_3 whose Cauchy transforms generate (after appropriate scalings, see Theorem 6.1) the three branches of Γ near ∞. Note that all three supports end only at the branching points of Γ.

As far as the measures μ_i are concerned, in this chapter we establish only some of their most important properties. To prove these, we use two facts about $\mathcal{C}_{\mu_i}(z)$. First, that $a_i^{-1}\mathcal{C}_{\mu_i}(z)$ satisfies the algebraic equation (6.2) and hence can be locally written as a finite sum $\sum_{\ell=1}^{r}\chi_\ell a_\ell^{-1}\Psi_\ell$, where the χ_ℓs are characteristic functions of certain sets. Second, that

$$\frac{\partial \mathcal{C}_{\mu_i}(z)}{\partial \bar{z}} \geq 0$$

in the sense of distributions. Using these facts we can develop a natural algebraic geometric setting and build on complex analytic techniques based on the main results of [36] in order to prove the next two theorems.

Theorem 6.3. *For each pencil T_λ of general type, there exists a real-analytic subset Γ_2 of $\mathbb{C}$ such that any limiting measure μ_i has properties (A)–(C) below in any sufficiently small neighborhood $\Omega(z_0)$ of any point $z_0 \in \mathbb{C}\backslash\Gamma_2$. In what follows Γ denotes the curve associated with the pencil T_λ, $\alpha_i \in \mathbb{C}$ is as in Proposition 6.1, and for any branch y_m of Γ, we define $A_m := \alpha_i y_m$ (clearly, A_m depends on i).*

(A) *The support $\mathcal{S}$ of the measure μ_i restricted to $\Omega(z_0)$ is a finite union of smooth curves $\mathcal{S}_r$, $r \in J$.*

(B) *For each $\mathcal{S}_r$ and any $\tilde{z} \in \mathcal{S}_r$ lying in $\Omega(z_0)$, one can always choose two branches $y_1(z)$ and $y_2(z)$ of Γ such that the tangent line $l(\tilde{z})$ to S_r is orthogonal to $\overline{A_1(\tilde{z})} - \overline{A_2(\tilde{z})}$.*

(C) *The density of the measure μ_i at $\tilde{z}$ equals $\frac{|A_1(\tilde{z})-A_2(\tilde{z})|ds}{2\pi}$, where ds is the length element along the curve $\mathcal{S}_r$.*

In fact for most pencils we can do better:

Theorem 6.4. *For a typical general type pencil T_λ, the set Γ_2 in the preceding theorem is finite.*

Remark 6.1. In the case of the usual spectral problem, strong results in this direction were obtained in Chapter 2, see [27].

We note that Theorem 6.1 also leads to a partial progress in the following intriguing question in potential theory. Let $Alg(i,j)$ denote the linear space of all polynomials in the variables (y,z) of bidegrees at most (i,j), i. e., each $P(y,z) \in Alg(i,j)$ has the form $\sum_{l=0}^{i} P_l(z)y^l$, where $\deg P_l(z) \le j$, $l = 1,\dots,i$. Abusing the notation, we identify each $P(y,z) \not\equiv 0$ with the algebraic function in the variable z defined by the equation $P(y,z) = 0$.

Problem 6.2. Describe the subset $PAlg(i,j) \subset Alg(i,j)$ consisting of all algebraic functions which have a branch near ∞ coinciding with the Cauchy transform of some probability measure compactly supported in $\mathbb{C}$.

Partial progress in this direction can be found in Chapter 7. We refer to such algebraic functions as *positive Cauchy transforms*. The only case when a complete answer to the above problem seems to be obvious is for bidegrees $(1,j)$, i. e., the case of rational functions. Namely, a rational function $r(z) = \frac{p(z)}{q(z)}$ is a positive Cauchy transform if and only if $r(z)$ is of the form

$$\sum_{l=1}^{j} \frac{c_l}{z - z_l} \quad \text{with } c_l > 0, 1 \le l \le j,\ z_k \ne z_l \text{ for } k \ne l,\ \sum_{l=1}^{j} c_l = 1.$$

Let us finally give yet another interpretation of Theorem 6.1 and Proposition 6.2.

Corollary 6.1. *Each branch near infinity of an algebraic function satisfying (6.2) is a positive Cauchy transform multiplied by an appropriate constant.*

The structure of the chapter is as follows. In § 6.2 we study the asymptotics of the eigenvalues to (6.1) and some simple properties of the corresponding eigenfunctions as well as the defining algebraic curve (6.2). In § 6.3 we solve the spectral problem defined by the operator (6.1) in formal power series near ∞ using the variable $y := z^{-1}$. In § 6.4 we show that the power series solution obtained in § 6.3 converges in some neighborhood of ∞. Based on these results we prove Theorem 6.2 as well as Theorem 6.1 and its corollaries in § 6.5. In § 6.6 we prove Theorem 6.3 and complete the proof of Theorem 6.4. Finally, in § 6.7 we propose a number of open problems and conjectures on the asymptotic behavior of polynomial functions for linear ordinary differential operators and place these in a wider context that encompasses both old and new literature on this and related topics.

6.2 Basic facts on asymptotics of eigenvalues and polynomial eigenfunctions

In this section we prove Proposition 6.1, that is, we describe the polynomial solutions of an *ES* spectral pencil, and we also prove the easy part of Theorem 6.1 saying that if the limit functions $\Psi_i(z)$, $1 \leq i \leq k$, exist locally and have derivatives of sufficiently high order, then they must satisfy equation (6.2).

Denote by $\mathcal{D}_n \subset ES_k$ the subset of all *ES*-operators T of order at most k such that the equation $Ty = 0$ has a polynomial solution of degree exactly n.

Lemma 6.1. *In the notation of Theorem 6.1 the closure $\overline{\mathcal{D}_n}$ of the discriminant $\mathcal{D}_n$ is a hyperplane in ES_k given by the equation*

$$\sum_{j=0}^{k} n(n-1)\cdots(n-j+1)q_{j,j} = \sum_{j=0}^{k} (n)_j q_{j,j} = 0. \tag{6.4}$$

Proof. Since any *ES*-operator T (non-strictly) preserves the infinite flag $\mathcal{P}_0 \subset \mathcal{P}_1 \subset \cdots \subset \cdots$ of linear spaces of polynomials of at most given degree, it follows that T has an eigenpolynomial of degree exactly n if and only if

(i) T restricted to $\mathcal{P}_n$ is degenerate.

(ii) The kernel of T restricted to $\mathcal{P}_n$ intersects $\mathcal{P}_n \setminus \mathcal{P}_{n-1}$.

The closure $\overline{\mathcal{D}_n}$ consists of all T having an eigenpolynomial of degree at most n. Now T is upper triangular in the standard monomial basis and the mth diagonal entry of T equals $\sum_{j=0}^{k}(m)_j q_{j,j}$. Therefore $\overline{\mathcal{D}_n}$ is given by equation (6.4). □

Remark 6.2. If $T \in \overline{\mathcal{D}_n}$, but $T \notin \overline{\mathcal{D}_l}$ for $0 \le l < n$, then T has a polynomial solution of degree exactly n. Otherwise T has at least a polynomial solution of degree equal to $\min\{l \in \{0, \dots, n-1\} : T \in \overline{\mathcal{D}_l}\}$ (and probably some other polynomial solutions as well).

We now turn to Proposition 6.1.

Proof. By Lemma 6.1 the homogenized spectral pencil T_λ has a polynomial solution of degree at most n if

$$\sum_{j=0}^{k} (n)_j q_{j,j} \lambda^{k-j} = 0. \tag{6.5}$$

This equation is of degree exactly k in λ since $q_{0,0} \neq 0$. Considering temporarily n as a complex variable, the equation defines an algebraic curve in $\mathbb{C}^2$ with k branches over $\mathbb{C}$. The behavior of the branches at infinity may be found by substituting $\widetilde{\lambda} = \lambda/n$ and dividing the left-hand side of (6.5) by n^k, which gives

$$\sum_{j=0}^{k} 1 \cdot \frac{(n-1)}{n} \cdot \frac{(n-2)}{n} \cdot \dots \cdot \frac{(n-i+1)}{n} \cdot q_{j,j} \cdot \widetilde{\lambda}^{k-j} = 0. \tag{6.6}$$

If $n \to \infty$ the latter family of equations tends coefficientwise to equation (6.3). Since the latter has exactly k different solutions, this is true for (6.6) as well if $n \in N$ for some simply connected neighborhood $N \subset \mathbb{CP}^1$ of infinity. Hence in this neighborhood N there are k different branches $\widetilde{\lambda}_{n,1}, \dots, \widetilde{\lambda}_{n,k}$ of the algebraic function $\widetilde{\lambda}$ defined as a function of the complex variable n by equation (6.6), and corresponding branches $\lambda_{n,1}, \dots, \lambda_{n,k}$ of the algebraic function λ. Then clearly one has

$$\lim_{n\to\infty} \lambda_{n,i}/n = \lim_{n\to\infty} \widetilde{\lambda}_{n,i} = \alpha_i, \quad i = 1, \dots, k.$$

This completes the proof of item (ii).

Further, to prove that for each branch $\lambda_{n,i}$, $i = 1, \dots, k$, the associated spectral pencil T_λ has a polynomial eigenfunction $p_{n,i}(z)$ of degree *exactly* n, we need to show that for a given λ and n that solve equation (6.5), there is no $m \neq n$ such that λ solves equation (6.5) for that m. This follows if we first prove that there are no solutions $n \neq m$ to $\lambda_{n,i} = \lambda_{m,i}$, $i = 1, \dots, k$, and secondly that if $i_1 \neq i_2$ then there are no solutions to $\lambda_{n,i_1} = \lambda_{m,i_2}$ (both statements in a possibly smaller neighborhood of infinity). The first statement is an immediate consequence of the assumptions since they imply that $\alpha_i \neq 0$, $i = 1, \dots, k$, and thus for each i, the function $n \mapsto \lambda_{n,i}$ is one-to-one in a neighborhood of infinity.

For the second statement, argue by contradiction as follows. If the statement is false, there are unbounded sequences of positive integers n_ℓ, m_ℓ, $\ell = 1, 2, \dots$ such that $\lambda_{n_\ell,i_1} = \lambda_{m_\ell,i_2}$ for all i. We may assume (by choosing a subsequence and possibly interchanging i_1 and i_2) that $\lim_{\ell\to\infty} m_\ell/n_\ell = r \in \mathbb{R}$. Hence

$$\alpha_{i_1} = \lim_{\ell\to\infty} \lambda_{n_\ell,i_1}/n_\ell = \lim_{\ell\to\infty} (\lambda_{m_\ell,i_2}/m_\ell)(m_\ell/n_\ell) = r\alpha_{i_2}$$

and so the arguments of α_{i_1} and α_{i_2} are equal, which contradicts the non-degeneracy assumptions of the proposition. This completes the proof. □

Remark 6.3. It is also possible to calculate the asymptotics for the eigenvalues even in the case when (6.3) has multiple or vanishing roots.

Let us now prove the easy part of Theorem 6.1.

Proposition 6.3. *Let $\{p_{n,i}(z)\}$ be a family of polynomial eigenfunctions of a homogenized spectral pencil T_λ with corresponding family of eigenvalues $\{\lambda_{n,i}\}$, i. e., $p_{n,i}(z)$ satisfies the equation $T_{\lambda_{n,i}} p_{n,i}(z) = 0$. Assume that the following holds:*
(i) $\lim_{n\to\infty} \lambda_{n,i} = \infty$.
(ii) *There exists an open set $\Omega \subseteq \mathbb{CP}^1$ where the normalized logarithmic derivatives $L_{n,i}(z) = \frac{p'_{n,i}(z)}{\lambda_{n,i} p_{n,i}(z)}$ are defined for all sufficiently large n and the sequence $\{L_{n,i}(z)\}$ converges in Ω to a function*

$$\Psi_i(z) := \lim_{n\to\infty} L_{n,i}(z).$$

(iii) *The $k-1$ first derivatives of the sequence $\{L_{n,i}(z)\}$ are uniformly bounded in Ω.*

Then if $\Psi_i(z)$ does not vanish identically, then it satisfies the equation

$$\sum_{j=0}^{k} Q_j(z)\Psi_i^j(z) = 0. \tag{6.7}$$

Proof. In order to simplify the notation in this proof, let us fix the value of the index $i \in \{1,\ldots,k\}$ and drop it.

Note that each $L_n(z) = \frac{p'_n(z)}{\lambda_n p_n(z)}$ is well defined and analytic in any disk $\mathbf{D}$ free from the zeros of $p_n(z)$. Choosing such a disk $\mathbf{D}$ and a branch of the logarithm such that $\log p_n(z)$ is defined in $\mathbf{D}$, let us consider a primitive function $M(z) = \lambda_n^{-1}\log p_n(z)$, which is also well defined and analytic in $\mathbf{D}$.

Straightforward calculations give: $e^{\lambda_n M(z)} = p_n(z)$, $p'_n(z) = p_n(z)\lambda_n L_n(z)$, and $p''_n(z) = p_n(z)(\lambda_n^2 L_n^2(z) + \lambda_n L'_n(z))$. More generally,

$$\frac{d^\ell}{dz^\ell}(p_n(z))p_n(z)(\lambda_n^\ell L_n^\ell(z) + \lambda_n^{\ell-1}F_\ell(L_n(z), L'_n(z),\ldots,L_n^{(\ell-1)}(z))),$$

where the second term

$$\lambda_n^{\ell-1}F_\ell(L_n, L'_n,\ldots,L_n^{(\ell-1)}) \tag{6.8}$$

is a polynomial of degree $\ell-1$ in λ_n. The equation $T_{\lambda_n}p_n(z) = 0$ gives

$$p_n(z)\left(\sum_{j=0}^{k} Q_j(z)\lambda_n^{k-j}(\lambda_n^j L_n^j(z) + \lambda_n^{j-1}F_j(L_n(z), L'_n(z),\ldots,L_n^{(j-1)}(z)))\right) = 0$$

or, equivalently,

$$\lambda_n^k \sum_{j=0}^{k} Q_j(z)(L_n^j(z) + \lambda_n^{-1} F_j(L_n(z), L_n'(z), \dots, L_n^{(j-1)}(z))) = 0. \tag{6.9}$$

Letting $n \to \infty$ and using the boundedness assumption for the first $k-1$ derivatives, we get the required equation (6.7). □

We end this section by establishing an important property of the curve Γ given by (6.2). Note that unless $Q_k(x) \equiv 0$ the curve Γ is a k-sheeted branched covering of the z-plane. We want to describe the behavior of Γ at infinity. Using coordinate $y := z^{-1}$, we can rewrite equation (6.2) as

$$\sum_{j=0}^{k} Q_j(z) w^j = \sum_{j=0}^{k} z^{-j} Q_j(z)(wz)^j = \sum_{j=0}^{k} P_j(y)\xi^j = 0,$$

where

$$P_j(y) = z^{-j} Q_j(z) = \sum_{\ell=0}^{j} q_{j\ell} y^{\ell-j}$$

and $\xi := wz = w/y$. Note also that at the point $y = 0$ one gets the reciprocal characteristic equation

$$q_{k,k}\xi^k + q_{k-1,k-1}\xi^{k-1} + \dots + q_{0,0} = 0. \tag{6.10}$$

Remark 6.4. The roots $\xi_1, \dots, \xi_k$ of (6.10) are the inverses of the roots $\alpha_1, \dots, \alpha_k$ of (6.3) respectively.

Using the above argument, we get the following simple statement.

Lemma 6.2. *If the roots $\xi_1, \dots, \xi_k$ of equation (6.10) are pairwise distinct, then there exist k branches $y_i(z)$, $i = 1, \dots, k$, of the curve Γ that are well defined in some common neighborhood of $z = \infty$. The ith branch $y_i(z)$ satisfies the normalization condition $\lim_{z\to\infty} zy_i(z) = \xi_i$.*

6.3 Solving equation (6.9) formally

This and the following two sections are completely devoted to the proof of Theorems 6.1 and 6.2. The sketch of the proof of Theorem 6.1 is as follows. In Proposition 6.3 of § 6.2 we have transformed the linear differential equation for the eigenpolynomials into a non-linear Riccati type equation for the logarithmic derivative. In this section we first closely analyze the terms of this new equation. We then describe the recursion scheme for solving this equation formally in a neighborhood of infinity and see how the solutions behave when $\lambda \to \infty$. Finally, in the next section we show that there is a neighborhood of $z = \infty$ where the formal solutions are analytic and we complete the proofs of Theorems 6.1 and 6.2 in § 6.5.

Throughout this section we use the variable $y := 1/z$ near $z = \infty$.

6.3.1 Differential algebra $A_{z,L}$

As the *first step* we describe the terms occurring in equation (6.9) more precisely. It is convenient to do this in a universal setting using the following infinitely generated free commutative $\mathbb{C}$-algebra (or rather the free differential algebra, cf. [96])

$$A_{z,L} := \mathbb{C}[\lambda, \lambda^{-1}, z, z^{-1}, L^{(0)}, L^{(1)}, L^{(2)}, \dots].$$

This algebra should be thought of as (a universal object) containing the terms in equation (6.9). Concrete instances can be obtained by specialization. In particular, the $L^{(i)}$s correspond to the normalized logarithmic derivatives in (6.9).

Note that the monomials $\lambda^i z^j L^I$ form a basis of $A_{z,L}$ considered as a vector space, where for any multi-index $I = (i_1, \dots, i_r)$, the symbol L^I denotes the product

$$L^I = \prod_{s=1}^{r} L^{(i_s)}.$$

It suffices to use multi-indices I that are finite non-decreasing sequences of non-negative integers. Denote the set of all such multi-indices by FS. (By definition, $L^{(0)} := L$.) Such index sequences may alternatively be thought of as finite multi-sets. In particular, we include the empty sequence $\emptyset$ in FS and interpret $L^{\emptyset} := 1$. For a given multi-index I, define its *modulus* by $|I| = \sum_{s=1}^{r} i_s$ and its *length* by $lng(I) = r$. (By definition, $|\emptyset| := 0$ and $lng(\emptyset) := 0$.) For a given $I \in FS$, denote by I_+ the sequence obtained from I by discarding all its 0 elements.

$A_{z,L}$ is equipped with a natural derivation D_z, which is a prototype of $\frac{d}{dz}$. Namely, D_z is uniquely defined by the relations: $D_z * \lambda = 0$, $D_z * z = 1$, and $D_z * L^{(i)} = L^{(i+1)}$. (We use the symbol "$*$" to denote the action of a differential operator as opposed to the product of differential operators. Note that the ring of differential operators generated by $A_{z,L}$ and D_z acts on $A_{z,L}$.)

6.3.2 The logarithmic derivative in the universal setting

Next we can use $A_{z,L}$ to describe the relation between the derivatives of an eigenpolynomial and its normalized logarithmic derivative. This is done by defining a differential $A_{z,L}$-module. Consider the free rank one $A_{z,L}$-module $A_{z,L}e^M$, where e^M is the generator. Define

$$D_z * e^M := \lambda L e^M$$

and extend the action of D_z to the whole $A_{z,L}e^M$ using the Leibnitz rule. Note that this action intuitively just says that L is the normalized logarithmic derivative $D_z * e^M/\lambda e^M$ of the generator e^M. The following lemma can be easily proved by induction.

Lemma 6.3. *In the above notation, one has* $(D_z)^m * e^M = R_m e^M$, *where* $R_0 = 1$, $R_1 = \lambda L$ *and* $R_{m+1} = (\lambda L + D_z) * R_m$ *for* $m \geq 1$. *In other words,*

$$R_m = (\lambda L + D_z)^m * 1 = (\lambda L + D_z)^{m-1} * \lambda L, \quad m \geq 1.$$

By universality, R_ms considered as polynomials $R_m(\lambda, L^{(0)}, \dots)$ have the following property.

Lemma 6.4. *Let* $g = \lambda^{-1} D \log f$, *where* f *is a non-vanishing analytic function in an open subset* $\Omega \subset \mathbb{C}$ *and* $\lambda \in \mathbb{C} \setminus \{0\}$ *(i. e., g is the logarithmic derivative of f normalized with respect to λ). Then in the above notation, one has*

$$f^{(m)} = R_m(\lambda, g, g^{(1)}, \dots) f, \quad m \geq 1.$$

6.3.3 Differential algebra $A_{z,L}$ at ∞

Next we rewrite (6.9) using the variable $y = z^{-1}$. Note that if $p(z)$ is a non-constant polynomial, then for any $\kappa \neq 0$, the logarithmic derivative $L(z) = p'(z)/\kappa p(z)$, rewritten using the variable $y = \frac{1}{z}$, has a simple zero at $y = 0$. Hence we should look for solutions of equation (6.9) in the form $L(z) = yN(y)$.

First we describe $A_{z,L}$ near $z = \infty$. For this we define an analogous free commutative algebra

$$B_{y,N} := \mathbb{C}[\lambda, \lambda^{-1}, y, y^{-1}, N^{(0)}, N^{(1)}, \dots].$$

As above, it has a natural derivation D_y, which is a prototype of $\frac{d}{dy}$ satisfying the relations: $D_y * \lambda = 0$, $D_y * y = 1$, and $D_y * N^{(m)} = N^{(m+1)}$. Note that $\frac{d}{dz} = -y^2 \frac{d}{dy}$ and recall that $L := L^{(0)}$ and $N := N^{(0)}$. Define an injection $\Theta : A_{z,L} \to B_{y,N}$ of algebras determined by

$$\Theta(\lambda) = \lambda, \quad \Theta(z) = y^{-1}, \quad \Theta(L) = yN, \quad \Theta(L^{(m)}) = (-y^2 D_y)^m * yN.$$

The following lemma describes the connection between $A_{z,L}$ and $B_{y,N}$.

Lemma 6.5. *The injection* Θ *has the property that for any* $a \in A_{z,L}$, *one has*

$$\Theta(D_z^m * a) = (-y^2 D_y)^m * \Theta(a).$$

Proof. By definition the above formula is valid in the case $m = 1$ for all the generators of $A_{z,L}$. Since D_z and $-y^2 D_y$ are derivations, the formula then works for all elements in this algebra and $m = 1$. Simple induction shows that the above formula is also valid for all $m > 1$. □

6.3.4 Main lemma on R_m

The preceding lemmas imply the following relation:

$$\lambda^{-m}y^{-m}\Theta(R_m) = y^{-m}(yN - y^2\lambda^{-1}D_y)^m * 1 = y^{-m}(yN - y^2\lambda^{-1}D_y)^{m-1}y * N. \tag{6.11}$$

We need to know which monomial terms of the form $\lambda^{\alpha_1}y^{\alpha_2}N^I$ occur in this equation. These monomials are contained in the subalgebra $B_0 \subset B_{y,N}$ defined as

$$B_0 := \mathbb{C}[\lambda^{-1}, y, N, \lambda^{-1}yN^{(1)}, \dots, (\lambda^{-1}y)^j N^{(j)}, \dots],$$

see Lemma 6.6 below. Define a two-sided ideal $J \subset B_0$ as follows:

$$J := \langle \lambda^{-1}, (\lambda^{-1}y)^2 N^{(1)}N^{(1)}, \dots, (\lambda^{-1}y)^{j_1+j_2}N^{(j_1)}N^{(j_2)} \dots \rangle,$$

where $j_1 \geq 1, j_2 \geq 1$. (As we will see, the ideal J contains all the terms that do not influence the asymptotic behavior of equation (6.9).)

One can easily show that the set of all monomials of the form

$$(\lambda^{-1})^{\beta_1} y^{\beta_2} (\lambda^{-1}y)^{|I|} N^I,$$

where $\beta_1, \beta_2 \in \mathbb{Z}_{\geq 0}$, and $I \in FS$ constitutes a basis of B_0 as a vector space. Such a monomial belongs to J if and only if either $\beta_1 \geq 1$ or $lng(I_+) \geq 2$.

Lemma 6.6. *For all $m \geq 0$, the following identity between elements in B_0*

$$\lambda^{-m}y^{-m}\Theta(R_m) = N^m + \sum_{j=1}^{m-1} \binom{m}{j+1} (-1)^j (\lambda^{-1}y)^j N^{m-1-j} N^{(j)} + h \tag{6.12}$$

holds for some unique $h \in J$. Moreover, if $m \geq 1$, the exponents of all non-vanishing monomials $(\lambda^{-1})^{\beta_1} y^{\beta_2} (\lambda^{-1}y)^{|I|} N^I$ occurring in the right-hand side of (6.12) *satisfy the inequality $\beta_1 + |I| \leq m - 1$.*

Proof. The case $m = 0$ is trivial, so we assume that $m \geq 1$.

Claim. The differential operator $L_m := y^{-(m+1)}(yN - y^2\lambda^{-1}D_y)^m y$ preserves both B_0 and J and satisfies the relation

$$L_m * N = N^{m+1} + \sum_{j=1}^{m} \binom{m+1}{j+1} (-1)^j N^{m-j} (\lambda^{-1}y)^j N^{(j)} \pmod{J}, \tag{6.13}$$

where "(mod J)" means that the expression is considered modulo the ideal $J \subset B_0$.

The lemma immediately follows from this claim. Indeed, note that by Lemma 6.3 one has

$$\lambda^{-m}y^{-m}\Theta(R_m) = y^{-m}\big(yN - y^2\lambda^{-1}D_y\big)^{m-1}y * N = L_{m-1} * N, \tag{6.14}$$

and so (6.12) is a consequence of (6.14) and (6.13). It thus remains to prove the three assertions in the above claim, which we do below by induction. As the base of induction, note that since $L_0 = 1$, it is obvious that L_0 preserves both B_0 and J and that it satisfies (6.13).

Steps of induction

We first show inductively that all operators L_m preserve both B_0 and J and then using this we check formula (6.13) also using induction. Now the following identity:

$$\begin{aligned} L_{m+1} * v &= y^{-(m+2)}(yN - y^2\lambda^{-1}D_y)y^{m+1}L_m * v \\ &= (N - (m+1)\lambda^{-1})L_m * v - y\lambda^{-1}D_y(L_m * v), \quad v \in B_0, \end{aligned} \tag{6.15}$$

is immediate. Hence in order to check that B_0 and J are preserved by L_{m+1}, it suffices (using the induction assumption) to show that both $(N - (m+1)\lambda^{-1})$ and $y\lambda^{-1}D_y$ preserve B_0 and J. Since N and λ^{-1} are contained in B_0, it is clear that they preserve both B_0 and J, so it is enough to verify that the same holds for $\lambda^{-1}yD_y$. Let us first show that the latter operator preserves the ideal J, under the assumption that it preserves B_0. For this we note that an arbitrary element $h \in J$ may be written as $h = \sum b_\ell h_\ell$, where h_ℓ are the generators given in the definition of J and $b_\ell \in B_0$. Thus it is enough to prove that $y\lambda^{-1}D_y * b_\ell h_\ell \in J$. As we will now explain, this property follows from the identity

$$y\lambda^{-1}D_y * b_\ell h_\ell = y\lambda^{-1}h_\ell(D_y * b_\ell) + y\lambda^{-1}b_\ell(D_y * h_\ell).$$

Indeed, we claim that its right-hand side belongs to J. To show this, note that the first term clearly belongs to J, so it suffices to check that $y\lambda^{-1}(D_y * h_\ell) \in J$. Now if $h_\ell = \lambda^{-1}$ then this is again obvious. For the other generators we simply use the identity

$$\begin{aligned} &y\lambda^{-1}D_y * (\lambda^{-1}y)^{j_1+j_2}N^{(j_1)}N^{(j_2)} \\ &\quad = (j_1+j_2)\lambda^{-1}(\lambda^{-1}y)^{j_1+j_2}N^{(j_1)}N^{(j_2)} + (\lambda^{-1}y)^{j_1+j_2+1}N^{(j_1+1)}N^{(j_2)} + (\lambda^{-1}y)^{j_1+j_2+1}N^{(j_1)}N^{(j_2+1)}. \end{aligned}$$

The terms in the right-hand side of the latter formula clearly belong to J, and so from (6.15) we get that L_{m+1} preserves J as soon as L_m does. The fact that $\lambda^{-1}yD_y$ preserves B_0 is proved in exactly the same fashion, namely by first checking that this is true for the generators of B_0 and then applying an inductive argument based on the observation that $\lambda^{-1}yD_y$ is a derivation.

Now that we established that both B_0 and J are preserved by the operators L_m, we use this information to check the induction step in formula (6.13). Since $\lambda^{-1} \in J$, equation (6.15) taken modulo J gives

$$L_{m+1} * v = (N - (m+1)\lambda^{-1})L_m * v - y\lambda^{-1}D_y(L_m * v) = NL_m * v - y\lambda^{-1}D_y(L_m * v) \quad (6.16)$$

for $v \in B_0$. Using the relations $y\lambda^{-1}D_y * N^{m+1} = (m+1)y\lambda^{-1}N^m N^{(1)}$ and

$$y\lambda^{-1}D_y * N^{m-j}y^j\lambda^{-j}N^{(j)} = N^{m-j}y^{j+1}\lambda^{-(j+1)}N^{(j+1)} \pmod{J},$$

we get from (6.16) and the induction assumption that

$$\begin{aligned} L_{m+1} * N &= NL_m * N - y\lambda^{-1}D_y * (L_m * N) \\ &= N^{m+2} + \sum_{j=1}^{m}\binom{m+1}{j+1}(-1)^j N^{m+1-j}y^j\lambda^{-j}N^{(j)} - (m+1)y\lambda^{-1}N^m N^{(1)} \\ &\quad - \sum_{j=1}^{m}\binom{m+1}{j+1}(-1)^j N^{m-j}y^{j+1}\lambda^{-(j+1)}N^{(j+1)} \pmod{J}. \end{aligned}$$

The usual properties of binomial coefficients now accomplish the proof of the step of induction, which completes the proof of formula (6.13).

The last statement in Lemma 6.6 follows trivially since the degree of $\lambda^{-m}y^{-m}\Theta(R_m)$ (considered as an element of $B_{y,N}$) in the variable λ^{-1} is precisely $m-1$, see (6.14), and the degree in λ^{-1} of the monomial $(\lambda^{-1})^{\beta_1}y^{\beta_2}(\lambda^{-1}y)^{|I|}N^I$ equals $\beta_1 + |I|$. □

6.3.5 Description of terms in equation (6.9)

Consider now the homogenized spectral pencil $T_\lambda = \sum_{j=0}^{k} Q_j(z)\lambda^{k-j}\frac{d^j}{dz^j}$. It acts (if we substitute D_z^j for $\frac{d^j}{dz^j}$) on the differential module $A_{z,L}e^M$ and satisfies the obvious relation

$$T_\lambda * e^M = \sum_{j=0}^{k} Q_j(z)\lambda^{k-j}R_j e^M.$$

Recall from Lemma 6.3 that each R_j is a polynomial $R_j(\lambda, L^{(0)}, \ldots)$. Now the non-linear equation $\sum_{j=0}^{k} Q_j(z)\lambda^{k-j}R_j = 0$ in L is the analog over $A_{z,L}$ of the non-linear equation (6.9). A change of variables $y = z^{-1}$ and division by λ^k transforms the previous equation into the equation

$$\lambda^{-k}S_\lambda(yN) := \Theta\left(\sum_{j=0}^{k} Q_j(z)\lambda^{-j}R_j\right) = 0. \quad (6.17)$$

In what follows we use the notation

$$\Theta(Q_j(z)) = \Theta(q_{j,j}z^j + q_{j,j-1}z^{j-1} + \cdots + q_{j,0}) = y^{-j}(q_{j,j} + q_{j,j-1}y + \cdots + q_{j,0}y^j) = y^{-j}P_j(y),$$

where $P_j(y)$ is a polynomial in y of degree at most j. Equation (6.17) has the form

$$\lambda^{-k}S_\lambda(yN) := \sum_{j=0}^{k} P_j(y)y^{-j}\lambda^{-j}\Theta(R_j) = 0. \tag{6.18}$$

Lemma 6.6 implies the following.

Lemma 6.7. *In the above notation, for some $b \in J$, one has*

$$\lambda^{-k}S_\lambda(yN) = \sum_j P_j(y)N^j + \sum_{j=0}^{k}\sum_{\ell=1}^{j-1} P_j(y)\binom{j}{\ell+1}(-1)^\ell N^{j-1-\ell}y^\ell\lambda^{-\ell}N^{(\ell)} + b. \tag{6.19}$$

The exponents of all non-zero monomials $(\lambda^{-1})^{\beta_1}y^{\beta_2}(\lambda^{-1}y)^{|I|}N^I$ occurring in the right-hand side satisfy the condition $\beta_1 + |I| \le k-1$.

6.3.6 Generalities on power series

Lemma 6.7 provides us with all necessary information about the explicit form of the terms in equation (6.9). The *second step* in solving (6.9) is to find a recurrence relation for the formal power series solutions of (6.18) using (6.19). To do this, we will use the ansatz $L(y) = yN(y) = \sum_{\ell=1}^{\infty}\epsilon_\ell y^\ell$. In algebraic terms this means that instead of an element of $B_{y,N}$ we consider its image under the map of differential rings $B_{y,N} \to \mathbf{C}[\epsilon_1,\epsilon_2,\ldots][[y]]$ given by $N \mapsto \sum_{\ell=0}^{\infty}\epsilon_{\ell+1}y^\ell$. (The latter ring is equipped with the usual derivation $\frac{d}{dy}$; in particular, $\frac{d}{dy} * \epsilon_\ell = 0$.)

We will repeatedly use some easily verified algebraic properties of the derivatives of formal power series similar to $N(y)$. These properties are summarized below.

Notation 6.3. If $A = \sum_{\ell=0}^{\infty}k_\ell y^\ell \in K[[y]]$ is a formal power series with coefficients in a field K, we define $[A]_m := k_m$. The notation "*EXP* $\mathrm{mod}(\epsilon_1,\ldots,\epsilon_m)$" means that expression *EXP* is taken modulo the vector space generated by all monomials in the indicated variables $(\epsilon_1,\ldots,\epsilon_m)$. (Note that this usage is different from the earlier used "(mod J)", where J is some ideal.)

Lemma 6.8. *Let $N(y) = \sum_{\ell=0}^{\infty}\epsilon_{\ell+1}y^\ell \in \mathbf{C}[\epsilon_1,\epsilon_2,\ldots][[y]]$. Then the following relations hold:*

(1) $[y^\ell N^{(\ell)}]_m = m(m-1)\cdots(m-\ell+1)\epsilon_{m+1}$. *If $0 \le m < \ell$, then $[y^\ell N^{(\ell)}]_m = 0$.*

(2) *One has*

$$[y^{|J|}N^J]_m = \sum_{c_1+\cdots+c_s=m}\prod_{\ell=1}^{s} c_\ell(c_\ell-1)\cdots(c_\ell-j_\ell+1)\epsilon_{c_\ell+1}.$$

(If $j_\ell = 0$, then abusing our notation we set $c_\ell(c_\ell-1)\cdots(c_\ell-j_\ell+1) = 1$.) If J contains strictly more than one non-zero element, then $[y^{|J|}N^J]_m = 0 \ \mathrm{mod}(\epsilon_1,\ldots,\epsilon_m)$.

(3) $[N^j]_m = j\epsilon_1^{j-1}\epsilon_{m+1} \ \mathrm{mod}(\epsilon_1,\ldots,\epsilon_m)$ *if $m \ge 1$ while $[N^j]_0 = \epsilon_1^j$.*

(4) $[y^\ell N^j N^{(\ell)}]_m = m(m-1)\cdots(m-\ell+1)\epsilon_1^j\epsilon_{m+1} \ \mathrm{mod}(\epsilon_1,\ldots,\epsilon_m)$ *if $\ell \ge 1$.*

(5) *If the monomial* $M = (\lambda^{-1})^{\beta_1} y^{\beta_2} (\lambda^{-1}y)^{|I|} N^I$ *belongs to* B_0, *then*

$$[M]_m = B(\epsilon_1, \dots, \epsilon_{m+1}, \lambda^{-1}, \lambda^{-1}m),$$

where $B(\epsilon_1, \dots, \epsilon_{m+1}, x, y)$ *is a polynomial of degree at most* $|I|$ *in* y *and at most* β_1 *in* x.

Proof. We prove items (2) and (5) leaving the rest as an exercise for the interested reader. Note first that

$$[y^{|J|} N^J]_m = \sum_{c_1+\dots+c_s=m} \prod_{\ell=1}^{s} [y^{j_i} N^{j_i}]_{c_\ell}.$$

By item (1), this gives the required relation in (2). Observe then that the only case when one has

$$\prod_{\ell=1}^{s} c_\ell (c_\ell - 1) \cdots (c_\ell - j_\ell + 1) \epsilon_{c_\ell+1} \neq 0 \bmod (\epsilon_1, \dots, \epsilon_m),$$

is if some $\epsilon_{c_\ell+1} = \epsilon_{m+1}$, and then $c_\ell = m$, and all other $c_l = 0, l \neq \ell$. But if for some such $l \neq \ell$ we have $j_l > 0$, then $c_l(c_l - 1) \cdots (c_l - j_l + 1) = 0$; hence at most one $j_\ell, 1 \leq \ell \leq s$ does not vanish. This finishes the proof of (2).

In order to prove item (5), notice that item (1) implies that the expression

$$\begin{aligned} A_\ell(\epsilon_{m+1}, \lambda^{-1}, \lambda^{-1}m) := [(\lambda^{-1}y)^\ell N^{(\ell)}]_m &= \lambda^{-\ell} m(m-1) \cdots (m - \ell + 1)\epsilon_{m+1} \\ &= (\lambda^{-1}m)(\lambda^{-1}m - \lambda^{-1}) \cdots (\lambda^{-1}m - (\ell - 1)\lambda^{-1})\epsilon_{m+1} \end{aligned}$$

is a polynomial whose degree in the last variable equals i. Therefore, the expression

$$A_\ell(\epsilon_1, \dots, \epsilon_{m+1}, \lambda^{-1}, \lambda^{-1}m) := [(\lambda^{-1}y)^{|J|} N^J]_m = \sum_{c_1+\dots+c_s=m} \prod_{\ell=1}^{s} [(\lambda^{-1}y)^{j_\ell} N^{j_\ell}]_{c_\ell}$$

is a polynomial whose degree in the last variable is at most $|J|$. This implies (5). □

The following description of the coefficients of the power series in the ideal J is an immediate consequence of Lemma 6.8.

Lemma 6.9. *Let* $M = (\lambda^{-1})^{\beta_1} y^{\beta_2} (\lambda^{-1}y)^{|I|} N^I$, *where* $\beta_1, \beta_2 \in \mathbb{Z}_{\geq 0}$, $I \in FS$, *be a monomial in* B_0. *Then*

(1) $[M]_0 \neq 0$ *if and only if* $M = (\lambda^{-1})^{\beta_1} N^j$ *for some* j. *If in addition* $M \in J$, *then* $\beta_1 \geq 1$.
(2) *If* $M \in J$ *and* $m \geq 1$ *then* $[M]_m \neq 0 \bmod (\epsilon_1, \dots, \epsilon_m)$ *implies that:*
   (a) $\beta_1 \geq 1$,
   (b) $\beta_2 = 0$,
   (c) $I = (0, \dots, 0, \sigma)$ *with* $lng(I) = j + 1$ *and contains at most one non-zero element* σ.

If $\sigma \geq 1$, then $M = (\lambda^{-1})^{\beta_1}(\lambda^{-1}y)^{\ell}N^jN^{(\sigma)}$ and

$$[M]_m = (\lambda^{-1})^{\beta_1+\sigma} m(m-1)\cdots(m-\sigma+1)\epsilon_1^j\epsilon_{m+1} \bmod(\epsilon_1,\ldots,\epsilon_m).$$

(3) *If $b \in J$, then $[b]_m = \lambda^{-1}E(\epsilon_1,\lambda^{-1},\lambda^{-1}m)\epsilon_{m+1} \bmod(\epsilon_1,\ldots,\epsilon_m)$, where E is a polynomial.*

Now we have all the tools needed to get an adequate picture of the recurrence relation corresponding to equation (6.9). We use the ansatz $N(y) = \sum_{\ell=0}^{\infty}\epsilon_{\ell+1}y^{\ell}$ in equation (6.18).

6.3.7 Constant term

The first step in solving our recurrence relation is to get an equation for ϵ_1.

Lemma 6.10. *Let $N(y) = \sum_{\ell=0}^{\infty}\epsilon_{\ell+1}y^{\ell}$. The values of λ for which $\lambda^{-k}S_\lambda(yN)$ has no constant term are precisely the solutions of $R(\epsilon_1,\lambda) = 0$. Here $R(\epsilon_1,\lambda)$ is given by*

$$R(\epsilon_1,\lambda) := [\lambda^{-k}S_\lambda(yN)]_0 = \sum_{j=0}^{k} q_{jj}\epsilon_1^j + \lambda^{-1}E(\epsilon_1,\lambda^{-1}) \tag{6.20}$$

for some $E(\epsilon_1,\lambda^{-1})$, which is a polynomial of degree at most k in ϵ_1 (and it has degree at most $k-2$ in λ^{-1}).

Proof. Item (1) of Lemma 6.9 and item (3) of Lemma 6.8 imply that $[P_j(y)N^j]_0 = q_{jj}\epsilon_1^j$ and $[P_j(y)N^{j-1-\ell}y^{\ell}\lambda^{-\ell}N^{(\ell)}]_0 = 0$ for all $\ell \geq 1$. Hence by Lemma 6.7 the constant term of $\lambda^{-k}S_\lambda(yN)$ is given by

$$\sum_j q_{jj}\epsilon_1^j + [b]_0.$$

By item 3 of Lemma 6.9, there is a polynomial E such that $[b]_0 = \lambda^{-1}E(\epsilon_1,\lambda^{-1})$. Item 1 of this same lemma gives that the only terms in b contributing a non-zero term of degree j in ϵ_1 to $[b]_0$ come from terms of the form $c(\lambda^{-1})^{\alpha_1}N^j$ with $\beta_1 \geq 1$ and $c \in \mathbb{C}$. It is clear that such a term will have $j \leq k$ since by (6.17) and (6.11) it stems from some $\lambda^{-m}y^{-m}\Theta(R_m) = y^{-m}(yN - y^2\lambda^{-1}D_y)^m * 1$ with $m \leq k$, which contains N at most in the power m. By the same observation, it follows that the degree in λ^{-1} of such a term is less than or equal to $k-1$. □

Hence the initial step in computing the formal solution to (6.17) is solving the equation

$$R(\epsilon_1,\lambda) = \sum_{j=0}^{k} a_{jj}\epsilon_1^j + \lambda^{-1}E(\epsilon_1,\lambda^{-1}) = 0. \tag{6.21}$$

Note that this equation tends to the equation $\sum_{j=0}^{k} a_{jj}\epsilon_1^j = 0$ as $\lambda \to \infty$ and that the latter coincides with equation (6.10). In particular, under the assumptions of Theorem 6.1, or equivalently, of Lemma 6.2, equation (6.21) has k distinct solutions for any sufficiently large value of λ. Choose one of the branches $\epsilon_1 = \epsilon_1(\lambda)$ that solves (6.21) in a neighborhood of $\lambda = \infty$ in $\mathbb{CP}^1$. This means that we have determined ϵ_1 for any large enough value of λ and we can start determining the other coefficients of $N(y) = \sum_{\ell=0}^{\infty} \epsilon_{\ell+1} y^\ell$ in terms of the chosen ϵ_1.

6.3.8 Recurrence relation

Assume now that the coefficients $\epsilon_1, \dots, \epsilon_m$ of N have already been defined. The recursive step that defines ϵ_{m+1} then corresponds to solving the equation $[\lambda^{-k} S_\lambda(yN)]_m = 0$.

Lemma 6.11. *One has*

$$[\lambda^{-k} S_\lambda(yN(y))]_m = \Phi_0(\epsilon_1, \lambda^{-1}, \lambda^{-1}m)\epsilon_{m+1} + B(\epsilon_1, \dots, \epsilon_m, \lambda^{-1}, \lambda^{-1}m),$$

where B and Φ_0 are polynomials and the latter satisfies

$$\Phi_0(\epsilon_1, \lambda^{-1}, \lambda^{-1}m) = -\sum_{j=0}^{k} a_{jj} \frac{(\epsilon_1 - \lambda^{-1}m)^j - \epsilon_1^j}{\lambda^{-1}m} + \lambda^{-1}E(\epsilon_1, \lambda^{-1}, \lambda^{-1}m), \tag{6.22}$$

where E is a polynomial whose degree in the third variable $\lambda^{-1}m$ is strictly less than $k-2$. The degree of B in the last variable $\lambda^{-1}m$ is less than or equal to $k-1$.

Proof. By equation (6.19) one has

$$[\lambda^{-k} S_\lambda(yN)]_m = \sum_j [P_j N^j]_m + \sum_{j=0}^{k}\sum_{\ell=1}^{j-1} \binom{j}{\ell+1} (-1)^\ell [P_j N^{j-1-\ell} y^\ell \lambda^{-\ell} N^{(\ell)}]_m + [b]_m. \tag{6.23}$$

By Lemma 6.8 the following two relations hold $\bmod(\epsilon_1, \dots, \epsilon_m)$:

$$[P_j N^j]_m = a_{jj} i \epsilon_1^{j-1} \epsilon_{m+1} = q_{jj} \binom{j}{1} \epsilon_1^{j-1} \epsilon_{m+1}$$

and

$$[P_j N^{j-1-\ell} y^\ell \lambda^{-\ell} N^{(\ell)}]_m = a_{jj} \lambda^{-\ell} m(m-1)\cdots(m-\ell+1)\epsilon_1^{j-1-\ell}\epsilon_{m+1}.$$

Finally, by Lemma 6.9 one has $[b]_m = \lambda^{-1}E_2(\epsilon_1, \lambda^{-1}, \lambda^{-1}m)\epsilon_{m+1}$, where E_2 is a polynomial. The identity

$$\lambda^{-\ell} m(m-1)\cdots(m-\ell+1) = \lambda^{-\ell} m^\ell + \lambda^{-1} F_j(\lambda^{-1}, \lambda^{-1}m),$$

where F_j is a polynomial, implies that mod$(\epsilon_1, \ldots, \epsilon_m)$ one further has

$$[\lambda^{-k} S_\lambda(yN)]_m = \left(\sum_{j=0}^{k} q_{jj} \sum_{\ell=0}^{j-1} \binom{j}{\ell+1} (-1)^\ell \epsilon_1^{j-1-\ell} \lambda^{-\ell} m^\ell + \lambda^{-1} E(\epsilon_1, \lambda^{-1}, \lambda^{-1} m) \right) \epsilon_{m+1}.$$

The sum over j in the latter expression equals $-\lambda m^{-1}((\epsilon_1 - \lambda^{-1}m)^i - \epsilon_1^i)$.

This completes the proof of the formula for Φ_0. The bounds for the degrees of the involved polynomials formulated in the lemma follow directly from the last item of Lemma 6.7 and item (5) of Lemma 6.8. □

Remark 6.5. Once we prove—which we will do in the next section—that there is an open set D containing $z = \infty$ where $\Phi_0(\epsilon_1, \lambda^{-1}, \lambda^{-1} m) \neq 0$ for all m and $\lambda \in \Omega$, then by Lemma 6.11 we immediately get a recursive determination of $\epsilon_{m+1}(\lambda)$. Indeed, the initial data is the choice of $\epsilon_1(\lambda)$. In (6.21) we saw that $\lim_{\lambda\to\infty} \epsilon_1(\lambda)$ is a root of (6.10), so in particular, it is bounded. Since in the right-hand side of (6.23) the only part not being a multiple of λ^{-1} is $\sum_\ell [P_\ell N^\ell]_m$, we get that $yN_\lambda(y) = \sum_{\ell=1}^{\infty} \epsilon_\ell(\lambda) y^\ell$ converges formally to a formal power series $yN(y) = \sum_{\ell=1}^{\infty} \epsilon_\ell y^\ell$, which is a formal solution to the equation $\sum_\ell P_\ell N^\ell = 0$. This shows that $yN(y)$ is a formal solution (at infinity) to the equation of the plane curve associated with the pencil T_λ.

Summarizing we have considered the ansatz

$$L(z) = yN(y) = \sum_{\ell=1}^{\infty} \epsilon_\ell y^\ell$$

for equation (6.9) at $z = \infty$ with respect to the variable $y = 1/z$ (cf. (6.17)). Using Lemma 6.10 we observed that $\epsilon_1 = \epsilon_1(\lambda)$ has to be a solution to equation (6.21) and that under the assumption of Theorem 6.1 it is always possible to find such a solution for large λ. Furthermore, Lemma 6.11 is immediately reformulated as the recurrence relation

$$\epsilon_{m+1} = -\Phi_0(\epsilon_1, \lambda^{-1}, \lambda^{-1} m)^{-1} B(\epsilon_1, \ldots, \epsilon_m, \lambda^{-1}, \lambda^{-1} m), \tag{6.24}$$

which formally solves equation (6.17) under the assumption that

$$\Phi_0(\epsilon_1, \lambda^{-1}, \lambda^{-1} m) \neq 0$$

for all $m = 1, 2, \ldots$. Clearly, the next step is to study this latter condition, which will be done in the next section.

Example 6.1. Consider $T_\lambda = \sum_{j=0}^{2} Q_j(z) \lambda^{2-j} \frac{d^j}{dz^j}$ with $Q_j(z) = \sum_{\ell=0}^{j} q_{j\ell} z^\ell$, $\ell = 0, 1, 2$. Then

$$\Phi_0(\epsilon_1, \lambda^{-1}, \lambda^{-1} m) = q_{22}(2\epsilon_1 - \lambda^{-1} m) + q_{11} - q_{22} \lambda^{-1}.$$

Remark 6.6. The tricky part in proving that there actually *exists* a formal solution to (6.9) is that the polynomial $\Phi_0(\epsilon_1, \lambda^{-1}, \lambda^{-1}m)$ might in fact vanish for some values of m. Moreover, proving the *convergence* of formal solutions is also non-trivial due to the possibility that $\Phi_0(\epsilon_1, \lambda^{-1}, \lambda^{-1}m) \to 0$ when $m \to \infty$. We can see from Lemma 6.11 and from the above example that the terms in $\Phi_0(\epsilon_1, \lambda^{-1}, \lambda^{-1}m)$ have degree in m, which is less than or equal to their total degree in λ^{-1}. This observation will be the crucial point in the proof of the convergence of solutions given by the formal power series.

6.4 Estimating the radius of convergence

We use the classical method of majorants (see, e. g., [83] or [92]) to show that under certain conditions the formal power series solutions of equation (6.17) actually represent analytic functions. The idea is to find an upper bound for the solutions to the studied recurrence relation in terms of a simpler one which we can analyze. In other words, we want to substitute the original recurrence relation by a simpler recurrence ensuring that during this process the absolute value of the solutions will not decrease. For λ fixed, the recurrence relation (6.24) has the form

$$\epsilon_{m+1} = G_m(\epsilon_1, \dots, \epsilon_m), \tag{6.25}$$

where $G_m(\epsilon_1, \dots, \epsilon_m) = \sum_I \delta_I^m \epsilon^I$, $m = 1, 2, \dots$ is a sequence of polynomials.

The following lemma is borrowed from [83].

Lemma 6.12. *Given a recurrence relation (6.25), assume that*

$$H_m(\epsilon_1, \dots, \epsilon_m) = \sum_I^m A_I \epsilon^I, \quad m = 1, 2, \dots,$$

is a sequence of polynomials with positive coefficients satisfying

$$|G_m(\epsilon_1, \dots, \epsilon_m)| \leq H_m(|\epsilon_1|, \dots, |\epsilon_m|) \tag{6.26}$$

for all $\epsilon_i \in \mathbb{C}$. (This is true, for example, if $A_I^m \geq |\delta_I^m|$.) Take $E_1 \geq |\epsilon_1|$ and define inductively $E_{m+1} = H_m(E_1, \dots, E_m)$. Then, for any m, one has that $E_m \geq |\epsilon_m|$, and hence the radius of convergence for the series $\sum_{\ell=1}^{\infty} \epsilon_i t^i$ is greater than or equal to the radius of convergence for the series $\sum_{\ell=1}^{\infty} E_i t^i$.

Lemma 6.12 implies the following statement.

Proposition 6.4. *Consider a set $\Lambda \subset \mathbb{CP}^1$ containing ∞ and such that $\lambda = 0$ is an interior point in $\mathbb{CP}^1 \setminus \Lambda$. For $\lambda \in \Lambda$, let $\epsilon_1 = \epsilon_1(\lambda)$ be a continuous solution to (6.21). If the polynomials $\Phi_0(\epsilon_1(\lambda), \lambda^{-1}, \lambda^{-1}m)$, $m = 1, 2, \dots$, satisfy the condition*

$$\sup\{|\Phi_0(\epsilon_1(\lambda), \lambda^{-1}, \lambda^{-1}m)^{-1}(\lambda^{-1}m)^r| : \lambda \in \Lambda,\ m \geq 0,\ r = 0, 1, \dots, k-1\} < \infty, \tag{6.27}$$

then there exists $D > 0$ such that the series $\sum_{i\geq 1} \epsilon_i y^i$ defined by the recurrence (6.24) converges for any $\lambda \in \Lambda$ in the disk $|y| < D$.

Proof. By Lemma 6.7 we know that

$$\lambda^{-k} S_\lambda(yN) = \sum c_{\beta_1\beta_2 J}(\lambda^{-1})^{\beta_1} y^{\beta_2}(\lambda^{-1}y)^{|J|} N^J, \tag{6.28}$$

where the sum is taken over a finite set of non-negative indices (β_1, β_2) and $J \in FS$ such that $\beta_1 + |J| \leq k - 1$.

The choice of ϵ_1 (depending on λ) determines $N(y) = \epsilon_1 + M(y)$, where $M(y) = \sum_{i=1}^{\infty} \epsilon_{i+1} y^i$ has no constant term. If J starts with precisely l zeros, i. e., $J = 0(l) \cup J_+$, then

$$N^J = (\epsilon_1 + M)^l M^{J_+} = \sum_{s=0}^{l} \binom{l}{s} \epsilon_1^{l-s} M^{0(s)\cup J_+} = \sum_{J_1 \in W} d_{J_1} M^{J_1},$$

where all J_1s that occur in the last sum satisfy the relation $|J_1| = |J|$. Using this and substituting $M(y)$ into (6.28), we get the equation

$$\sum d_{\beta_1\beta_2 I}(\lambda^{-1})^{\beta_1} y^{\beta_2}(\lambda^{-1}y)^{|I|} M^I = 0, \tag{6.29}$$

where the relation $\beta_1 + |I| \leq k - 1$ is still valid. Note that (6.29) contains no non-zero $d_{\beta_1\beta_2 J}$ such that $\beta_2 = 0$ and $|J| = 0$ since this case will contribute to a non-trivial constant term (i. e., a term of degree 0 in y). But the constant term was already eliminated by the explicit choice of ϵ_1. The coefficients in (6.29) are polynomials in $\epsilon_1(\lambda)$. From the study of the asymptotic behavior of equation (6.21) for $\epsilon_1(\lambda)$ we already know that $\lim_{\lambda\to\infty} \epsilon_1(\lambda)$ is finite. Therefore these coefficients are bounded in the domain Λ.

Now we will analyze how the recurrence formula (6.24) is affected by our choice of a branch $\epsilon_1(\lambda)$. We will do this in a way similar to the arguments in Lemma 6.11. Since by Lemma 6.8 items (2)–(4) the constant term of M vanishes, one gets that only the terms $(\lambda^{-1})^{\beta_1} y^{\beta_2}(\lambda^{-1}y)^{|I|} M^I$ for which $I = (l)$ has length 1 and $\beta_2 = 0$ will satisfy the condition that the mth coefficient is non-zero $\mathrm{mod}(\epsilon_1, \ldots, \epsilon_m)$. For these terms one has

$$[(\lambda^{-1})^{\beta_1}(\lambda^{-1}y)^l M^{(l)}]_m = (\lambda^{-1})^{\beta_1} m(m-1)\cdots(m-l+1)\epsilon_{m+1}.$$

Hence we can split the terms in the left-hand side of equation (6.29) into two groups, namely, the sum $\sum_{\beta_1,l} d_{\beta_1 0(l)}(\lambda^{-1})^{\beta_1}(\lambda^{-1}y)^l M^{(l)}$ and the remaining terms $\sum_V d_{\beta_1\beta_2 I}(\lambda^{-1})^{\beta_1} y^{\beta_2}(\lambda^{-1}y)^{|I|} M^I$. These remaining terms correspond to a certain index set V. The first sum has the mth coefficient equal to $\Phi_0(\epsilon_1, \lambda^{-1}, \lambda^{-1}m)\epsilon_{m+1}$, while the mth coefficient in the second sum is a polynomial in the preceding coefficients. By the hypothesis of Proposition 6.4 we have $\Phi_0(\epsilon_1, \lambda^{-1}, \lambda^{-1}m) \neq 0$ if $\lambda \in \Lambda$. Hence the recurrence relation expressing ϵ_{m+1} can be rewritten as

$$\epsilon_{m+1} = -\Phi_0(\epsilon_1, \lambda^{-1}, \lambda^{-1}m)^{-1} \sum_V d_{\beta_1\beta_2 I}[(\lambda^{-1})^{\beta_1} y^{\beta_2}(\lambda^{-1}y)^{|I|} M^I]_m. \tag{6.30}$$

Interpret its right-hand side as a polynomial $G_m(\epsilon_2, \dots, \epsilon_m)$ with coefficients depending on λ. If $K = \max\{|d_{\beta_1\beta_2 I}|, (\beta_1, \beta_2, I) \in V\}$, then clearly

$$|G_m(\epsilon_2, \dots, \epsilon_m)| \leq K \sum_V \left|\Phi_0(\epsilon_1, \lambda^{-1}, \lambda^{-1}m)^{-1}[(\lambda^{-1})^{\beta_1} y^{\beta_2} (\lambda^{-1}y)^{|I|} M^I]_m\right|. \tag{6.31}$$

We can find an even simpler upper bound for these terms. Set $J = (j_1, j_2, \dots, j_s)$ and recall that

$$[y^{|J|}M^J]_m = \sum_{c_1+\dots+c_s=m} \prod_{i=1}^{s} c_i(c_i-1)\cdots(c_i-j_i+1)\epsilon_{c_i}$$

by Lemma 6.8. Clearly, $\prod_{i=1}^{s} c_i(c_i-1)\cdots(c_i-j_i+1) \leq m^{|J|}$, and thus we get

$$|[y^{|J|}M^J]_m| \leq m^{|J|} \sum_{c_1+\dots+c_s=m} \prod_{i=1}^{s} |\epsilon_{c_i}|.$$

Setting $M_{\mathrm{abs}} := \sum_{\ell=2}^{\infty} |\epsilon_\ell| t^i$ and noting that $s = lng(J)$, we obtain

$$\sum_{c_1+\dots+c_s=m} \prod_{i=1}^{s} |\epsilon_{c_i}| = [M_{\mathrm{abs}}^{lng(J)}]_m,$$

and so we have the bound

$$|[y^{|J|}M^J]_m| \leq m^{|J|}[M_{\mathrm{abs}}^{lng(J)}]_m.$$

This implies the inequality

$$|G_m(\epsilon_2, \dots, \epsilon_m)| \leq K \sum_V |\Phi_0(\epsilon_1, \lambda^{-1}, \lambda^{-1}m)^{-1} \lambda^{-\beta_1} (\lambda^{-1}m)^{|I|}| [y^{\beta_2} M_{\mathrm{abs}}^{lng(I)}]_m. \tag{6.32}$$

As it was mentioned in the beginning of the proof, for each term in this sum one has $\beta_1 + |I| \leq k-1$. Hence both inequalities $\beta_1 \leq k-1$ and $|I| \leq k-1$ hold. The assumption that 0 is an interior point of the complement of Λ implies that there is a positive real number D such that $|\lambda^{-i}| \leq D$ for $\lambda \in \Lambda$ and $i = 1, \dots, k-1$. By the second assumption of the proposition, there is a constant C such that

$$|\Phi_0(\epsilon_1, \lambda^{-1}, \lambda^{-1}m)^{-1}(\lambda^{-1}m)^{|I|}| \leq C$$

for $\lambda \in \Lambda$ and any I occurring in (6.32). So

$$|G_m(\epsilon_2, \dots, \epsilon_m)| \leq KCD \sum_V [y^{\beta_2} M_{\mathrm{abs}}^{lng(I)}]_m \leq L \sum_{i,j} [y^i M_{\mathrm{abs}}^j]_m =: H_m(|\epsilon_2|, \dots, |\epsilon_m|)$$

for some positive real number $L := KCDr$, where r is the maximal number of indices in V (the index set introduced above) that correspond to some $i = \beta_2$ and $j = lng(I)$. Note that L is independent of m since the index set V just depends on k. The summation in H_m is taken over all pairs of non-negative integers (i,j) except $(0,0)$ and $(0,1)$. (The fact that $(0,0)$ is not needed follows from the observation that there are no terms in (6.29) with $\beta_2 = 0$ and $lng(I) = 0$; the fact that $(0,1)$ is not needed is implied by the observation that the terms with $\beta_2 = 0$ and $lng(I) = 1$ were excluded from V by the splitting of terms in (6.29) for the construction of V.) This is an infinite sum, but we still get a well-defined polynomial since M contains no constant term.

The recurrence formulae $\epsilon_{m+1} = H_m(\epsilon_2, \dots, \epsilon_m)$, $m \geq 1$, correspond to equating the mth coefficients in both sides of the relation

$$M = L\sum_{\ell,j} y^\ell M^j = L\left(\frac{1}{(1-y)(1-M)} - 1 - M\right)$$

defining the power series $M(y) = \sum_{\ell=1}^{\infty} \epsilon_{\ell+1} y^\ell$. (Note that for $m = 1$ we simply get $\epsilon_2 = H_1 \in \mathbb{R}$.) The above equation is equivalent to the quadratic equation

$$(1+L)M^2 - M + Ly/(1-y) = AM^2 - M + C = 0$$

and its solutions are $\frac{1}{2A} \pm \frac{\sqrt{1-4CA}}{2A}$. These solutions are univalent analytic functions in case $|4CA| < 1$. The latter condition is easily seen to be satisfied for $|y| < \frac{1}{(1+2L)^2}$. These solutions are developable in power series in this disk and their power series will satisfy our recurrence formulae. This accomplishes the proof of Proposition 6.4. Indeed, we have found a sequence of polynomials that satisfy the assumptions of Lemma 6.12 and proved that the formal power series that they define is analytically convergent in some disk around $y = 0$. (This argument mimics the elegant proof that an algebraic function has a convergent power series expansion given in [83, pp. 34–35].) □

Remark 6.7. It is worth mentioning that if we know the supremum in Proposition 6.4 we can easily get the estimate $\frac{1}{(1+2L)^2}$ for the radius of convergence of the solutions to equation (6.19) at least for small k.

6.4.1 Convergence of polynomial solutions

We will now show that the condition on the supremum in Proposition 6.4 is generically satisfied for the sequences of polynomial eigenfunctions used in Propositions 6.1 and 6.3. This means that we assume that the corresponding pencil is of general type. If $p_n(z)$ is any monic degree n polynomial, then its normalized logarithmic derivative $L_n(z) = \frac{p_n'(z)}{\lambda_n p_n(z)}$ expanded at infinity using the variable $y = 1/z$ satisfies the relation

$$L_n(z) = yN(y) = \frac{n}{\lambda_n} y + \cdots = \sum_{\ell=1}^{\infty} \epsilon_{\ell,n} y^\ell.$$

In particular, $\epsilon_{1,n} = \frac{n}{\lambda_n}$. (Note that so far $\lambda_n \neq 0$ is any constant.)

Now let $\{\lambda_n\}$ be one of the sequences of eigenvalues in Propositions 6.1. In a certain neighborhood Λ of ∞ and for n so large that $\lambda_n \in \Lambda$, there exists a unique eigenpolynomial $p_n(z)$ of degree n satisfying the equation $T_{\lambda_n} p_n(z) = 0$.

By Proposition 6.1 for large n, one has $\lambda_n \sim an$, where a is one of the k distinct roots of (6.3). This gives

$$\lambda_n^{-1} n = \epsilon_{1,n} \to 1/a$$

as $n \to \infty$. Note that $1/a$ is a root of the reciprocal characteristic equation (6.10)

$$q_{k,k}\xi^k + q_{k-1,k-1}\xi^{k-1} + \cdots + q_{0,0} = 0. \tag{6.33}$$

Next we show that the assumption in Proposition 6.4 ensuring the convergence of our formal power series solutions is satisfied for a large class of cases.

Proposition 6.5. *Assume that*
(i) *The polynomial $\sum_{j=0}^{k} q_{jj} t^j$ has degree k (i. e., $q_{kk} \neq 0$).*
(ii) *The root a^{-1} of $\sum_{j=0}^{k} q_{jj} t^j = 0$ is simple.*
(iii) *For any $r > 0$, the number $(1 - r)a^{-1}$ is never a root of this same polynomial (in particular, taking $r = 1$ implies that 0 is not a root of the above polynomial, which gives $q_{00} \neq 0$).*

Then there exists a neighborhood $\tilde{\Lambda} \in \mathbb{C}P^1$ of ∞ with the following property. If $n \in \mathbb{Z}_{\geq 0}$ is such that $\lambda_n \in \tilde{\Lambda}$ and $\epsilon_{1,n} = \frac{n}{\lambda_n}$, then

$$\sup\{|\Phi_0(\epsilon_{1,n}, \lambda_n^{-1}, \lambda_n^{-1} m)^{-1} (\lambda_n^{-1} m)^r| : m \geq 0,\ r = 0, 1, \ldots, k-1\} < \infty.$$

In this case the condition in Proposition 6.4 is valid.

Proof. Recall first that by Lemma 6.11 we have

$$\Phi_0(\epsilon_1, \lambda^{-1}, \lambda^{-1} m) = -\sum_{j=0}^{k} q_{jj} \frac{(\epsilon_1 - \lambda^{-1} m)^j - \epsilon_1^j}{\lambda^{-1} m} + \lambda^{-1} E(\epsilon_1, \lambda^{-1}, \lambda^{-1} m). \tag{6.34}$$

We need to establish the existence of a finite supremum in the previous proposition. Let us argue by contradiction. If the proposition is not true then there exist sequences of values m_ℓ and n_ℓ, $\ell = 1, 2, \ldots$, such that
- $\lim_{\ell \to \infty} n_\ell = \infty$.
- For some $j \leq k - 1$ one has $(\lambda_{n_\ell}^{-1} m_\ell)^{-i} \Phi_0(\epsilon_{1,n_\ell}, \lambda_{n_\ell}^{-1}, \lambda_{n_\ell}^{-1} m_\ell) \to 0$ as $\ell \to \infty$.

In what follows we use the notation $n = n_\ell$, $m = m_\ell$, $\epsilon_1 = \epsilon_{1,n} = \frac{n}{\lambda_n}$, and $\lambda = \lambda_n \in \Lambda$, where $n \in \mathbb{Z}_{\geq 0}$, and thus suppress indices in the sequence. All limits will be taken as $\ell \to \infty$.

By taking a subsequence we see that there exist two possibilities: either there is a sequence as above for which m/n converges to a finite limit $r \in \mathbb{R}$, or there is such a sequence for which $m/n \to \infty$. In the second case we first note that also $\lambda^{-1}m \to \infty$ (since $\lim \lambda^{-1}m = \lim \lambda^{-1}n \cdot \lim(m/n) = a^{-1} \lim m/n$). Hence by our assumption also $(\lambda^{-1}m)^{-(k-1)}\Phi_0(\epsilon_1, \lambda^{-1}, \lambda^{-1}m) \to 0$. To get a contradiction we use the fact that the degree in $\lambda^{-1}m$ of E is less than $k-2$ by Lemma 6.11, and that $\lambda^{-1} \to 0$ for the chosen sequence (recall that λ depends on n_j and $n_j \to \infty$). It follows that $\lim(\lambda^{-1}m)^{-(k-1)}E(\epsilon_1, \lambda^{-1}, \lambda^{-1}m) = 0$. Then the hypothesis that $q_{kk} \neq 0$ and the explicit description in (6.34) imply that $\lim(\lambda^{-1}m)^{-(k-1)}\Phi_0(\epsilon_1, \lambda^{-1}, \lambda^{-1}m) = (-1)^{k-1}q_{kk} \neq 0$, which is the desired contradiction.

In the first case, the hypothesis implies that

$$\lambda^{-1}m = (\lambda^{-1}n)(m/n) \to r/a$$

for our sequence of m and n. Furthermore

$$\lim \Phi_0(\epsilon_1, \lambda^{-1}, \lambda^{-1}m)(\lambda^{-1}m)^{-i} = (r/a)^i \lim((\lambda^{-1}m)^{-i}\Phi_0(\epsilon_1, \lambda^{-1}, \lambda^{-1}m)) = 0. \tag{6.35}$$

Now we also have $\lambda \to \infty$, which implies that $\lim \lambda^{-1}E(\epsilon_1, \lambda^{-1}, \lambda^{-1}m) = 0$, and thus (6.35) yields

$$\lim \sum_{j=0}^{k} q_{jj} \frac{(\epsilon_1 - \lambda^{-1}m)^j - \epsilon_1^j}{\lambda^{-1}m} = 0. \tag{6.36}$$

If $r \neq 0$, this immediately implies that both $a^{-1} - ra^{-1}$ and $a^{-1} = \lim \epsilon_1$ solve the same equation $\sum_{j=0}^{k} q_{jj}t^j = 0$. This contradicts the assumptions put on this equation. On the other hand, if $r = 0$ then a calculation shows that the left-hand side of (6.36) is $\frac{d}{dt}(\sum_{j=0}^{k} q_{jj}t^j)|_{t=a^{-1}}$ which, in its turn, contradicts the fact that a^{-1} is a simple root and finishes the proof. □

6.5 Proving Theorems 6.1 and 6.2

We start with Theorem 6.2. Consider a pencil T_λ of general type as defined in § 6.1. By Proposition 6.1 its polynomial eigenfunctions split into k distinct families $p_{n,i}$, $i = 1, \ldots, k$. Fix one of these families $p_n := p_{n,i}$ with the corresponding eigenvalues $\lambda_n \sim \alpha_i n$ as $n \to \infty$.

Equation (6.9) holds for each λ. Among its solutions in an open set Ω are all normalized logarithmic derivatives $\frac{p'}{\lambda p}$ of eigenfunctions p (i. e., $T_\lambda * p = 0$) that are defined and non-vanishing in Ω. The recurrence relation (6.24) studied in the last two sections solves equation (6.9) by means of the power series expansion at $y^{-1} = z = \infty$ of a special solution of $yN(y) = \sum_{\ell=0}^{\infty} \epsilon_{\ell+1} y^{\ell+1}$ (see Lemma 6.11 in § 6.3). In Lemma 6.10 it was shown that for each λ large enough, there are k possible initial values of $\epsilon_1 = \epsilon_1(\lambda)$. Let $\lambda = \lambda_n$

be large enough for this to be true. Since T_λ is of general type the condition in Proposition 6.5 is satisfied. Hence by Proposition 6.4 there exists a non-trivial disk D around ∞ where for any n, all k formal solutions to the recurrence relation (6.24) converge to analytic functions.

Consider now $L_n(z) = \frac{p_n'(z)}{\lambda_n p_n(z)}$ for an arbitrarily fixed n. This is a solution to (6.9) and has the required zero at infinity, so there is a neighborhood of ∞ where $L_n(z)$ can be expanded as a convergent power series $yN_n(y)$. In particular, this power series will be one of the k solutions of the recurrence relation, and thus it will actually converge in D. It follows that the zeros of the polynomials p_n are contained in the compact region of the complex plane, i. e., in the complement of D. This fact immediately implies Theorem 6.2.

Let us now concentrate on Theorem 6.1 and start by proving the last part of this theorem. Recall that by Remark 6.5 the recurrence formulas (6.24) converge when $\lambda \to \infty$ to recursive formulas determining formal solutions to the algebraic equation (6.2). This implies that the formal power series at ∞ for $L_n(z)$ converges formally (i. e., coefficientwise) to the formal power series at ∞ of the algebraic function $y(z)$ for which $\lim_{z\to\infty} zy(z) = \alpha^{-1}$ (see Lemma 6.2 in § 6.2). Since the convergence radius has a uniform non-zero lower bound, this gives that the convergence is uniform in a neighborhood of ∞. Distinct branches of the algebraic function $y(z)$ at ∞ correspond to distinct values of α, and this makes it clear that they all occur (cf. Lemma 6.2). This proves the last part of Theorem 6.1.

To settle the first part of Theorem 6.1, we need some basic properties of probability measures. If K is a compact set in $\mathbb{C}$, denote by $M(K)$ the space of all probability measures supported in K equipped with the weak topology. It is known that $M(K)$ is a sequentially compact Hausdorff space, which allows one to choose a convergent subsequence from any sequence of measures belonging to $M(K)$ (which is the statement of Helly's theorem).

Lemma 6.13 (cf. Lemma 8 of [27]). *Let $\{q_m(z)\}$ be a sequence of polynomials with $\deg q_m(z) \to \infty$ as $m \to \infty$. Denote by μ_m and μ_m' the root-counting measures of $q_m(z)$ and $q_m'(z)$, respectively, and assume that there exists a compact set K containing the supports of all measures μ_m and therefore also the supports of all measures μ_m'. If $\mu_m \to \mu$ and $\mu_m' \to \mu'$ as $m \to \infty$ and u and u' are the logarithmic potentials of μ and μ', respectively, then $u' \le u$ in $\mathbb{C}$ with equality in the unbounded component of $\mathbb{C} \setminus supp(\mu)$.*

Example 6.2. Consider the polynomial sequence $\{z^m - 1\}$. The measure μ is then the uniform distribution on the unit circle of total mass 1. Its logarithmic potential $u(z)$ equals $\log|z|$ if $|z| \ge 1$ and 0 in the disk $|z| \le 1$. On the other hand, the sequence of derivatives is given by $\{mz^{m-1}\}$, and the corresponding (limiting) logarithmic potential $u'(z)$ equals $\log|z|$ in $\mathbb{C} \setminus \{0\}$. Obviously, $u(z) = u'(z)$ in $|z| \ge 1$ and $u'(z) < u(z)$ in $|z| < 1$.

In the notation of Theorem 6.1, consider the sequence of eigenpolynomials $\{p_{n,i}(z)\}$ for some arbitrarily fixed value of the index $i = 1, \dots, k$. Assume that N_i is a sequence of the natural numbers such that for $\ell = 0, \dots, k$, there exists a limit

$$\mu_i^{(\ell)} := \lim_{n\to\infty,\, n\in N_i} \mu_{n,i}^{(\ell)} \tag{6.37}$$

where $\mu_{n,i}^{(\ell)}$ denotes the root-counting measure of $p_{n,i}^{(\ell)}(z)$. The existence of such N_i follows from the above remark on $M(K)$ (Helly's theorem). Notice that for each ℓ, the logarithmic potential $u_i^{(\ell)}$ of $\mu_i^{(\ell)}$ satisfies a. e. the identity

$$u_i^{(\ell)}(z) - u_i^{(0)}(z) = \lim_{n\to\infty,\, n\in N_i} \frac{1}{n} \log\left|\frac{p_{n,i}^{(\ell)}(z)}{(n)_i p_{n,i}(z)}\right|.$$

The next proposition completes the proof of Theorem 6.1 and also shows the remarkable property that if one considers a sequence of eigenpolynomials for some spectral pencil, then the situation $u'(z) < u(z)$ occurring in Example 6.2 can never happen.

Proposition 6.6. *The measures $\mu_i^{(\ell)}$, $\ell = 0,\dots,k$, are all equal and the scalar multiple $\widetilde{\Psi}_i = C_\mu/a_i$ of the Cauchy transform of this common measure μ_i satisfies equation* (6.2) *almost everywhere.*

Proof. For $n \in N_i$, one has

$$\frac{p_{n,i}^{(\ell+1)}(z)}{(n-\ell)p_{n,i}^{(\ell)}(z)} \to C^{(\ell+1)}(z) := \int_{\mathbb{C}} \frac{d\mu_i^{(\ell)}(\zeta)}{z-\zeta} \quad \text{as } n \to \infty$$

with convergence in L^1_{loc}. The well-known property of convergence in L^1_{loc} implies that passing to a subsequence one can assume that the above convergence is actually the pointwise convergence almost everywhere in $\mathbb{C}$. It follows that

$$\frac{p_{n,i}^{(\ell)}(z)}{n^\ell p_{n,i}(z)} \to C^{(1)}(z)\cdots C^{(\ell)}(z) \tag{6.38}$$

pointwise almost everywhere in $\mathbb{C}$. We claim that this limit is non-zero a. e. Given this, notice that

$$u_i^{(k)}(z) - u_i^{(0)}(z) = \lim_{n\to\infty,\, n\in N_i} \frac{1}{n} \log\left|\frac{p_{n,i}^{(k)}(z)}{(n)_k p_{n,j}(z)}\right| = 0$$

almost everywhere in $\mathbb{C}$. On the other hand, $u_i^0 \geq u_i^{(1)} \geq \cdots \geq u_i^{(k)}$ by Lemma 6.13. Hence the potentials $u_i^{(\ell)}$ are all equal and the corresponding measures $\mu_i^{(\ell)} = \Delta u_i^{(\ell)}/2\pi$ are equal as well.

It remains to settle the claim. Recall that $p_{n,i}(z)$ satisfies the differential equation $T_{\lambda_{n,i}} p_{n,i}(z) = 0$, i. e.,

$$Q_k(z)p_{n,i}^{(k)}(z) + \lambda_{n,j} Q_{k-1}(z)p_{n,i}^{(k-1)}(z) + \cdots + \lambda_{n,i}^k Q_0(z)p_{n,i}(z) = 0. \tag{6.39}$$

Therefore,

$$Q_k(z)\frac{p_{n,i}^{(k)}(z)}{n^k p_{n,i}(z)} + \frac{\lambda_{n,i}}{n} Q_{k-1}(z)\frac{p_{n,i}^{(k-1)}(z)}{n^{k-1} p_{n,i}(z)} + \cdots + \frac{\lambda_{n,i}^k}{n^k} Q_0(z) = 0.$$

Using the asymptotics $\lambda_{n,i} \sim a_i n$ and the pointwise convergence (6.38) a. e., we get

$$Q_k(z)\mathcal{C}^{(1)}(z)\cdots\mathcal{C}^{(k)}(z) + a_i Q_{k-1}(z)\mathcal{C}^{(1)}(z)\cdots\mathcal{C}^{(k-1)}(z) + \cdots + a_i^k Q_0(z) = 0. \tag{6.40}$$

Using the hypothesis that $Q_0 \neq 0 \neq a_i$, we conclude that $\mathcal{C}^{(1)}(z) \neq 0$ a. e.

To prove that $\mathcal{C}^{(2)}(z)$ also is non-zero a. e., we consider the differential equation satisfied by $p_{n,i}^{(1)}(z)$, obtained by differentiating (6.39).

$$Q_k(z)p_{n,i}^{(k+1)}(z) + (Q_k^{(1)} + \lambda_{n,i}Q_{k-1}(z))p_{n,i}^{(k)}(z) + \cdots + (\lambda_{n,i}^{k-1}Q_1^{(1)}(z) + \lambda_{n,i}^k Q_0(z))p_{n,i}^{(1)}(z) = 0.$$

Repeating the previous analysis, we get

$$Q_k(z)\frac{p_{n,i}^{(k+1)}(z)}{n^k p_{n,i}^{(1)}(z)} + \frac{(Q_k^{(1)} + \lambda_{n,i}Q_{k-1}(z))}{n}\frac{p_{n,i}^{(k)}(z)}{n^{k-1}p_{n,i}^{(1)}(z)} + \cdots + \frac{(\lambda_{n,i}^{k-1}Q_1^{(1)}(z) + \lambda_{n,i}^k Q_0(z))}{n^k} = 0.$$

Hence taking the limit we obtain

$$Q_k(z)\mathcal{C}^{(2)}(z)\cdots\mathcal{C}^{(k+1)}(z) + a_i Q_{k-1}(z)\mathcal{C}^{(2)}(z)\cdots\mathcal{C}^{(k)}(z) + \cdots + a_i^k Q_0(z) = 0.$$

From this we see that $\mathcal{C}^{(2)}(z)$ is again non-zero a. e. Similarly, for $\ell \geq 3$, $\mathcal{C}^{(\ell)}(z)$ is non-zero a. e., which settles the claim. Note that the polynomials $p_n(z) := z^n - 1$ in Example 6.2 satisfy the differential equation $zp_n^{(2)}(z) - (n-1)p_n^{(1)} = 0$, which has $Q_0 = 0$.

The fact that the multiple $\widetilde{\Psi}_i = \mathcal{C}_\mu / a_i$ of the Cauchy transform of this common measure μ_j satisfies equation (6.2) almost everywhere follows by (6.40), since the equality of the measures implies that $\mathcal{C}_\mu = \mathcal{C}^{(1)} = \mathcal{C}^{(2)} = \cdots$. Thus

$$Q_k(z)(\mathcal{C}_\mu(z))^k + a_i Q_{k-1}(z)(\mathcal{C}_\mu(z))^{k-1} + \cdots + a_i^k Q_0(z) = 0,$$

which is equivalent to the desired equation. □

The proof of Theorem 6.1 is now finished. For the sake of completeness, let us prove the result on L^1_{loc}-convergence used in the preceding proof.

Lemma 6.14. *Suppose that μ_n, $n = 1, 2, \ldots$, are probability measures that converge as distributions to the measure μ. Then $z^{-1} * \mu_n$ converges to $z^{-1} * \mu$ in L^1_{loc}.*

Proof. That $z^{-1} * \mu_n \in L^1_{\text{loc}}$ is a consequence of the fact that $z^{-1} \in L^1_{\text{loc}}$ combined with Fubini's theorem (see [69]). Write $z^{-1} = f_\epsilon + g_\epsilon$, where g_ϵ is a test function with compact support and $\|f_\epsilon\|_1 \leq \epsilon$. Then by Fubini's theorem one has $\|f_\epsilon * \mu_n\|_1 \leq \epsilon$. Since the functions $g_\epsilon * \mu_i$ converge to $g_\epsilon * \mu$ uniformly, they also converge in L^1_{loc} the lemma follows. □

6.6 Proving Theorems 6.3 and 6.4

Theorems 6.3 and 6.4 require a rather elaborate mixture of ideas from algebraic geometry and complex analysis. The latter are based on the main results of [36].

Definition 6.1. A function Φ defined in a domain $U \subset \mathbb{C}$ is called *piecewise analytic*, or a *PA-function* for short, if there exists a finite family of pairwise distinct analytic functions $\{A_m(z)\}_{m=1}^r$ in U and an associated family of disjoint open sets $\{M_m\}_{m=1}^r$ with $M_m \subset \Omega := \{z \in U \mid A_s(z) \neq A_t(z), 1 \le s \neq t \le r\}$, $1 \le m \le r$, such that $\{M_m\}_{m=1}^r$ forms a covering of U up to a set of Lebesgue measure 0 and

$$\Phi(z) = \sum_{m=1}^{r} A_m(z)\chi_m(z) \quad \text{in } U,$$

where χ_m is the characteristic function of the set M_m, $1 \le m \le r$.

Note that *PA*-functions need not be continuous—this will certainly not be the case in our situation. The following result singles out the properties of the limit $\Phi(z) := \mathcal{C}_{\mu_i}(z)$ that we will need, see Theorem 6.1.

Proposition 6.7. *$\Phi(z)$ is a PA-function in any simply-connected open set U. More precisely, we have that $\Phi(z) = \sum_{m=1}^{r} A_m(z)\chi_m(z)$, where χ_m are the characteristic functions of certain subsets of U (defined in the proof below) and $A_1, \dots, A_r$ are distinct (analytic) branches of the equation*

$$\sum_{j=0}^{k} Q_j(z)\alpha_i^{-j} w^j = 0$$

in U. Moreover, $\mu_i = \partial\Phi(z)/\partial\bar{z} \ge 0$ is a distribution supported in U.

Proof. This is an immediate corollary of Theorem 6.1, which implies that $\alpha_i^{-1}\mathcal{C}_{\mu_i}(z)$ satisfies an algebraic equation (6.2) almost everywhere in $\mathbb{C}$. Therefore in any simply-connected U one has that $\prod_{s=1}^{k}(\mathcal{C}_{\mu_i}(z) - A_s(z)) = 0$ almost everywhere. Let M_s (with characteristic function χ_s) be the subset of U where $\mathcal{C}_{\mu_i}(z) = A_s(z)$. Then we immediately get that $\Phi(z) := \mathcal{C}_{\mu_i}(z) = \sum_{s=1}^{r} A_s(z)\chi_s(z)$ and the subsets M_s cover U up to a set of Lebesgue measure zero. □

Using some of the main results of [36], we can derive several consequences for the support of the measure μ_i and prove Theorem 6.3. If $\Phi(z)$ is any PA-function with $r \ge 3$, we associate to each triple (s, t, w) of distinct indices in $\{1, \dots, r\}$ the following set:

$$\Gamma_{s,t,q} = \{z \in U \mid \exists c \in \mathbb{R} \text{ such that } A_s(z) - A_q(z) = c(A_t(z) - A_q(z))\}.$$

It is easy to see that this definition only depends on the set $\{s, t, q\}$. Clearly, if $r \le 2$, these sets are empty.

Theorem 6.5 (See Theorem 1 and Corollary 2 of [36]). *In the above notation, assume that $\partial\Phi/\partial\bar{z} \geq 0$ as a distribution supported in U and let $p \in U$ be such that any sufficiently small neighborhood N of p intersects each M_s, $1 \leq s \leq r$, in a set of a positive Lebesgue measure. Define $H_s(z) = \Re[\int_p^z A_s(w)dw]$ and suppose that*
(i) *$A_s(p) \neq A_t(p)$ for $1 \leq s \neq t \leq r$, i. e., p is a non-singular point of $H_s - H_t$.*
(ii) *There is at most one $\Gamma_{s,t,q}$ that contains p.*

Then there exists a neighborhood of p where $\partial\Phi/\partial\bar{z}$ is supported on a union of segments of the zero loci of the functions $H_s - H_t$, $1 \leq s \neq t \leq r$.

By Proposition 6.7 one can apply Theorem 6.5 to $\mathcal{C}_{\mu_i}(z)$. Let Γ_2 be the set of points where conditions (i) and (ii) of the above theorem fail to hold. In other words, Γ_2 is the set of points p, which are either one of the finite number of branch points where $A_s(p) = A_q(p)$ for a pair (s,q) of distinct indices, or satisfy $A_s(p) - A_q(p) \in \mathbb{R}(A_t(p) - A_q(p))$ for at least two distinct triples $\{s,t,q\}$ in $\{1,\ldots,r\}$. Notice that the original measure μ_i can be restored from its Cauchy transform by

$$\mu_i = \frac{1}{\pi}\frac{\partial\mathcal{C}_{\mu_i}(z)}{\partial\bar{z}},$$

where $\frac{1}{\pi}\frac{\partial\mathcal{C}_{\mu_i}(z)}{\partial\bar{z}}$ is considered as a distribution, see, e. g., [69, Chapter 2]. Then Theorem 6.5 implies that in a neighborhood of any point $p \notin \Gamma_2$ the support of μ_i coincides with a smooth zero locus of some function $\Re[\int_p^z A_s(w)dw - \int_p^z A_t(w)dw]$, $s \neq t$, which proves items A and B of Theorem 6.3. Item C follows easily from [36, Theorem 2].

We turn to proving Theorem 6.4, i. e., to showing that generically, the set Γ_2 is finite. The plan is to describe the locus L of curves such that Γ_2 is finite. Now L is a semi-algebraic set, that is, L is given by real polynomial equalities and inequalities. We cannot describe these explicitly since even simple cases become quite involved, as shown by Example 6.3 below. Instead we will prove that L contains an open subset (in the Zariski topology and hence also in the Euclidean topology) essentially by analyzing the construction of $\Gamma_{s,t,q}$ in real-algebraic geometry terms, as opposed to semi-algebraic. It will then suffice to find just one example of a curve in L in order to show that L is non-empty and therefore also dense, which we do in Example 6.3.

Clearly, $\Gamma_{s,t,q}$ is either a real analytic curve or else there exists $c \in \mathbb{R}$ such that $A_s(z) - A_q(z) = c(A_t(z) - A_q(z))$ for all $z \in U$. This latter case is certainly non-generic, and we assume that all the $\Gamma_{s,t,q}$ are real analytic curves.

First define the variety Y, which parametrizes roots of equations of degree n. Namely, set $P = \sum_{j=0}^k Q_j y^j \in \mathbb{C}[Q_0,Q_1,\ldots,Q_k,y]$ and let $Y \subset \mathbb{C}^{k+1}$ be defined by $P = 0$ and $Q_k \neq 0$. Denote by $p_1 : \mathbb{C}^{k+2} \to \mathbb{C}^{k+1}$ the projection on the $k+1$ first coordinates $p_1 : (Q_0,Q_1,Q_2,\ldots,Q_k,y) \mapsto (Q_0,Q_1,Q_2,\ldots,Q_k)$ and by p_2 the projection on the last coordinate. The image of Y under p_1 is the open set $X \subset \mathbb{C}^{k+1}$ where $Q_k \neq 0$. The reason for imposing this last condition is to guarantee that $p_1 : Y \to X$ is a finite map of affine varieties.

We can also parametrize triples of roots (so we tacitly assume that $n \geq 3$) by means of the variety Y. Indeed, use p_1 to form the fiber product $Y \times_X Y \times_X Y \subset (\mathbb{C}^{k+1} \setminus V(Q_k)) \times \mathbb{C}^3$ and let

$$V := \{(y_1, y_2, y_3) \in Y \times_X Y \times_X Y \mid y_i \neq y_j \text{ if } i \neq j\}.$$

The Zariski closure $\bar{V}$ of V is a closed subset of $Y \times_X Y \times_X Y$. The permutation group S_3 acts on the last three coordinates of $\bar{V} \subset (\mathbb{C}^{k+1} \setminus V(Q_k)) \times \mathbb{C}^3$. The orbit space $Z := \bar{V}//S_3$ can be embedded as a closed subset of $(\mathbb{C}^{k+1} \setminus V(Q_k)) \times \mathbb{C}^3$ since $\mathbb{C}^3//S_3 \cong \mathbb{C}^3$. Hence Z is a quasi-affine variety. There is a natural projection from Z to X induced by $p_1 : Y \to X$, which we denote by

$$p : Z \subset (Y \times_X Y \times_X Y)//S_3 \to X.$$

This map is finite. The fiber of p over a point $(Q_0, Q_1, \ldots, Q_k) \in X$ consists of the orbits of the finite set $\{A_{s,a} \times A_{t,a} \times A_{q,a} \mid 1 \leq s, t, q \leq k\}$, where $A_{s,a}$, $1 \leq i \leq k$, are the distinct solutions to the equation $P(Q, y) = 0$ under permutation of coordinates.

Any affine subvariety A of $\mathbb{C}^n = \mathbb{R}^{2n}$ can be considered as the set of $\mathbb{R}$-rational points of a real algebraic variety (cf. [33, Chapter 3.1]) by taking the real and imaginary parts of the complex coordinates as new coordinates. For example, $\mathbb{C}$ itself with the ring of functions $\mathbb{C}[z]$ is the set of $\mathbb{R}$-rational points of the real-algebraic variety $\mathbb{R}^2$ with ring of functions $\mathbb{R}[x, y]$ and the correspondence is given by $z = x + iy$. For a closed subvariety A of $\mathbb{C}^n$ the corresponding ideal of $\mathbb{R}[x_1, \ldots, x_n, y_1, \ldots, y_n]$ is

$$I_R(A) := \langle \Re P(x + iy), \Im P(x + iy) \mid P \in \mathbb{C}[z],\ P(z) = 0 \text{ if } z \in A \rangle.$$

Denote by $O_R(A) = \mathbb{R}[x_1, \ldots, x_n, y_1, \ldots, y_n]/I_R(A)$ the ring of real-algebraic functions on A. An open subset of $\mathbb{C}^n$ defined by the non-vanishing of a polynomial f may be considered as the set of $\mathbb{R}$-rational points of the real-algebraic variety defined by $\mathbb{R}[x_1, \ldots, x_n, y_1, \ldots, y_n]_{\tilde{f}}$, where $\tilde{f} = (\Re f)^2 + (\Im f)^2$. More generally, any quasi-affine subvariety in $\mathbb{C}^n$ can be thought of as the set of $\mathbb{R}$-rational points of a quasi-affine real-algebraic variety. We stress this since we will need the fact that an algebraic map of complex quasi-affine varieties $A \to B$ is finite is preserved when turning A and B into real algebraic varieties (in the sense of algebraic geometry, meaning that $O_R(A)$ is a finite module over $O_R(B)$).

We now return to the global description of the curves $\Gamma_{s,t,q}$. By the above we can consider $\bar{V}$ and its orbit space Z as real algebraic varieties, and as such Z contains the real algebraic subset $\tilde{\Gamma}$ defined by

$$\tilde{\Gamma} = \{(y_1, y_2, y_3) \in Y \times_X Y \times_X Y \mid y_\ell = (a, z, y_\ell) \text{ and } \Im(p_2(y_1) - p_2(y_2))\overline{(p_2(y_3) - p_2(y_2))} = 0\}.$$

Notation 6.4. Consider the algebraic curve in $\mathbb{C}^2$ given by

$$P(z,y) = \sum_{i=0}^{k} Q_j(z)y^j = 0, \tag{6.41}$$

where $\deg_z Q_j \le j$ for $j < k$ and $\deg_z Q_k = k$. For $Q_j(z) = \sum_{0\le\ell\le j} q_{j\ell} z^\ell$, $0 \le j \le k$, set $\alpha = \{(q_{j\ell})\}$ and denote by L_α the curve (6.41) corresponding to $\alpha = (q_{j\ell})$.

Without loss of generality we can assume that $q_{k,k} = 1$. Then the polynomial $P(z,y)$ is parametrized by the r-tuple

$$\alpha = \{(q_{j,\ell})\} \in \mathbb{C}^r := \mathbb{C}^k \times \mathbb{C}^k \times \mathbb{C}^{k-1} \times \cdots \times \mathbb{C}.$$

In addition we assume that the constant polynomial $Q_0 = q_{0,0} \ne 0$. In particular, the set $V(Q_k, Q_{k-1}, \ldots, Q_0) \subset \mathbb{C}^r$ where all the Q_is vanish is empty. Let $c_\alpha : \mathbb{C} \to \mathbb{C}^k$ be the map $c_\alpha(z) = (Q_1(z), \ldots, Q_k(z))$. Then the pullback under c_α of $p_1 : \mathbb{C}^{k+1} \to \mathbb{C}^k$ is the curve $L_\alpha \subset \mathbb{C}^2 \mapsto \mathbb{C} \setminus V(Q_k)$, where $V(Q_k)$ is the zero locus ("support") of Q_k. Hence the pullback of $Y \times_X Y \times_X Y \to X$ over a point $z \in \mathbb{C}$ will locally have the finite fiber

$$\{(z, A_s(z) \times A_t(z) \times A_q(z)) \mid 1 \le s, t, q \le k\},$$

where A_s, $1 \le s \le k$, are the distinct branches of the algebraic curve L_α.

Define the real-algebraic variety Γ as the image of $\tilde{\Gamma}$ under the finite map of real-algebraic varieties $\tilde{V} \to Z$.

Lemma 6.15. *In the above notation, one has:*
(i) *$\tilde{\Gamma}$ is preserved by the action of S_3.*
(ii) *Let $\alpha = (a_{s,t})$ and*

$$c_\alpha : \mathbb{C} \to \mathbb{C}^{r+1}$$

be the map such that $c_\alpha(z) = (Q_1(z), \ldots, Q_k(z), z)$. Then the pullback $c_\alpha^{-1}(p(\Gamma)) \subset \mathbb{C}$ is locally the same as $\bigcup \Gamma_{s,t,q}$, where the union is taken over all indices s, t, q for the curve L_α.

Proof. The identity $\mathfrak{I}(u-w)\overline{(v-w)} = \mathfrak{I}(u\bar{v} - u\bar{w} - w\bar{v})$ gives the first part. Now $\mathfrak{I}(u - w)\overline{(v-w)} = 0$ if and only if $\frac{u-w}{v-w} \in \mathbb{R}$, which together with the definitions yields statement (ii). □

Let R_Γ be the $\mathbb{R}$-algebra whose subset of $\mathbb{R}$-rational points coincides with Γ, and let R_X be the $\mathbb{R}$-algebra that corresponds to X in similar fashion. Then the map p corresponds to the map $p^* : R_X \to R_\Gamma$ and the kernel J of p^* defines the image of p. Furthermore, since p^* is finite, the R_X-annihilator $J_2 \supset J$ of $R_\Gamma/p^*(R_X)$, that is,

$$J_2 = \{r \in R_X \mid p(r)v = 0,\ v \in R_\Gamma\}$$

exists and defines the *support* $V(J_2)$ of $R_\Gamma/p^*(R_X)$. It also defines the locus of prime ideals m in $\operatorname{Spec} R_X$ such that the fiber $R_\Gamma/p^*(m)$ is not equal to R_X/m. Thus $V(J_2)$ defines the locus where the fiber has multiplicity greater than one. (The existence of such an ideal was the whole point of our excursion into algebraic geometry.) Note that the fact that the multiplicity of the fiber over a $\mathbb{R}$-rational point in X is greater than one does not guarantee the existence of $\mathbb{R}$-rational points in the fiber.

Now we can reduce the proof of Theorem 6.4 to finding an explicit example. Consider the map $c : \mathbb{C}^r \times \mathbb{C} \to \mathbb{C}^k$ defined by $c(a,z) = (Q_1(z), \ldots, Q_k(z))$. (The relation between a and Q_i is explained above.) Set $A = \mathbb{C}^r \times \mathbb{C} \setminus V(Q_k)$. Take the map $c^* : R_X \to R_A$ corresponding to c. The ideal $c^*(I_2)R_A$ describes the points in $\operatorname{Spec} R_A$ such that the fiber of the pullback has multiplicity greater than one. Denote the real coordinates in $\mathbb{C}^r \times \mathbb{C}$ by x_{st}, y_{st}, x, y such that $a_{st} = x_{st} + iy_{st}$ and $z = x + iy$. A point $a = (a_{st}) \in \mathbb{C}^r$ corresponds to a maximal ideal m_a. By Lemma 6.15 the ring

$$S_a := \mathbb{R}[x,y]/m_a c^*(I_2)R_A$$

describes the set of points in $\mathbb{C}$ that belongs to two distinct $\Gamma_{i,j,l}$s.

The following lemma (whose simple proof is included here for completeness) shows that the property of a that S_a is finite as a $\mathbb{R}$-vector space is generic.

Lemma 6.16. *Suppose that $c : T \to S$ is a map of finitely generated $\mathbb{R}$-algebras, and let U be the set of $\mathbb{R}$-rational points m in* Spec *T such that $S/c(m)$ is a finite $\mathbb{R}$-algebra. Then U is open.*

Proof. Let V be a finite-dimensional vector space that generates S and containing $1 \in S$. Let V^n denote the set of sums of products of at most n elements from V with coefficients in $c(T)$. The fact that V generates S implies that $S = \bigcup_{n\geq 0} V^n$. Now $S/c(m)$ is a finite $\mathbb{R}$-algebra if and only if there exists a non-negative integer n such that

$$T/m \otimes_T (V^{n+1}/V^n) = 0.$$

Let $J_k := \operatorname{Ann}_T V^{k+1}/V^k$, $k \geq 0$. Since T is a ring, one has that $J_k \subset J_{k+1}$. Set $J = \bigcup_{k\geq 0} J_k$. Then U consists of all $\mathbb{R}$-rational points m in Spec T that are not contained in the support $V(J)$. Since the latter is obviously closed, this proves the lemma. □

Applied to our situation this means that there exists a set of algebraic equations in the variables x_{st}, y_{st} such that when some of them are violated the number of points in $\mathbb{C}$ that lie on two distinct $\Gamma_{i,j,l}$s is finite. Thus to conclude the proof of Theorem 6.4 it only remains to find an example, which we do next.

Example 6.3. Let $b_1, \ldots, b_k$ be complex numbers. Consider the reducible plane curve L given by

$$\prod_{m=1}^{k} ((b_m - z)y - 1) = 0.$$

This equation has the solutions $y = \frac{1}{b_m - z}$, $m = 1, \ldots, k$, and it satisfies the conditions of (6.41). To describe the sets $\Gamma_{s,t,q}$, we use the following notation. Let α_i, β_i, $i = 1, 2, 3$, be our variables and define an action of the symmetric group S_3 on polynomial functions in these variables by permuting the pairs (α_i, β_i). For a monomial M, let η_M denote the alternating monomial function

$$\eta_M = \sum_{\sigma \in S_3} (-1)^{\operatorname{sign}(\sigma)} \sigma M.$$

(For example, $\eta_{\alpha_1\beta_2} = \alpha_1\beta_2 - \alpha_2\beta_1 - \alpha_1\beta_3 + \alpha_2\beta_3 + \alpha_3\beta_1 - \alpha_3\beta_2$.) The set Γ for the curve L is reducible and consists of one component for each $\{s, t, q\}$. Simple calculations show that the set $\Gamma_{1,2,3}$ is a circle in $\mathbb{C}$ consisting of all points $z = u + iv$ satisfying the equation

$$p_{1,2,3} = -\eta_{\alpha_1\beta_2}(u^2 + v^2) + (\eta_{\alpha_1\beta_2^2} + \eta_{\alpha_1\alpha_2^2})v + (\eta_{\beta_1^2\beta_2} + \eta_{\alpha_1^2\beta_2})u + (\eta_{\alpha_1\beta_2\beta_3^2} + \eta_{\alpha_1\alpha_3^2\beta_2}) = 0.$$

The center of this circle is given by

$$-\left(\frac{\eta_{\beta_1^2\beta_2} + \eta_{\alpha_1^2\beta_2}}{2\eta_{\alpha_1\beta_2}}, \frac{\eta_{\alpha_1\beta_2^2} + \eta_{\alpha_1\alpha_2^2}}{2\eta_{\alpha_1\beta_2}}\right),$$

where $b_j = \alpha_j + i\beta_j$, $j = 1, 2, 3$, and $\Gamma_{1,2,3}$ is non-empty (which is generically true). Hence for a generic set of complex numbers $b_1, \ldots, b_k$, all sets $\Gamma_{i,j,k}$ are circles with different centers, so that in particular the intersection of two sets $\Gamma_{i,j,k}$ corresponding to two different index sets $\{i, j, k\}$ is finite. S_a is the direct sum of the algebras $\mathbb{R}[u, v]/I$, where I is the ideal generated by two different quadrics $p_{i,j,k}$ and $p_{r,s,t}$, and it is therefore finite. This shows that the set U in Lemma 6.16 is non-empty.

The proof of Theorem 6.4 is now complete.

Remark 6.8. The starting point of our study was a number of rather surprising numerical experiments in calculating the zeros of polynomial eigenfunctions for the usual spectral problem, which led to several conjectures presented in [122]. Most of these conjectures were later proved in [27], see Chapter 2. The idea to consider the homogenized spectral problem comes mainly from [68, 145], and references [7, 8].

Remark 6.9. By carefully going through the proofs of the above theorems, one can see that actually these apply to a wider class of *ES*-pencils. Namely, consider *ES*-pencils of the form $T_\lambda = \sum_{j=0}^{k} Q_j(z, \lambda)\frac{d^j}{dz^j}$, where $Q_j(z, \lambda)$ are polynomials in z and λ. Assume that $\deg_\lambda Q_j(z, \lambda) \leq k - j$ and that $\deg_z Q_j(z, \lambda) \leq j$. Decompose $Q_j(z, \lambda) = Q_{j,0}(z) + \cdots + Q_{j,k-j}(z)\lambda^{k-j}$. If the homogenized pencil $\tilde{T}_\lambda = \sum_{j=0}^{k} Q_{j,k-j}(z)\lambda^{k-j}\frac{d^j}{dz^j}$ is of general type, then all the results in this chapter apply. However, for the sake of brevity, we have chosen to focus only on the homogeneous case.

6.7 Open problems in Chapter 6

6.1. One of the major objects in the classical asymptotic theory is the *global Stokes line* of a given linear differential equation, see, e. g., [7, 8]. The theory of *global Stokes lines* is well understood in the physically important case of second order linear odes. By contrast with the latter case, the development of this theory for higher order equations experiences serious difficulties. These are mostly due to the appearance of new Stokes lines (discovered in [29]) originating from the intersection points of the initial Stokes lines, which emanate from the turning points of the considered equation. The supports of the asymptotic measures μ_j studied in the present chapter show many similar features to the global Stokes line of a linear differential operator (6.1), which leads to the following conjecture.

Conjecture 6.2. *The union of the supports of the measures $\mu_1, \dots, \mu_k$ is a subset of the global Stokes line for the differential operator (*6.1*).*

6.2. The main open questions related to the results of the present chapter are Conjecture 6.1 (see § 6.1) and a complete description of the topological properties of the support of the limiting measures μ_i, $i = 1, \dots, k$. (We strongly believe that μ_i is unique for each i.) In particular, the following problem appears to be of fundamental importance, and all the numerical evidence we obtained so far is in favor of this conjecture.

Conjecture 6.3. *For each j, the set $\mathbb{C} \setminus \operatorname{supp}(\mu_i)$ is connected.*

Problem 6.3. Is it true that the support of each μ_i is topologically a tree?

In general, it would be highly desirable to describe the stratification of the space ES_k according to the topology of the supports of the k-tuple of measures $\mu_1, \dots, \mu_k$.

6.3. A further natural question is as follows.

Problem 6.4. Which of the above results hold for operators (6.1) of non-general type?

Numerical experiments carried out in [26] show the existence of asymptotic root-counting measures even for such operators. However, these measures usually have non-compact supports, which leads to considerable additional complications since the arguments in all our proofs do not extend *mutatis mutandis* to such situations.

6.4. As we already mentioned in the introduction to this chapter, Problem 6.2 is completely solved only for rational functions.

Problem 6.5. Characterize algebraic germs of real (respectively, positive) Cauchy transforms.

(Partial information about this problem can be found in Chapter 7.)

7 Algebraic equations satisfied by Cauchy transforms of positive measures

(The material of this chapter is based on [44].) The results and questions posed in Chapter 6 motivate and clarify the content of this chapter.

7.1 Preliminaries

The study of local and global properties of the Cauchy transform and the Cauchy–Stieltjes integral was initiated by A. Cauchy and T. Stieltjes in the middle of the 19th century. Large numbers of papers and several books are partially or completely devoted to this area, which is closely connected with the potential theory in the complex plane and, especially, to the inverse problem and to the inverse moment problem, see, e. g., [22, 52, 69, 139, 199].

As above, for a finite compactly supported complex measure μ on the complex plane $\mathbb{C}$, we define the *logarithmic potential* of μ as

$$u_\mu(z) := \int_{\mathbb{C}} \ln|z-\xi| d\mu(\xi)$$

and the *Cauchy transform* of μ as

$$\mathcal{C}_\mu(z) := \int_{\mathbb{C}} \frac{d\mu(\xi)}{z-\xi}.$$

Standard facts about the Cauchy transform include:

- $\mathcal{C}_\mu$ is locally integrable; in particular, it defines a distribution on $\mathbb{C}$ and therefore can be acted upon by $\frac{\partial}{\partial z}$ and $\frac{\partial}{\partial \bar{z}}$.
- $\mathcal{C}_\mu$ is analytic in the complement to the support of μ in $\mathbb{C}P^1 \simeq \mathbb{C}\cup\{\infty\}$. For example, if μ is supported on the unit circle (which is the most classical case), then $\mathcal{C}_\mu$ is analytic both inside the open unit disc and outside the closed unit disc.
- The relations between μ, $\mathcal{C}_\mu$, and u_μ are

$$\mathcal{C}_\mu = \frac{\partial u_\mu}{\partial z} \quad \text{and} \quad \mu = \frac{1}{\pi}\frac{\partial \mathcal{C}_\mu}{\partial \bar{z}} = \frac{1}{\pi}\frac{\partial^2 u_\mu}{\partial z \partial \bar{z}}$$

  which should be understood as equalities of distributions.
- The Laurent series of $\mathcal{C}_\mu$ around ∞ is given by

$$\mathcal{C}_\mu(z) = \frac{m_0(\mu)}{z} + \frac{m_1(\mu)}{z^2} + \frac{m_2(\mu)}{z^3} + \cdots,$$

  where

$$m_k(\mu) = \int_{\mathbb{C}} z^k d\mu(z), \quad k = 0, 1, \dots$$

https://doi.org/10.1515/9783112207741-008

are the harmonic moments of measure μ. (These moments are important conserved quantities in the theory of Hele–Shaw flow, [110].)

During the last decades the notion of a *motherbody* of a solid domain or, more generally, of a positive Borel measure was discussed both in geophysics and mathematics, see, e. g., [173, 159, 80, 200]. It was apparently pioneered in the 1960s by a Bulgarian geophysicist D. Zidarov [200] and later mathematically developed by B. Gustafsson [80]. Although a number of interesting results about motherbodies was obtained in several special cases [159, 80, 200], there is still no consensus about its appropriate general definition. In particular, no general existence and/or uniqueness results are known at present.

Below we use one possible definition of a motherbody (measure) and study a natural exterior inverse motherbody problem in the complex plane. (In what follows we will only use Borel measures.)

Problem 7.1 (Main problem). Given a germ $f(z) = a_0/z + \sum_{n\geq 2}^{\infty} a_n/z^n$, $a_0 \in \mathbb{R}$ of an algebraic (or, more generally, analytic) function near ∞, is it possible to find a compactly supported signed measure whose Cauchy transform coincides with (a branch of) the analytic continuation of $f(z)$ a. e. in $\mathbb{C}$? Additionally, for which $f(z)$ it is possible to find a positive measure with the above properties?

If such a signed measure exists and its support is a finite union of compact semi-analytic curves and isolated points, we will call it *a real motherbody measure* of the germ $f(z)$, see Definition 7.1 below. If the corresponding measure is positive, then we call it *a positive motherbody measure* of $f(z)$.

Remark 7.1. Observe that for some germs of analytic functions near ∞, the singularities of their analytic continuation to $\mathbb{C}$ might be very ill-behaved, which implies that a motherbody measure of such a germ does not exist for this reason. However, since we consider only algebraic functions, this phenomenon never occurs in our situation.

An obvious necessary condition for the existence of a positive motherbody measure is $a_0 > 0$ since a_0 equals the total mass. A germ (branch) of an analytic function $f(z) = a_0/z + \sum_{n\geq 2}^{\infty} a_n/z^n$ with $a_0 > 0$ will be called *positive*. If $a_0 = 1$, then $f(z)$ is called a *probability* branch. (Necessary and sufficient conditions for the existence of a probability branch of an algebraic function are given in Lemma 7.1 below.)

The formal definition of a motherbody measure that we are using is as follows.

Definition 7.1. Given a germ $f(z) = a_0/z + \sum_{n\geq 2}^{\infty} a_n/z^n$, $a_0 > 0$ of an analytic function near ∞, we say that a signed measure μ_f is its *motherbody* if:

(i) Its support $supp(\mu_f)$ is the union of finitely many points and finitely many compact semi-analytic curves in $\mathbb{C}$.
(ii) The Cauchy transform of μ_f coincides in each connected component of the complement $\mathbb{C} \setminus supp(\mu_f)$ with some branch of the analytic continuation of $f(z)$.

(Here by a compact semi-analytic curve we mean a compact segment of a real-analytic curve in $\mathbb{C} \simeq \mathbb{R}^2$.)

Remark 7.2. Notice that by Theorem 1 of [32], if the Cauchy transform of a positive measure coincides a. e. in $\mathbb{C}$ with an algebraic function $f(z)$, then the support of this measure is a finite union of semi-analytic curves and isolated points. Therefore it is a motherbody measure according to the above definition. Whether the latter result extends to signed measures is unknown at present, which motivates the following question.

Problem 7.2. Is it true that if there exists a signed measure whose Cauchy transform satisfies an irreducible algebraic equation a. e. in $\mathbb{C}$, then there exists, in general, another signed measure whose Cauchy transform satisfies a. e. in $\mathbb{C}$ the same algebraic equation and whose support is a finite union of compact semi-analytic curves and isolated points? Does there exist such a measure with a singularity on each connected component of its support?

Classically the inverse problem in potential theory deals with the question of how to restore a solid body or a (positive) measure from the information about its potential near infinity. The main efforts in this inverse problem since the foundational paper of P. S. Novikov [142] were concentrated around the question about the uniqueness of a (solid) body with a given exterior potential, see, e. g., recent [70] and [71] and the references therein. P. S. Novikov (whose main mathematical contributions are in the areas of mathematical logic and group theory) proved uniqueness for convex and star-shaped bodies with a common point of star-shapeness. The question about the uniqueness of contractible domains in $\mathbb{C}$ with a given sequence of holomorphic moments was later posed anew by H. S. Shapiro, see Problem 1, p. 193 of [20] and answered negatively by M. Sakai in [158]. A similar non-uniqueness example for non-convex plane polygons was reported by geophysicists in [46], see also [148].

It turns out that the existence of a compactly supported positive measure with a given Cauchy transform $f(z)$ near ∞ imposes no conditions on a germ except for the obvious $a_0 > 0$, see Theorem 7.1 below.

On the other hand, the requirement that the Cauchy transform coincides with (the analytic continuation) of a given germ $f(z)$ a. e. in $\mathbb{C}$ often leads to additional restrictions on the germ $f(z)$, which are not easy to describe in terms of the defining algebraic equation, see § 7.4.

Below we study two classes of algebraic functions of very different origin, and our results for these two cases are very different as well. For the first class, the obvious necessary condition $a_0 = 1$ seems to be sufficient for the existence of a positive motherbody measure. (At present we can prove this fact only under certain additional restrictions.) For the second class, the set of admissible germs has a quite complicated structure. These results together with a number of conjectures seem to indicate that it is quite difficult, in general, to answer when a given algebraic germ $f(z)$ admits a motherbody measure, and if it does, then how many.

Our interest in probability measures whose Cauchy transforms are algebraic functions a. e. in $\mathbb{C}$ was sparked by the pioneering work [27], see Chapter 2. Since then the class of interesting examples where such situation occurs has been substantially broadened, see [39, 84, 168]. Some general local results when one considers a collection of locally analytic functions instead of a global algebraic function were obtained in [36] and later extended in [32].

7.2 Some general facts

The first rather simple result of the present chapter.

Theorem 7.1. *Given an arbitrary germ $f(z) = a_0/z + \sum_{n\geq 2}^{\infty} a_n/z^n$, $a_0 > 0$ of an analytic function near ∞, there exist (families of) positive compactly supported in $\mathbb{C}$ measures whose Cauchy transform near ∞ coincides with $f(z)$.*

Proof of Theorem 7.1. Given a branch $f(z) = a_0/z + \sum_{n\geq 2} a_n/z^n$ of an analytic function near ∞, we first take a germ of its "logarithmic potential", i. e., a germ $h(z)$ of harmonic function such that $h(z) = a_0 \log|z| + \cdots$ satisfying the relation $\partial h/\partial z = f(z)$ in a punctured neighborhood of ∞. Here . . . stands for a germ of harmonic function near ∞. For any sufficiently large positive v, the connected component γ_v of the level curve $h(z) = v$ near infinity is closed and simple. It makes one turn around ∞. To get a required positive measure whose Cauchy transform coincides with $f(z)$ near ∞, take γ_v for any v large enough and consider the complement $\Omega_v = \mathbb{C}P^1 \setminus Int(\gamma_v)$, where $Int(\gamma_v)$ is the interior of γ_v containing ∞. Consider the equilibrium measure of mass a_0 supported on Ω_v. By Frostman's theorem, this measure is in fact supported on γ_v (since γ_v has no polar subsets), its potential is constant on γ_v and it is harmonic in the complement to the support. (For the definition and properties of equilibrium measures as well as Frostman's theorem, consult [155].) Thus it should coincide with $h(z)$ in $Int(\gamma_v)$ since the total mass is correctly chosen. Then its Cauchy transform coincides with $f(z)$ in $Int(\gamma_v)$. □

Example 7.1. If we choose $f(z) = 1/z$ as our branch at ∞, then $h(z) = \log|z|$ and γ_v is the circle $|z| = e^v$. The equilibrium measure is the uniform probability measure on this circle, and its Cauchy transform $C(z)$ vanishes inside the circle and equals $1/z$ outside. Since the constant 0 is not the analytic continuation of $1/z$, the uniform measure on the circle is not a motherbody measure of the germ $1/z$. However, the Cauchy transform of the unit point mass placed at the origin equals $1/z$ in $\mathbb{C} \setminus 0$. Therefore the unit mass at the origin is a motherbody measure for the germ $1/z$ at ∞.

We now give a necessary condition and a slightly stronger sufficient condition for an algebraic curve given by the equation

$$P(\mathcal{C}, z) = \sum_{(i,j)\in S(P)} a_{i,j} \mathcal{C}^i z^j = 0 \tag{7.1}$$

to have a probability branch at ∞. Here every $a_{i,j} \neq 0$ and $S(P)$ is an arbitrary finite subset of pairs of non-negative integers, i. e., an arbitrary finite set of monomials in 2 variables. In other words, $S(P)$ is the set of all monomials appearing in P with non-vanishing coefficients. (In what follows, $\mathcal{C}$ stands for the variable corresponding to the Cauchy transform.)

The following group of results consisting of Lemmas 7.1 and 7.2 as well as Corollary 7.1 are very straightforward although we could not find them in the literature.

Lemma 7.1. *If the algebraic curve given by equation* (7.1) *has a probability branch at* ∞, *then*

$$\sum_i a_{i,i-m(P)} = 0 \quad \text{where } m(P) := \min_{(i,j)\in S(P)} i-j. \tag{7.2}$$

In particular, there should be at least two distinct monomials in $S(P)$ *whose difference of indices* $i-j$ *equals* $m(P)$.

If equation (7.2) *is satisfied, and additionally*

$$\sum_i i a_{i,i-m(P)} \neq 0, \tag{7.3}$$

then there is a unique probability branch at ∞ *satisfying equation* (7.1).

Proof. Substituting $w = 1/z$ in (7.1), we get $P(\mathcal{C}, w) = \sum_{(i,j)\in S(P)} a_{i,j}\mathcal{C}^i/w^j$. Assuming that the algebraic curve given by $P(\mathcal{C}, w) = 0$ has a branch

$$\mathcal{C} = w + \sum_{n=2}^{\infty} a_n w^n,$$

where a_l, $l = 2, 3, \dots$ are undetermined coefficients, we substitute the latter expression in the above equation and obtain the expression

$$\sum_{(i,j)\in S(P)} a_{i,j}\left(w + \sum_{n=2}^{\infty} a_n w^n\right)^i / w^j = 0. \tag{7.4}$$

Collecting the entries of the minimal total degree $m(P) := \min_{(i,j)\in S(P)} i-j$, we obtain an obvious necessary condition for the solvability of (7.4) given by

$$\sum a_{i,i-m(P)} = 0. \tag{7.5}$$

Let us now show that (7.2), together with (7.3), is sufficient for the solvability of (7.4) with respect to the sequence of coefficients $a_2, a_3, \dots$. Indeed, due to algebraicity of $P(\mathcal{C}, z)$, it suffices to prove the formal solvability of (7.4). Let $\mathcal{C} = \mathcal{D}w$ and rewrite the equation as

$$\tilde{P}(\mathcal{D}, w) = \sum_{(i,j)\in S(P)} a_{i,j}\mathcal{D}^i w^{i-j-m(P)} = 0,$$

which is now a polynomial in w and $\mathcal{D}$. Assume that $d_r = 1 + a_2 w + \cdots + a_{r+1} w^r$ satisfies

$$\tilde{P}(d_r, w) \equiv 0 \bmod z^{r+1}. \tag{7.6}$$

The fact that this equation holds for $d_0 = 1$ is exactly relation (7.5), which gives the basis of the inductive construction of $d_1, d_2, \ldots$, as follows. Letting $\tilde{P}_1'(\mathcal{C}, w)$ be the partial derivative of $\tilde{P}(\mathcal{C}, w)$ with respect to the first variable, we have the following relation for the undetermined coefficient a_{r+2}:

$$\begin{aligned}\tilde{P}(d_r + a_{r+2} w^{r+1}, w) &\equiv \tilde{P}(d_r, w) + \tilde{P}_1'(d_r, w) a_{r+2} w^{r+1} \\ &\equiv \tilde{P}(d_r, w) + \tilde{P}_1'(1, 0) a_{r+2} w^{r+1} \bmod w^{r+2}.\end{aligned}$$

Since we have assumed that $\tilde{P}_1'(1,0) = \sum_i i a_{i,i-m(P)}$ is non-zero, and that, by the induction assumption, $\tilde{P}(d_r, w) \equiv b w^{r+1} \bmod w^{r+2}$, $b \in \mathbb{C}$, we can solve the latter equation for a_{r+2}. Thus, by induction, we have proven that there is a formal series solution of (7.4). Therefore conditions (7.2) and (7.3) are sufficient for the existence of a probability branch (which is also unique in this case). □

Remark 7.3. Note that for an irreducible algebraic curve defined by (7.1), the second condition (7.3) in Lemma 7.1 guarantees that $z = \infty$ is not its branch point. It is clearly possible, though cumbersome, to give necessary and sufficient conditions for the existence of a probability branch in terms of algebraic relations between the coefficients of the equation. An example of an equation that does not satisfy both conditions in Lemma 7.1, but still has a probability branch as a solution, is $P(\mathcal{C}, z) = (\mathcal{C}z - 1)^2$.

For the polynomials whose Newton polygons (i. e., the convex hulls of the set of monomials) are shown in Figure 7.1, the necessary condition of Lemma 7.1 says that the sum of the coefficients of the monomials marked by the letters a, b, ... should vanish.

Since we are working with irreducible algebraic curves (obtained as the analytic continuation of given branches at ∞), we will need the following statement. Let S be an arbitrary set of monomials in 2 variables. Denote by $\mathcal{N}_S$ the *Newton polygon* of S, i. e.,

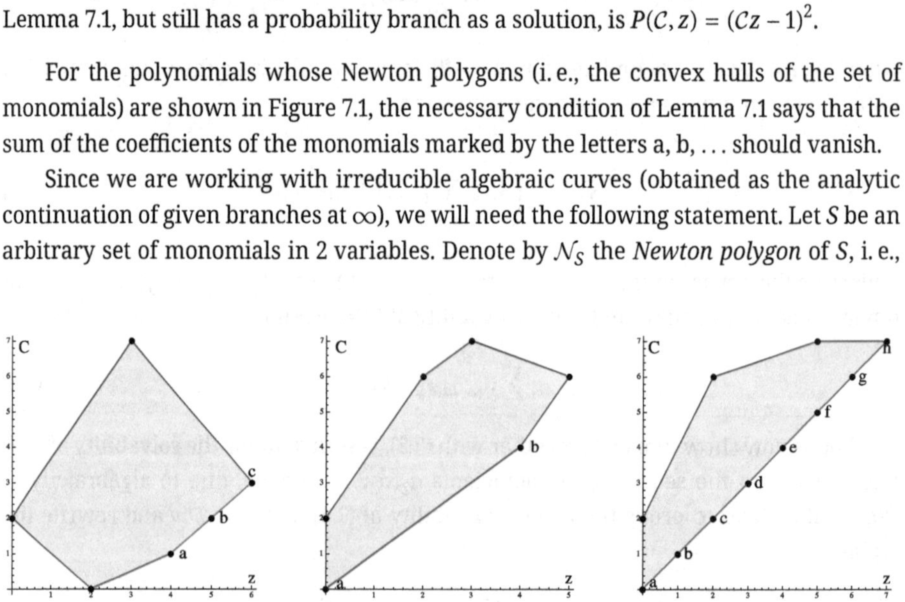

Figure 7.1: Three examples of Newton polygons.

the convex hull of S in the plane of exponents of monomials. Denote by Pol_S the linear span of all monomials in S with complex coefficients.

Lemma 7.2. *A generic polynomial in Pol_S is irreducible if and only if:*
(i) *$\mathcal{N}_S$ is two-dimensional, i. e., not all monomials in S lie on the same affine line.*
(ii) *S contains vertices on both coordinate axes, i. e., pure powers of both variables.*

Notice that (ii) is satisfied for both coordinate axes if S contains the origin (i. e., polynomials in Pol_S might have a non-vanishing constant term).

Proof. Observe that the property that a generic polynomial from Pol_S is irreducible is inherited. In other words, if S contains a proper subset S' with the same property, then it automatically holds for S. The necessity of both conditions (i) and (ii) is obvious. If (i) is violated, then any polynomial in Pol_S can be represented as a polynomial in one variable after an appropriate change of variables. If (ii) is violated, then any polynomial in Pol_S is divisible by a (power of a) variable.

To prove sufficiency, we have to consider several cases. If S contains $\mathcal{C}^i$ and z^j where both i and j are positive, then already a generic curve of the form $\alpha\mathcal{C}^i + \beta\mathcal{C}^j z^k + \gamma z^l = 0$ is irreducible unless $\mathcal{C}^i$, $\mathcal{C}^j z^k$ and z^l lie on the same line. If for some positive i, S contains $\mathcal{C}^i$ and 1 and no other pure powers of z (or, similarly, z^k and 1 for some positive k and no other pure powers of $\mathcal{C}$), then S should contain another monomial $\mathcal{C}^l z^k$ with l and k positive. But then a generic curve of the form $A + B\mathcal{C}^i + C\mathcal{C}^l z^k = 0$ is irreducible. Finally, if the only monomial on the union of the axes is 1, but there exist two monomials $\mathcal{C}^{i_1} z^{j_1}$ and $\mathcal{C}^{i_2} z^{j_2}$ with $i_1/j_1 \neq i_2/j_2$, then a generic curve of the form $A + B\mathcal{C}^{i_1} z^{j_1} + C\mathcal{C}^{i_2} z^{j_2} = 0$ is irreducible. □

Corollary 7.1. *An irreducible polynomial $P(\mathcal{C}, z)$ having a probability branch has a non-negative $M(P) := \max_{(i,j)\in S(P)}(i - j)$ as well as a non-positive $m(P) := \min_{(i,j)\in S(P)}(i - j)$.*

If $m(P) = 0$, then the set $S(P)$ of monomials of $P(\mathcal{C}, z)$ must contain the origin, i. e., must have a non-vanishing constant term.

If we denote by $\mathcal{N}_P = \mathcal{N}_{S(P)}$ the Newton polygon of $P(\mathcal{C}, z)$, (i. e., the convex hull of its monomials), then the geometric interpretation of the latter corollary is that $\mathcal{N}_P$ should (non-strictly) intersect the bisector of the first quadrant in the plane with coordinates $(z, \mathcal{C})$.

The case $m(P) = 0$ will be of special interest to us, see § 7.3. We say that a polynomial $P(\mathcal{C}, z)$ (and the algebraic function given by $P(\mathcal{C}, z) = 0$) is *balanced* if it satisfies the condition $m(P) = 0$. For polynomials with $m(P) < 0$ having a positive branch, the problem of existence of a motherbody measure seems to be dramatically more complicated than for $m(P) = 0$, see § 7.4.

A rather simple situation occurs, when $\mathcal{C}$ is a rational function.

Lemma 7.3. *A (germ at ∞ of a) rational function $\mathcal{C}(z) = \frac{z^n+\cdots}{z^{n+1}+\cdots}$ with coprime numerator and denominator admits a motherbody measure if and only if it has all simple poles* *with real residues. If all residues are positive, the corresponding motherbody measure is positive.*

Proof. Recall the classical relation

$$\mu = \frac{1}{\pi}\frac{\partial \mathcal{C}_\mu(z)}{\partial \bar{z}}$$

between a measure μ and its Cauchy transform $\mathcal{C}_\mu$. Here the derivative is taken in the sense of distributions. By assumption, $\mathcal{C}_\mu(z)$ coincides almost everywhere with a given rational function. Therefore, by the above formula, μ is the measure supported at the poles of the rational function. Moreover the measure concentrated at each pole coincides (up to a factor π) with the residue of $\mathcal{C}_\mu(z)$ at this pole. Since by assumption μ has to be a real measure, this implies that all poles of the rational function should be simple and with real residues. □

The above statement implies that the set of rational functions of degree n admitting a motherbody measure has dimension equal to half of the dimension of the space of all rational functions of degree n. The case when $\mathcal{C}_\mu$ satisfies a quadratic equation is considered in more detail in § 7.4.

7.3 Balanced algebraic functions

In the case of balanced algebraic functions, i. e., for bivariate polynomials (7.1) with $m(P) := \min_{(i,j)\in S(P)}(i-j) = 0$ and satisfying the necessary condition (7.2), our main conjecture is as follows.

Conjecture 7.1. *An arbitrary balanced irreducible polynomial $P(\mathcal{C}, z)$ with a positive branch admits a positive motherbody measure.*

Appropriate Newton polygons are shown on the central and the rightmost pictures in Figure 7.1. The next result essentially proven in Chapter 6 and [39] strongly supports the above conjecture. Further supporting statements can be found in [26].

Theorem 7.2. *An arbitrary balanced irreducible polynomial $P(\mathcal{C}, z)$ with a unique probability branch which additionally satisfies the requirements:*
(i) *$S(P)$ contains a diagonal monomial $\mathcal{C}^n z^n$, which is lexicographically larger than any other monomial in $S(P)$;*
(ii) *The probability branch is the only positive branch of $P(\mathcal{C}, z)$;*

admits a probability motherbody measure.

By "lexicographically larger" we mean that the pair (n, n) is coordinate-wise not smaller than any other pair of indices in $S(P)$, see the rightmost picture in Figure 7.1. Condition (i) is the only essential restriction in Theorem 7.2 compared to Conjecture 7.1 since condition (ii) is generically satisfied. Observe also that an irreducible balanced polynomial $P(C, z)$ must necessarily have a non-vanishing constant term, see Lemma 7.2. Balanced polynomials satisfying assumptions of Theorem 7.2 are called *excellent balanced* polynomials.

An interesting detail about the latter theorem is that its proof has hardly anything to do with potential theory. Our proof of Theorem 7.2 is based on certain properties of eigenpolynomials of the ES-operators, see, e. g., [39]. We will construct the required motherbody measure as the weak limit of a sequence of root-counting measures of these eigenpolynomials.

Definition 7.2. For a linear ordinary differential operator

$$T = \sum_{j=1}^{k} Q_j(z) \frac{d^j}{dz^j}, \tag{7.7}$$

its symbol Sm_T is, by definition, the bivariate polynomial

$$Sm_T(C, z) = \sum_{j=1}^{k} Q_j(z) C^i. \tag{7.8}$$

Observe that T is exactly solvable if and only if Sm_T is balanced. (Notice that Sm_T, by definition, has no constant term.)

Given an exactly solvable $T = \sum_{j=1}^{k} Q_j(z) \frac{d^j}{dz^j}$, consider the following *homogenized spectral problem*:

$$Q_k(z)p^{(k)} + \lambda Q_{k-1}(z)p^{(k-1)} + \lambda^2 Q_{k-2}(z)p^{(k-2)} + \cdots + \lambda^{k-1} Q_1(z)p' = \lambda^k p, \tag{7.9}$$

where λ is called the *homogenized spectral parameter*, which we discussed in Chapter 6. Given a positive integer n, we want to determine all values λ_n of the spectral parameter such that equation (7.9) has a polynomial solution $p_n(z)$ of degree n.

Using notation $Q_j(z) = q_{j,j}z^j + q_{j,j-1}z^{j-1} + \cdots + q_{j,0}$, $i = 1, \dots, k$, one can easily check that the eigenvalues λ_n satisfy the equation:

$$q_{k,k}n(n-1)\dots(n-k+1) + \lambda_n q_{k-1,k-1}n(n-1)\dots(n-k+2) + \cdots + \lambda_n^{k-1} n q_{1,1} = \lambda_n^k. \tag{7.10}$$

(If T is non-degenerate, i. e., $q_{k,k} \neq 0$, then there are typically k distinct values of λ_n for n large enough.)

Introducing the normalized eigenvalues $\tilde{\lambda}_n = \lambda_n/n$, we get that $\tilde{\lambda}_n$ satisfy the equation

$$q_{k,k}\frac{(n)_k}{n^k} + \tilde{\lambda}_n q_{k-1,k-1}\frac{(n)_{k-1}}{n^{k-1}} + \cdots + \tilde{\lambda}_n^{k-1} q_{1,1} = \tilde{\lambda}_n^k.$$

If $\tilde{\lambda} = \lim_{n\to\infty} \tilde{\lambda}_n$ exists, then it should satisfy the relation

$$q_{k,k} + q_{k-1,k-1}\tilde{\lambda} + \cdots + q_{1,1}\tilde{\lambda}^{k-1} = \tilde{\lambda}^k. \tag{7.11}$$

If T is a degenerate ES-operator, then let j_0 be the maximal i such that $\deg Q_i(z) = i$. By definition, $q_{j,j}$ vanish for all $j > j_0$. Thus, the first $k - j_0$ terms in (7.10) vanish as well, implying that (7.10) has j_0 non-vanishing and $(k - j_0)$ vanishing roots.

Lemma 7.4. *Given an ES-operator T as above,*
(i) *For the sequence of vanishing "eigenvalues" $\lambda_n = 0$, there exists a finite upper bound on the degree of a non-vanishing eigenpolynomial $\{p_n(z)\}$.*
(ii) *For any sequence $\{\lambda_n\}$ of eigenvalues such that $\lim_{n\to\infty} \frac{\lambda_n}{n} = \widetilde{\lambda}_l$, where $\widetilde{\lambda}_l$ is some non-vanishing simple root of (7.11), the sequence of their eigenpolynomials is well defined for all $n > N_0$, i. e., for all sufficiently large n.*

Proof. Item (i) is obvious since when $\lambda = 0$, (7.9) reduces to $Q_k(z)p^{(k)} = 0$, implying that $p^{(k)} = 0$. But this is impossible for any polynomial of degree exceeding $k - 1$.

To prove (ii), notice that (7.10) defines $\widetilde{\lambda}$ as an algebraic function of the (complex) variable n. Then (7.11) defines the value of $\widetilde{\lambda}$ at $n = \infty$. By assumption there is a unique branch with the value $\widetilde{\lambda}_l$. Additionally, the corresponding branch of the algebraic function will intersect other branches in at most finite number of points. Hence for n very large, we can identify $\widetilde{\lambda}_l(n)$ as belonging to this particular branch. □

The next result is central in our consideration.

Theorem 7.3 (See Chapter 6 and [39, Theorem 2]). *For any non-degenerate exactly solvable operator T, such that equation (7.11) has no double roots, there exists N_0 such that the roots of all eigenpolynomials of the homogenized problem (7.9) whose degree exceeds N_0 are bounded, i. e., there exists a disk centered at the origin containing all of them at once.*

The next local result is a keystone in the proof of Theorem 7.2, comp. Proposition 3 of [39].

Theorem 7.4. *Given an ES-operator $T = \sum_{j=1}^{k} Q_j(z)\frac{d^j}{dz^j}$, consider a sequence $\{p_n(z)\}$, $\deg p_n(z) = n$ of the eigenpolynomials of (7.9) such that the sequence $\{\lambda_n\}$ of their eigenvalues satisfies the condition $\lim_{n\to\infty} \frac{\lambda_n}{n} = \widetilde{\lambda}_l$, where $\widetilde{\lambda}_l$ is some non-vanishing root of (7.11). Let $L_n(z) = \frac{p_n'(z)}{\lambda_n p_n(z)}$ be the normalized logarithmic derivative of $p_n(z)$. If the sequence $\{L_n(z)\}$ converges to a function $L(z)$ in some open domain $\Omega \subset \mathbb{C}$, and the derivatives of $L_n(z)$ up to order k are uniformly bounded in Ω, then in the domain Ω the function $L(z)$ satisfies the symbol equation*

$$Q_k(z)L^k(z) + Q_{k-1}(z)L^{k-1}(z) + \cdots + Q_1(z)L(z) = 1. \tag{7.12}$$

Proof. Note that each $L_n(z) = \frac{p_n'(z)}{\lambda_n p_n(z)}$ is well defined and analytic in any open domain Ω free from the zeros of $p_n(z)$. Choosing such a domain Ω and an appropriate branch of the logarithm such that $\log p_n(z)$ is defined in Ω, consider a primitive function $M(z) = \lambda_n^{-1} \log p_n(z)$. The latter function is also well defined and analytic in Ω.

Straightforward calculations give $e^{\lambda_n M(z)} = p_n(z)$, $p_n'(z) = p_n(z)\lambda_n L_n(z)$, and $p_n''(z) = p_n(z)(\lambda_n^2 L_n^2(z) + \lambda_n L_n'(z))$. More generally,

$$\frac{d^j}{dz^j}(p_n(z)) = p_n(z)(\lambda_n^j L_n^j(z) + \lambda_n^{j-1} F_j(L_n(z), L_n'(z), \dots, L_n^{(j-1)}(z))),$$

where the second term

$$\lambda_n^{j-1} F_j(L_n, L_n', \dots, L_n^{(j-1)}) \tag{7.13}$$

is a polynomial in λ_n of degree $j-1$. The equation $T(p_n) + \lambda_n p_n = 0$ gives us

$$p_n(z)\left(\sum_{j=0}^{k} Q_j(z)\lambda_n^{k-j}(\lambda_n^j L_n^j(z) + \lambda_n^{j-1} F_j(L_n(z), L_n'(z), \dots, L_n^{(j-1)}(z)))\right) = 0$$

or, equivalently,

$$\lambda_n^k \sum_{j=0}^{k} Q_j(z)(L_n^j(z) + \lambda_n^{-1} F_j(L_n(z), L_n'(z), \dots, L_n^{(j-1)}(z))) = 0. \tag{7.14}$$

Letting n tend to ∞ and using the boundedness assumption for the first $k-1$ derivatives, we get the required equation (7.12). □

Sketch of proof of Theorem 7.2. Take an excellent balanced irreducible polynomial $P(\mathcal{C}, z) = \sum_{j=1}^{k} Q_j(z)\mathcal{C}^j$ having a probability branch at ∞. Since $P(\mathcal{C}, z)$ has a non-vanishing constant term, we can, without loss of generality, assume that it is equal to -1. Consider its corresponding differential operator $T_P(z) = \sum_{j=1}^{k} Q_j(z)\frac{d^j}{dz^j}$, i. e., the operator whose symbol Sm_T equals $P(\mathcal{C}, z) + 1$. Notice that since $P(\mathcal{C}, z)$ has a probability branch at ∞, then there exists a root of (7.11) that is equal to 1. Therefore, there exists a sequence $\{\lambda_n\}$ of eigenvalues of $\mathfrak{d}_P(z)$ satisfying the condition $\lim_{n\to\infty} \frac{\lambda_n}{n} = 1$.

To settle Theorem 7.2, notice that by Theorem 7.3 for any $T_P(z)$ as above, the union of all roots of all its eigenpolynomials of all sufficiently large degrees is bounded. Choose a sequence $\{\lambda_n\}$ of eigenvalues of $T_P(z)$ satisfying the condition $\lim_{n\to\infty} \frac{\lambda_n}{n} = 1$. (Such subsequences exist for any excellent $P(\mathcal{C}, z)$.) Consider the corresponding sequence $\{p_n(z)\}$ of eigenpolynomials of $\mathfrak{d}_P(z)$, and the sequence $\{\mu_n\}$ of the root-counting measures of these eigenpolynomials, together with the sequence $\{\mathcal{C}_n(z)\}$ of their Cauchy transforms. (Observe that the Cauchy transform of a polynomial P of degree n equals $\frac{P'}{nP}$.) By Theorem 7.3, all zeros of all these $p_n(z)$ are contained in some fixed disk. Therefore the supports of all μ_n are bounded implying that there exist weakly converging subsequences $\{\mu_{i_n}\}$ of the sequence $\{\mu_n\}$. Choosing a further appropriate subsequence, we can guarantee that the corresponding subsequence $\{\mathcal{C}_{i_n}(z)\}$ converges almost everywhere in $\mathbb{C}$. (The rigorous argument for the latter claim is rather technical, see details in [39].) Taking a further subsequence, if necessary, we get that the subsequences $\{C'_{i_n}(z)\}, \{C''_{i_n}(z)\}, \dots, \{C^{(k)}_{i_n}(z)\}$ will be bounded a. e. on any compact subset of $\mathbb{C}$. Finally, notice that if the limit $\lim_{n\to\infty} C_{i_n}(z) = C(z)$ exists in some domain Ω, then also $\lim_{n\to\infty} L_{i_n}(z)$ exists in Ω and equals $C(z)$. Here $L_n(z) = \frac{p'_n(z)}{\lambda_n p_n(z)}$ and the sequence $\{\lambda_n\}$ is chosen as above. Since it satisfies the condition $\lim_{n\to\infty} \frac{\lambda_n}{n} = 1$, we have settled the existence of a probability measure whose Cauchy transform coincides a. e. in $\mathbb{C}$ with a branch of the algebraic function given by the equation $\sum_{j=1}^{k} Q_j(z)C^j = 1$. The latter equation coincides with $P(C, z) = 0$. □

7.4 Quadratic differentials and Cauchy transforms satisfying quadratic equations

In this section we discuss which quadratic equations of the form

$$P(z)C^2 + Q(z)C + R(z) = 0, \tag{7.15}$$

with $\deg P = n + 2$, $\deg Q \le n + 1$, $\deg R \le n$ admit motherbody measure(s).

For the subclass of (7.15) with $Q(z)$ identically vanishing, such results were in large obtained in [168]. Very close statements were independently and simultaneously obtained in [132] and [133]. To go further, we need to recall some basic facts about quadratic differentials.

7.4.1 Basics on quadratic differentials

The following material can be easily found in the classical sources [179] and [95], see also [174, 175].

A (meromorphic) quadratic differential Ψ on a compact orientable Riemann surface Y without boundary is a (meromorphic) section of the tensor square $(T^*_{\mathbb{C}}Y)^{\otimes 2}$ of the holomorphic cotangent bundle $T^*_{\mathbb{C}}Y$. The zeros and the poles of Ψ constitute the set of *critical points* of Ψ denoted by $Sing_\Psi$. Zeros and poles are called *finite critical points.* (Non-critical points of Ψ are usually called *regular*.)

If Ψ is locally represented in two intersecting charts by $h(z)dz^2$ and by $\tilde{h}(\tilde{z})d\tilde{z}^2$ resp. with a transition function $\tilde{z}(z)$, then $h(z) = \tilde{h}(\tilde{z})(d\tilde{z}/dz)^2$. Any quadratic differential induces a canonical metric on its Riemann surface Y, whose length element in local coordinates is given by

$$|dw| = |h(z)|^{\frac{1}{2}}|dz|.$$

The above canonical metric $|dw| = |h(z)|^{\frac{1}{2}}|dz|$ on Y is closely related to two distinguished line fields given by the condition that $h(z)dz^2 > 0$ and $h(z)dz^2 < 0$ resp. The integral curves of the first field are called *horizontal trajectories* of Ψ, while the integral curves of the second field are called *vertical trajectories* of Ψ. In what follows we will mostly use horizontal trajectories of quadratic differentials and reserve the term *trajectories* for the horizontal ones.

Here we only consider rational quadratic differentials, i. e., $Y = \mathbb{CP}^1$. Any such quadratic differential Ψ will be globally given in $\mathbb{C}$ by $\phi(z)dz^2$, where $\phi(z)$ is a complex-valued rational function.

Trajectories of Ψ can be naturally parameterized by their arclength. In fact, in a neighborhood of a regular point z_0 on $\mathbb{C}$, one can introduce a local coordinate called a *canonical parameter* and given by

$$w(z) := \int_{z_0}^{z} \sqrt{\phi(\xi)}d\xi.$$

One can easily check that $dw^2 = \phi(z)dz^2$ implying that horizontal trajectories in the z-plane correspond to horizontal straight lines in the w-plane, i. e., they are defined by the condition $\Im w = \text{const}$.

A trajectory of a meromorphic quadratic differential Ψ given on a compact Y without boundary is called *critical* if there exists a finite critical point of Ψ belonging to its closure. For a given quadratic differential Ψ on a compact surface Y, denote by $K_\Psi \subset Y$ the union of all its critical trajectories and critical points. In general, K_Ψ can be very complicated. In particular, it can be dense in some subdomains of Y.

We say that a critical trajectory is *finite* if it approaches finite critical points in both directions, i. e., its both endpoints are finite critical points. We denote by $DK_\Psi \subseteq K_\Psi$ (the closure of) the set of finite critical trajectories of (7.16). (One can easily show that DK_Ψ is an imbedded (multi)graph in Y. Here by a *multigraph* on a surface we mean a graph with possibly multiple edges and loops.) Finally, denote by $DK_\Psi^0 \subseteq DK_\Psi$ the subgraph of DK_Ψ consisting of (the closure of) the set of finite critical trajectories whose both ends are zeros of Ψ.

A non-critical trajectory $\gamma_{z_0}(t)$ of a meromorphic Ψ is called *closed* if $\exists T > 0$ such that $\gamma_{z_0}(t+T) = \gamma_{z_0}(t)$ for all $t \in \mathbb{R}$. The least such T is called the *period* of γ_{z_0}. A quadratic differential Ψ on a compact Riemann surface Y without boundary is called *Strebel* if the set of its closed trajectories covers Y up to a set of Lebesgue measure zero.

7.4.2 Cauchy transforms satisfying quadratic equations

In this subsection we relate the question for which triples of polynomials (P, Q, R) equation (7.15) admits a real motherbody measure to a certain problem about rational quadratic differentials. We start with the following necessary condition.

Proposition 7.1. *Assume that equation* (7.15) *admits a real motherbody measure μ. Then the following two conditions hold:*

(i) *Any connected curve in the support of μ coincides with a horizontal trajectory of the quadratic differential*

$$\Theta = -\frac{D(z)}{P^2(z)} dz^2, \tag{7.16}$$

where $D(z) = 4P(z)R(z) - Q^2(z)$ is the usual discriminant of equation (7.15).

(ii) *the support of μ should include all the branch points of* (7.15).

Remark 7.4. Observe that if $P(z)$ and $Q(z)$ are coprime, the set of all branch points coincides with the set of all zeros of $D(z) = 4P(z)R(z) - Q^2(z)$. In particular, requirement (ii) of Proposition 7.1 implies that the set DK_Θ^0 for the differential Θ should contain all zeros of $D(z)$.

Proof. The fact that every curve in supp(μ) should coincide with some horizontal trajectory of (7.16) is well known and follows from the Plemelj–Sokhotsky formula, see, e. g., [154]. It is based on the local observation that if a real measure $\mu = \frac{1}{\pi}\frac{\partial \mathcal{C}}{\partial \bar{z}}$ is supported on a smooth curve γ, then the tangent to γ at any point $z_0 \in \gamma$ should be perpendicular to $\overline{\mathcal{C}_1(z_0)} - \overline{\mathcal{C}_2(z_0)}$, where $\mathcal{C}_1$ and $\mathcal{C}_2$ are the one-sided limits of $\mathcal{C}$ when $z \to z_0$, see, e. g., [27]. (Here $^-$ stands for the usual complex conjugation.) Solutions of (7.15) are given by

$$\mathcal{C}_{1,2} = \frac{-Q(z) \pm \sqrt{Q^2(z) - 4P(z)R(z)}}{2P(z)},$$

their difference being

$$\mathcal{C}_1 - \mathcal{C}_2 = \frac{\sqrt{Q^2(z) - 4P(z)R(z)}}{P(z)}.$$

Being orthogonal to $\overline{\mathcal{C}_1(z_0)} - \overline{\mathcal{C}_2(z_0)}$, the tangent line to the support of the real motherbody measure μ satisfying (7.15) at its (arbitrary) smooth point z_0, is given by the condition $\frac{4P(z_0)R(z_0) - Q^2(z_0)}{P^2(z_0)} dz^2 > 0$. The latter condition is exactly the one defining the horizontal trajectory of Θ at z_0.

Finally, the observation that supp μ should contain all branch points of (7.15) follows immediately from the fact that $\mathcal{C}_\mu$ is a well-defined univalent function in $\mathbb{C} \setminus \operatorname{supp} \mu$. □

In many special cases statements similar to Proposition 7.1 can be found in the literature, see, e. g., [16] and the references therein.

Proposition 7.1 allows us, under mild non-degeneracy assumptions, to formulate necessary and sufficient conditions for the existence of a motherbody measure for (7.15), which, however, are difficult to verify. Namely, let $\Gamma \subset \mathbb{CP}^1_{\mathcal{C}} \times \mathbb{CP}^1_z$ with coordinates $(\mathcal{C}, z)$ be the algebraic curve given by (the projectivization of) equation (7.15). Γ has bidegree

$(2, n+2)$ and is hyperelliptic. Let $\pi_z : \Gamma \to \mathbb{C}$ be the projection of Γ along the $\mathcal{C}$-coordinate onto the z-plane $\mathbb{CP}^1$. From (7.15) we observe that π_z induces a branched double covering of $\mathbb{CP}^1$ by Γ. If $P(z)$ and $Q(z)$ are coprime and if $\deg D(z) = 2n+2$, the set of all branch points of $\pi_z : \Gamma \to \mathbb{CP}^1$ coincides with the set of all zeros of $D(z)$. (If $\deg D(z) < 2n+2$, then ∞ is also a branch point of π_z of multiplicity $2n+2-\deg D(z)$.) We need the following lemma.

Lemma 7.5. *If $P(z)$ and $Q(z)$ are coprime, then at each pole of* (7.15), *i. e., at each zero of $P(z)$, only one of two branches of Γ goes to ∞. Additionally, the residue of this branch at this zero equals that of $-\frac{Q(z)}{P(z)}$.*

Proof. Indeed if $P(z)$ and $Q(z)$ are coprime, then no zero z_0 of $P(z)$ can be a branch point of (7.15) since $D(z_0) \neq 0$. Therefore only one of two branches of Γ goes to ∞ at z_0. More exactly, the branch $\mathcal{C}_1 = \frac{-Q(z)+\sqrt{Q^2(z)-4P(z)R(z)}}{2P(z)}$ attains a finite value at z_0 while the branch $\mathcal{C}_2 = \frac{-Q(z)-\sqrt{Q^2(z)-4P(z)R(z)}}{2P(z)}$ goes to ∞. (Here we use the agreement that $\lim_{z\to z_0} \sqrt{Q^2 - 4P(z)R(z)} = Q(z_0)$.) Now consider the residue of the branch $\mathcal{C}_2$ at z_0. Since residues depend continuously on the coefficients $(P(z), Q(z), R(z))$, it suffices to consider only the case when z_0 is a simple zero of $P(z)$. Further, if z_0 is a simple zero of $P(z)$, then

$$Res(\mathcal{C}_2, z_0) = \frac{-2Q(z_0)}{2P'(z_0)} = Res\left(-\frac{Q(z)}{P(z)}, z_0\right). \qquad \square$$

Besides the obvious condition that (7.15) has a real branch near ∞ with the asymptotics $\frac{\alpha}{z}$ for some $\alpha \in \mathbb{R}$, by Proposition 7.1 the necessary condition for (7.15) to admit a motherbody measure is that the set DK_Θ^0 for the differential (7.16) contains all branch points of (7.15), i. e., all zeros of $D(z)$. Consider $\Gamma_{cut} = \Gamma \setminus \pi_z^{-1}(DK_\Theta^0)$. Since DK_Θ^0 contains all branch points of π_z, Γ_{cut} consists of some number of open sheets each projecting diffeomorphically on its image in $\mathbb{CP}^1 \setminus DK_\Theta^0$. (The number of sheets in Γ_{cut} equals twice the number of connected components in $\mathbb{C} \setminus DK_\Theta^0$.) Observe that since we have chosen a real branch of (7.15) at infinity with the asymptotics $\frac{\alpha}{z}$, we have a marked point $p_{br} \in \Gamma$ over ∞. If we additionally assume that $\deg D(z) = 2n+2$, then ∞ is not a branch point of π_z, and therefore $p_{br} \in \Gamma_{cut}$.

Lemma 7.6. *If $\deg D(z) = 2n+2$, then any choice of a spanning (multi)subgraph $G \subset DK_\Theta^0$ without isolated vertices induces the unique choice of the section Sc_G of the curve Γ over $\mathbb{CP}^1 \setminus G$ which*

a) *Contains p_{br}.*
b) *Is discontinuous at any point of G.*
c) *Is projected by π_z diffeomorphically onto $\mathbb{CP}^1 \setminus G$.*

Here by a spanning subgraph we mean a subgraph containing all the vertices of the ambient graph. By a *section* of Γ over $\mathbb{CP}^1 \setminus G$ we mean a choice of one of two possible values of Γ at each point in $\mathbb{CP}^1 \setminus G$.

Proof. Obvious. □

Observe that the section Sc_G might attain the value ∞ at some points, i. e., contain some poles of (7.15). Denote the set of poles of Sc_G by $Poles_G$. Now we can formulate our necessary and sufficient conditions.

Theorem 7.5. *Assume that the following conditions are valid:*

(i) *Equation* (7.15) *has a real branch near ∞ with the asymptotic behavior $\frac{\alpha}{z}$ for some $\alpha \in \mathbb{R}$, comp. Lemma* 7.1.
(ii) *$P(z)$ and $Q(z)$ are coprime, and the discriminant $D(z) = Q^2(z) - 4P(z)R(z)$ of equation* (7.15) *has degree* $2n + 2$.
(iii) *The set DK_Θ^0 for quadratic differential Θ given by* (7.16) *contains all zeros of $D(z)$.*
(iv) *Θ has no closed horizontal trajectories.*

Then (7.15) *admits a real motherbody measure if and only if there exists a spanning (multi)subgraph $G \subseteq DK_\Theta^0$ with no isolated vertices such that all poles in $Poles_g$ are simple and all their residues are real, see notation above.*

Proof. Indeed assume that (7.15) satisfying (ii) admits a real motherbody measure μ. Assumption (i) is obviously necessary for the existence of a real motherbody measure. The necessity of assumption (iii) follows from Proposition 7.1 if (ii) is satisfied. The support of μ consists of a finite number of curves and possibly a finite number of isolated points. Since each curve in the support of μ is a trajectory of Θ and Θ has no closed trajectories, then the whole support of μ consists of finite critical trajectories of Θ connecting its zeros, i. e., the support belongs to DK_Θ^0. Moreover the support of μ should contain sufficiently many finite critical trajectories of Θ such that they include all the branch points of (7.15). By (ii) these are exactly all zeros of $D(z)$. Therefore the union of finite critical trajectories of Θ belonging to the support of μ is a spanning (multi)graph of DK_Θ^0 without isolated vertices. The isolated points in the support of μ are necessarily the poles of (7.15). Observe that the Cauchy transform of any (complex-valued) measure can only have simple poles (as opposed to the Cauchy transform of a more general distribution). Since μ is real the residue of its Cauchy transform at each pole must be real as well. Therefore the existence of a real motherbody under assumptions (i)–(iv) implies the existence of a spanning (multi)graph G with the above properties. The converse is also immediate. □

Remark 7.5. Observe that if (i) is valid, then assumptions (ii) and (iv) are generically satisfied. Notice, however, that (iv) is violated in the special case when $Q(z)$ is absent considered in § 7.4.3. Additionally, if (iv) is satisfied, then the number of possible motherbody measures is finite. On the other hand, it is assumption (iii) that imposes severe additional restrictions on admissible triples $(P(z), Q(z), R(z))$. At the moment there is no information about possible cardinalities of the sets $Poles_G$ introduced above. Thus it is difficult to estimate the number of conditions required for (7.15) to admit a motherbody measure. Theorem 7.5, however, leads to the following sufficient condition for the existence of a real motherbody measure for (7.15).

Corollary 7.2. *If, additionally to assumptions (i)–(iii) of Theorem 7.5, one assumes that all roots of $P(z)$ are simple and all residues of $\frac{Q(z)}{P(z)}$ are real, then (7.15) admits a real motherbody measure.*

Proof. Indeed if all roots of $P(z)$ are simple and all residues of $\frac{Q(z)}{P(z)}$ are real, then all poles of (7.15) are simple with real residues. In this case for any choice of a spanning (multi)subgraph G of DK^0_Θ, there exists a real motherbody measure whose support coincides with G plus possibly some poles of (7.15). Observe that if all roots of $P(z)$ are simple and all residues of $\frac{Q(z)}{P(z)}$ are real, one can omit assumption (iv). In case when Θ has no closed trajectories, all possible real motherbody measures are in a bijective correspondence with all spanning (multi)subgraphs of DK^0_Θ without isolated vertices. In the opposite case such measures are in a bijective correspondence with the unions of a spanning (multi)subgraph of DK^0_Θ and an arbitrary (possibly empty) finite collection of closed trajectories. □

Although we at present do not have rigorous results about the structure of the set of general equations (7.15) admitting a real motherbody measure, we think that generalizing our previous methods and statements from [39, 84], and [168], one would be able to settle the following conjecture.

Conjecture 7.2. *Fix any monic polynomial $P(z)$ of degree $n+2$ and an arbitrary polynomial $Q(z)$ of degree at most $n+1$. Let $\Omega_{P,Q}$ denote the set of all polynomials $R(z)$ of degree at most n such that (7.15) admits a probability measure, i. e., a positive motherbody measure of mass 1. Then $\Omega_{P,Q}$ is a real semi-analytic variety of (real) dimension n.*

7.4.3 Case $Q = 0$ and Strebel differentials

In this subsection we present in more detail the situation when the middle term in (7.15) is vanishing, i. e., $Q(z) = 0$. In this case one can obtain complete information about the number of possible signed motherbody measures and also a criterion of the existence of a positive motherbody measure.

It is known that for a meromorphic Strebel differential Ψ given on a compact Riemann surface Y without boundary, the set K_Ψ has several nice properties. In particular, it is a finite embedded multi-graph on Y whose edges are finite critical trajectories. In other words, for a Strebel quadratic differential Ψ, one gets $K_\Psi = DK_\Psi$.

Our next result relates a Strebel differential Ψ on $\mathbb{CP}^1$ with a double pole at ∞ to real-valued measures supported on K_Ψ.

Theorem 7.6. (i) Given two coprime polynomials $P(z)$ and $R(z)$ of degrees $n+2$ and n respectively where $P(z)$ is monic and $R(z)$ has a negative leading coefficient, the algebraic function given by the equation

$$P(z)\mathcal{C}^2 + R(z) = 0 \tag{7.17}$$

admits a motherbody measure μ_C if and only if the quadratic differential $\Psi = R(z)dz^2/P(z)$ is Strebel.

(ii) Such motherbody measures are, in general, non-unique. If we additionally require that the support of each such measure consists only of finite critical trajectories, i. e., is a spanning subgraph of $K_\Psi = DK_\Psi$, then for any Ψ as above, there exists exactly 2^{d-1} real measures where d is the total number of connected components in $\mathbb{CP}^1 \setminus K_\Psi$. These measures are in 1 – 1-correspondence with 2^{d-1} possible choices of the branches of $\sqrt{-R(z)/P(z)}$ in the union of $(d-1)$ bounded components of $\mathbb{CP}^1 \setminus K_\Psi$, see Figure 7.2.

Concerning possible positive measures, we can formulate an exact criterion of the existence of a positive measure for a rational quadratic differential $\Psi = R(z)dz^2/P(z)$ in terms of rather simple topological properties of K_Ψ. To do this, we need a few definitions. Notice that K_Ψ is a planar multi-graph with the following properties. The vertices of K_Ψ are the affine (i. e., different from ∞) critical points of Ψ, and its edges are critical trajectories connecting these critical points. Each (open) connected component of $\mathbb{C} \setminus K_\Psi$ is homeomorphic to an (open) annulus. K_Ψ might have isolated vertices that are the affine double poles of Ψ. Vertices of K_Ψ having valency 1 (i. e., hanging vertices) are exactly the simple poles of Ψ. Vertices different from the isolated and hanging vertices are the zeros of Ψ. The number of edges adjacent to a given vertex minus 2 is equal to the order of the zero of Ψ at this point. Finally, the sum of the multiplicities of all poles (including the one at ∞) minus the sum of the multiplicities of all zeros equals 4.

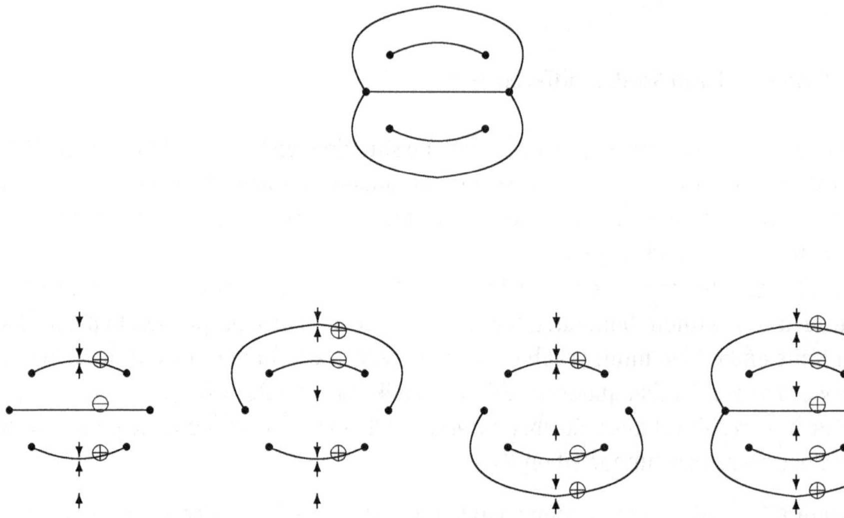

1st measure 2nd measure 3rd measure 4th measure

Figure 7.2: The set $K_\Psi = DK_\Psi$ of a Strebel differential Ψ and four related real measures.

By a *simple cycle* in a planar multi-graph K_Ψ, we mean any closed non-selfintersecting curve formed by the edges of K_Ψ. (Obviously, any simple cycle bounds an open domain homeomorphic to a disk, which we call the *interior of the cycle*.)

Proposition 7.2. *A Strebel differential* $\Psi = R(z)dz^2/P(z)$ *admits a positive motherbody measure if and only if no edge of* K_Ψ *is attached to a simple cycle from inside. In other words, for any simple cycle in* K_Ψ *and any edge not in the cycle, but adjacent to some vertex in the cycle, this edge does not belong to its interior. The support of the positive measure coincides with the forest obtained from* K_Ψ *after the removal of all its simple cycles.*

Remark. Notice that under the assumptions of Proposition 7.2, all simple cycles of K_Ψ are pairwise non-intersecting and, therefore, their removal is well defined in an unambiguous way.

In particular, the compact set shown on the rightmost picture of Figure 7.3 admits no positive measure since it contains an edge cutting a simple cycle (the outer boundary) in two smaller cycles. The left picture on Figure 7.3 has no such edges and, therefore, admits a positive measure (whose support consists of the four horizontal edges of K_Ψ).

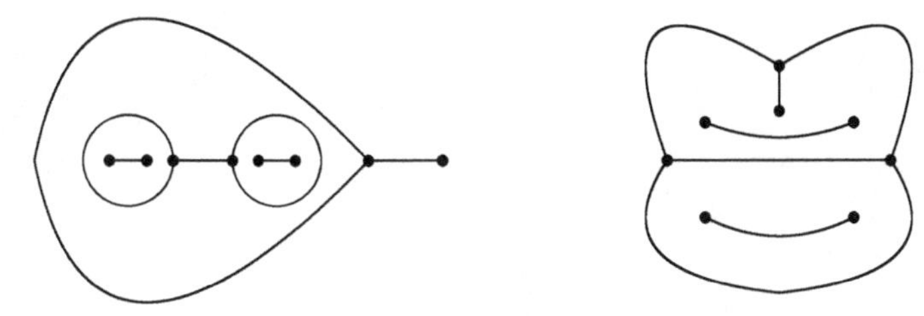

Figure 7.3: Two examples of K_Ψ admitting and not admitting a positive measure.

Let us finish the chapter with the following important observation.

Proposition 7.3. *For any monic* $P(z)$ *of degree* $n+2$, *the set of all polynomials* $R(z)$ *of degree* n *and with leading coefficient* -1 *such that the differential* $\frac{R(z)dz^2}{P(z)}$ *is Strebel, is dense in the space of all polynomials of degree* n *with leading coefficient* -1. *In fact, this set is the countable union of real semi-analytic varieties of positive codimension.*

This circumstance illustrates the difficulty of the general problem to determine for which algebraic equations a real motherbody measure exists.

7.5 Open problems in Chapter 7

7.1. The natural question of which algebraic functions of degree exceeding 2 with Newton polygon intersecting the diagonal in the $(\mathcal{C}, z)$-plane non-trivially admit a real motherbody measure is hard to answer. Some steps in this direction can be found in [84].

This topic is apparently closely related to the (non-existing) notion of Strebel differential of higher order, which we approach in Chapter 12. In any case, it is clear that no results similar to Theorem 7.2 can be true, and one needs to impose highly non-trivial additional conditions on such functions to ensure the existence of a probability measure.

7.2. Already very basic questions about the set of polynomials admitting a motherbody measure are widely open. We state the simplest of them below.

Problem 7.3. Given a finite set S of monomials satisfying the assumptions of Lemma 7.1 and Lemma 7.2, consider the linear space Pol_S of all polynomials $P(\mathcal{C}, z)$ whose Newton polygon is contained in S. What is the (Hausdorff) dimension of the subset $MPol_S \subseteq Pol_S$ of polynomials admitting a motherbody measure?

Part II: **Heine–Stieltjes problem**

8 Classical and generalized Heine–Stieltjes problem

In Part II of this book we consider polynomial solutions that occur for discrete values of parameters in special families of linear differential operators with polynomial coefficients. Such families were first considered in the middle of the 19th century by E. Heine and T. J. Stieltjes. Below we generalize many of their results from the case of second order operators to operators of arbitrary order. However, there are still many possible further directions in which their set-up/results can be extended and which are left outside our scope.

(The material of this chapter is based on [162].)

8.1 Preliminaries

As we mentioned in Chapter 1, the algebraic form of the classical Lamé equation is

$$Q(z)\frac{d^2S}{dz^2} + \frac{1}{2}Q'(z)\frac{dS}{dz} + V(z)S = 0, \tag{8.1}$$

where $Q_l(z)$ is a real polynomial of degree l with all real and distinct roots, and $V(z)$ is a polynomial of degree at most $l - 2$ whose choice depends on what type of solution to (8.1) we are looking for. In what follows, we will concentrate on the usual polynomial solutions of (8.1).

A *generalized Lamé equation* (see [197]) is the second order differential equation given by

$$Q_2(z)\frac{d^2S}{dz^2} + Q_1(z)\frac{dS}{dz} + V(z)S = 0, \tag{8.2}$$

where $Q_2(z)$ is a complex polynomial of degree l and $Q_1(z)$ is a complex polynomial of degree at most $l - 1$. The special case $l = 3$ is widely known as the Heun equation.

The following fundamental result announced in [82], and provided there with not quite satisfactory proof, was undoubtedly the starting point of the classical Heine–Stieltjes theory.

Theorem 8.1 (Heine). *If the coefficients of $Q_1(z)$ and $Q_2(z)$ are algebraically independent, then for any integer $n > 0$, there exists exactly $\binom{n+l-2}{n}$ polynomials $V(z)$ of degree exactly $(l - 2)$ such that equation (8.2) has a unique (up to a constant factor) polynomial solution S of degree exactly n.*

Remark 8.1. Notice that throughout this chapter, we count polynomials $V(z)$ per se and polynomials $S(z)$ up to a constant factor since they solve the homogeneous linear differential equation.

https://doi.org/10.1515/9783112207741-010

Later on, T. Stieltjes and his followers considered a special case of (8.2), which is important in physics and is closely related to the original equation (8.1). Namely, they studied the equation of the form

$$\prod_{i=1}^{l}(z-\alpha_i)\frac{d^2S}{dz^2}+\sum_{j=1}^{l}\beta_j\prod_{j\neq i}(z-\alpha_i)\frac{dS}{dz}+V(z)S=0, \tag{8.3}$$

with real and distinct $\alpha_1, \alpha_2, \ldots, \alpha_l$ ordered as $\alpha_1 < \alpha_2 < \cdots < \alpha_l$ and real positive $\beta_1, \ldots, \beta_l$. In particular, they proved the following important result.

Theorem 8.2 (Stieltjes–Van Vleck–Bôcher [176, 191, 43], and [181]). *Under the assumptions of* (8.3)*, for any integer* $n > 0$*, one has the following:*

1. *There exist exactly* $\binom{n+l-2}{n}$ *distinct polynomials* V *of degree* $(l-2)$ *such that equation* (8.3) *has a polynomial solution* S *of degree exactly* n*.*
2. *Every root of each* V *and each* S *is real and simple, and belongs to the interval* (α_1, α_l)*.*
3. *None of the roots of* S *can coincide with some of the* α_i*s. Moreover,* $\binom{n+l-2}{n}$ *polynomials* S *are in 1–1-correspondence with* $\binom{n+l-2}{n}$ *possible ways to distribute* n *roots of* S *into the* $(l-1)$ *open intervals* $(\alpha_1, \alpha_2), (\alpha_2, \alpha_3), \ldots, (\alpha_{l-1}, \alpha_l)$*.*

The polynomials V and the corresponding polynomial solutions S of equation (8.2) (or, equivalently, of (8.3)) are called *Van Vleck* and *Stieltjes* polynomials, respectively.

The case when α_is and/or β_js are not real is substantially less studied, see [129] and [135]. One nice result in this case is as follows (see [150]).

Theorem 8.3 (Pólya). *If in the notation of* (8.3) *not all* α_i*s lie on the same line and all* β_j*s are positive, then all roots of each* V *and each* S *belong to the convex hull* $Conv_{Q_2}$ *of the root locus* $\{\alpha_1, \ldots, \alpha_l\}$ *of* $Q_2(z)$*.*

Remark 8.2. The situation when all the numbers β_j are negative (for example, when $Q_1(z) = -Q_2'(z)$) or have different signs seems to differ drastically from the latter case, see, e. g., [59] and [64]. Further interesting results on the distribution of the zeros of Van Vleck and Stieltjes polynomials under weaker assumptions on α_is and β_js were obtained in [124, 125, 3, 4, 198].

In this chapter, we extend the above three fundamental results about generalized Lamé equations of the second order to the case of higher orders and/or complex coefficients. Namely, consider an arbitrary linear ordinary differential operator

$$T=\sum_{i=1}^{k}Q_i(z)\frac{d^i}{dz^i}, \tag{8.4}$$

with polynomial coefficients. As above the number $r = \max_{i=1,\ldots,k}(\deg Q_i(z) - i)$ will be called the *Fuchs index* of T. The operator T is called a *higher Lamé operator* if its Fuchs index r is non-negative. Notice that if $r = 0$, then T is exactly solvable; the case we considered in Part I of this book.

The operator T is called *non-degenerate* if $\deg Q_k(z) = k + r$. Notice that non-degeneracy of T is a quite natural condition equivalent to the requirement that T has either a regular or a regular singular point at ∞.

Given a higher Lamé operator T, consider the following multi-parameter spectral problem.

Problem 8.1. For each positive integer n, find all polynomials $V(z)$ of degree at most r such that the equation

$$TS(z) + V(z)S(z) = 0 \tag{8.5}$$

has a polynomial solution $S(z)$ of degree n.

Following the classical terminology, we call (8.5) a *higher Heine–Stieltjes spectral problem*, $V(z)$ is called a *higher Van Vleck polynomial*, and the corresponding polynomial $S(z)$ is called a *higher Stieltjes polynomial*. Below, we will often skip mentioning "higher". In what follows, we will always assume that the leading coefficient of T is a monic polynomial.

8.1.1 Generalizations of Heine's theorems, degeneracies, and non-resonance condition

We start with a number of generalizations of Heine's Theorem 8.1. Following Heine's original proof, one can obtain the following straightforward generalization.

Theorem 8.4. *For any non-degenerate higher Lamé operator T with algebraically independent coefficients of its polynomial coefficients $Q_i(z)$, $i = 1, \ldots, k$ and for any $n \geq 0$, there exist exactly $\binom{n+r}{r}$ distinct Van Vleck polynomials $V(z)$ whose corresponding Stieltjes polynomials $S(z)$ are unique (up to a constant factor) and have degree n. As above, r is the Fuchs index of T.*

Our next result, obtained by a linear-algebraic interpretation of (8.5), has no genericity assumptions and is crucial in the problem concerning the existence of solutions to (8.5) (compare [135, Problem 1]).

Theorem 8.5. *For any non-degenerate operator T with the Fuchs index $r \geq 0$ and any positive integer n, the total number of Van Vleck polynomials $V(z)$ (counted with natural multiplicities) that have a Stieltjes polynomial $S(z)$ of degree at most n equals $\binom{n+r+1}{r+1}$.*

Remark 8.3. Note that in Theorem 8.5, we do not require that there is a unique (up to a constant factor) Stieltjes polynomial corresponding to a given Van Vleck polynomial. For the exact (rather lengthy) description of the notion of the *natural multiplicity* of a Van Vleck polynomial consult Definition 8.2 in § 8.2.

8.1.2 More on degeneracies

Notice that Theorem 8.4 claims that for any positive n, a generic operator T has exactly $\binom{n+r}{r}$ distinct Van Vleck polynomials, each of which has a unique Stieltjes polynomial and this polynomial is of degree exactly n. If we drop the genericity assumptions on T, then the question about possible degeneracies occurring in Problems (8.2) and (8.5) is quite delicate. It may happen that Van Vleck polynomials attain non-trivial multiplicities or that the corresponding Stieltjes polynomials form linear spaces of dimension bigger than one, or Stieltjes polynomials of certain degrees do not exist at all. In particular, for any polynomial $Q(z)$ of degree l, no choice of a polynomial $V(z)$ of degree at most $l-2$ will supply the equation

$$Q(z)\frac{d^2S}{dz^2} - Q'(z)\frac{dS}{dz} + V(z)S = 0$$

with a polynomial solution S of degree $l+1$. (This follows from Proposition 5 and Lemma 4 of [64].) The fact that (8.2) can admit families (i. e., linear spaces of dimension at least 2) of polynomial solutions S corresponding to one and the same V was already mentioned by Heine in his original proof. More exact information is available nowadays. For example, a result of Varchenko–Scherbak gives necessary and sufficient condition for a Fuchsian second order equation to have two independent polynomial solutions, see [161] and [64]. Finally, high multiplicity of Van Vleck polynomials occurs, for example, in the case $Q_2(z) = z^l$, $Q_1(z) = 0$. Then one can easily show that for all $n \geq 2$, there exists just one and only polynomial $V_n(z) = -n(n-1)z^{l-2}$ solving the above problem; its corresponding Stieltjes polynomial equals $S_n(z) = z^n$. The multiplicity of the latter Van Vleck polynomial $V_n(z)$ is $\binom{n+l-2}{n}$.

To formulate necessary and sufficient conditions under which the conclusion of Heine's theorem holds for all positive integers n is apparently an impossible task. Heine himself mentions that for the validity of his result for a given fixed positive integer n, one has to avoid a certain discriminantal hypersurface (similar to the usual discriminant of univariate polynomials), but this hypersurface is difficult to obtain explicitly.

Below, we formulate a simple sufficient condition that allows us to avoid many of the above degeneracies and guarantees the existence of Stieltjes polynomials of a given degree. Namely, consider an arbitrary non-degenerate operator T of the form (8.4) with the Fuchs index r. Denote by $A_k, A_{k-1}, \ldots, A_1$ the coefficients at the highest possible degrees $k+r, k+r-1, \ldots, r+1$ in the polynomials $Q_k(z), Q_{k-1}(z), \ldots, Q_1(z)$ resp. (Notice that any subset of A_js can vanish but $A_k \neq 0$ due to the non-degeneracy of T.) Recall that

$$(j)_i = j(j-1)(j-2)\cdots(j-i+1),$$

stands for the Pochhammer symbol where j is a non-negative and i is a positive integer. In case $j = i$ one has $(j)_i = j!$ and in case $j < i$ one gets $(j)_i = 0$. For any non-negative n, we call the expression

$$\mathbb{L}_n = (n)_k A_k + (n)_{k-1} A_{k-1} + \cdots + (n)_1 A_1 \tag{8.6}$$

the nth diagonal coefficient.

Proposition 8.1. *Let n be a positive integer. If in the above notation, the nth non-resonance condition holds, that is,*

$$\mathbb{L}_n \neq \mathbb{L}_j, \quad j = 0, 1, \ldots, n-1, \tag{8.7}$$

then there exist Van Vleck polynomials that possess Stieltjes polynomials of degree exactly n and no other Stieltjes polynomials of degrees smaller than n. In this case, the total number of such Van Vleck polynomials (counted with natural multiplicities) equals $\binom{n+r}{r}$.

Remark 8.4. The above non-resonance condition is quite natural. It says that if equation (8.5) has a polynomial solution of degree n, then it has no polynomial solutions of smaller degrees. Another way to express this fact is that if the indicial equation of (8.5) at ∞ has $-n$ as its root, then it has no roots among non-positive integers $0, -1, -2, \ldots, 1-n$, see, e. g., [153, Chapter V].

The explicit formula (8.6) for $\mathbb{L}_n$ immediately shows that Theorem 8.5 and Proposition 8.1 are valid for any non-degenerate T and all sufficiently large n.

Corollary 8.1. *For any non-degenerate higher Lamé operator T and all sufficiently large n, the nth non-resonance condition holds. In particular, for any problem (8.5), there exist finitely many (up to a scalar multiple) Stieltjes polynomials of any sufficiently large degree.*

Remark 8.5. Notice that for an arbitrary non-degenerate operator T and a given integer n, it is difficult to find explicitly all Van Vleck polynomials that possess a Stieltjes polynomial of degree at most n. By this we mean that in order to do this one has, in general, to solve an overdetermined system of algebraic equations in the coefficients of V since the set of Van Vleck polynomials under consideration is not a complete intersection. (In other words, this system of determinantal equations contains many more equations than variables.) However, one consequence of Heine's recipe to prove Theorem 8.1 is as follows. For a given non-degenerate operator T with Fuchs index r and a positive integer n, denote by $\mathfrak{V}_n \subset Pol_r$ the set of all its Van Vleck polynomials possessing a Stieltjes polynomial of degree exactly n.

Theorem 8.6. *If in the above notation the nth non-resonance condition (8.7) holds, then $\mathfrak{V}_n$ is a complete intersection and the corresponding system of equations can be obtained explicitly in each concrete case.*

An explicit example of the defining system of r algebraic equations in r variables can be found in § 8.2, see Example 8.1. In a purely linear algebraic setting, this result and further information about relevant discriminants can be found in [165].

8.1.3 Generalizations of Stieltjes' theorem

We continue with a conceptually new generalization of Theorem 8.2. It was conjectured by the author after extensive computer experiments and was later proved by P. Bränden in [48]. Here we only quote a somewhat simplified version of his result. For further details consult loc. cit.

Definition 8.1. A differential operator $T = \sum_{i=m}^{k} Q_i(z)\frac{d^i}{dz^i}$, $1 \le m \le k$, where all $Q_i(z)$s are polynomials with real coefficients, is called a *strict hyperbolicity preserver* if for any real polynomial $P(z)$ with all real and simple roots, the image $\mathfrak{d}(P(z))$ either vanishes identically or is a polynomial with only real and simple roots.

Theorem 8.7. *For any strict hyperbolicity preserving non-degenerate Lamé operator T with the Fuchs index r as above and any integer $n \ge m$, we have:*
1. *There exist exactly $\binom{n+r}{n}$ distinct polynomials $V(z)$ of degree exactly r such that equation (8.5) has a polynomial solution $S(z)$ of degree exactly n.*
2. *All roots of each such $V(z)$ and $S(z)$ are real, simple, and coprime.*
3. *$\binom{n+r}{n}$ polynomials $S(z)$ are in 1–1-correspondence with $\binom{n+r}{n}$ possible arrangements of r real roots of polynomials V(z) and n real roots of the corresponding polynomials S(z).*

Using Theorem 8.5, one immediately sees that the latter result describes the set of all possible pairs (V, S) with $m \le n = \deg S$ for any hyperbolicity preserver T.

Remark 8.6. The interested reader can check that the sum of the first two terms in (8.3) is indeed a strict hyperbolicity preserver. It looks very tempting and important to find an analog of the electrostatic interpretation of the roots of classical Heine–Stieltjes and classical Van Vleck polynomials (alias 'Bethe ansatz') in the case of higher Heine–Stieltjes and Van Vleck polynomials (compare [123]).

Remark 8.7. Notice that the converse to the above theorem is false. Namely, one can show that the ES-operator $T(f) = f' + z(z+1)f''$ has all hyperbolic eigenpolynomials but is not a hyperbolicity preserver.

A straightforward application of Theorem 8.7 implies the following. Consider a differential equation

$$\prod_{i=1}^{l}(z-\alpha_i)\frac{d^kS}{dz^k} + \sum_{j=1}^{l}\beta_j\prod_{j\neq i}(z-\alpha_i)\frac{d^{k-1}S}{dz^{k-1}} + V(z)S = 0, \tag{8.8}$$

where $2 \le k \le l$, $\alpha_1 < \alpha_2 < \cdots < \alpha_l$ and $\beta_1, \ldots, \beta_l$ are positive.

Corollary 8.2. *Under the above assumptions concerning (8.8) and for any $n \ge k-1$:*
1. *There exist exactly $\binom{n+l-k}{n}$ polynomials $V(z)$ of degree $(l-k)$ such that equation (8.8) has a polynomial solution $S(z)$ of degree exactly n.*

2. *All roots of each $V(z)$ and $S(z)$ are real, simple, coprime and belong to the interval (α_1, α_l).*
3. *$\binom{n+l-k}{n}$ polynomials $S(z)$ are in 1–1-correspondence with $\binom{n+l-k}{n}$ possible arrangements of $(l-k)$ real roots of a polynomial $V(z)$ and n real roots of the corresponding polynomial $S(z)$ on the interval (α_1, α_l).*

It seems that Theorem 8.7 and Corollary 8.2 give a new interpretation of Theorem 8.2 even in the classical case (8.3). However, the following two statements provén by G. Shah explain this mystery (see Theorem 3 of [162] and Theorem 3 of [163]).

Proposition 8.2. *Under the assumptions of Theorem 8.2, the roots of any Van Vleck polynomial $V(z)$ and its corresponding $S(z)$ are coprime.*

Moreover, the following result is valid.

Proposition 8.3. *If $v_1 < v_2 < \cdots < v_r$, $r = l-2$ denote the roots of some Van Vleck polynomial $V(z)$ in the classical situation (8.3), then for each $i = 2, \ldots, l-1$, the interval (v_i, α_{i+1}) contains no roots of the corresponding $S(z)$. Therefore, for each polynomial $S(z)$, the distribution of its n roots into $(l-1)$ intervals $(\alpha_1, \alpha_2), \ldots, (\alpha_{l-1}, \alpha_l)$ coincides with the distribution of these roots defined by the roots of its Van Vleck polynomial $V(z)$.*

Remark 8.8. Note that we do not claim $v_i < \alpha_{i+1}$, i. e., the endpoints of the interval (v_i, α_{i+1}) may be placed in the wrong order or may coincide.

8.1.4 Generalizations of Pólya's theorem

We start with a simple-minded extension of Pólya's Theorem 8.3 [150].

Theorem 8.8. *If the constants $\beta_1, \ldots, \beta_l$ in (8.8) are non-negative, then all the roots of Vs and Ss lie in the (closed) convex hull $Conv_{Q_k}$ of the roots $(\alpha_1, \ldots, \alpha_l)$ of the polynomial $Q_k(z)$. (No assumptions on $(\alpha_1, \ldots, \alpha_l)$ are required.)*

The next Theorem 8.9 is far more general. It shows that Theorem 8.8 is *asymptotically* true for any non-degenerate Lamé operator. The question for which operators T the roots of all its Van Vleck and Stieltjes polynomials lie exactly (and not just asymptotically) in the convex hull of its leading coefficient seems to be very hard even in the classical case of equation (8.2).

Theorem 8.9. *For any non-degenerate higher Lamé operator T and any $\epsilon > 0$, there exists a positive integer N_ϵ such that the zeros of all Van Vleck polynomials $V(z)$ possessing a Heine–Stieltjes polynomial $S(z)$ of degree $n \geq N_\epsilon$ as well as all zeros of these Stieltjes polynomials belong to $Conv^\epsilon_{Q_k}$. Here $Conv_{Q_k}$ is the convex hull of all zeros of the leading coefficient Q_k and $Conv^\epsilon_{Q_k}$ is its ϵ-neighborhood in the usual Euclidean distance on $\mathbb{C}$.*

The latter theorem is closely related to the next somewhat simpler localization result having independent interest.

Proposition 8.4. *For any non-degenerate higher Lamé operator T, there exist a positive integer N_0 and a positive number R_0 such that all zeros of all Van Vleck polynomials $V(z)$ possessing a Stieltjes polynomial $S(z)$ of degree $n \geq N_0$ as well as all zeros of these Stieltjes polynomials lie in the disk $|z| \leq R_0$.*

Remark 8.9. Notice that the roots of all Van Vleck polynomials (and not just those whose Stieltjes polynomials are of sufficiently large degree) of any non-degenerate higher Lamé operator T lie in some disk. But this is no longer true for Stieltjes polynomials. If the set of all Stieltjes polynomials is discrete (up to a scalar multiple), then their roots are bounded. But as soon as some Van Vleck polynomial admits at least a 2-dimensional linear space of Stieltjes polynomials, then these roots become unbounded for obvious reasons. However, for sufficiently large n, no Van Vleck polynomial admits such families and the localization result holds, see Corollary 8.1.

In Figures 8.1–8.2 we show typical behavior of the zeros of Van Vleck polynomials and the corresponding Stieltjes polynomials obtained in numerical experiments. Below we consider an example of the operator $T = Q(z)\frac{d^3}{dz^3}$ with $Q(z) = (z^2+1)(z-3+i)(z-1-2i)$. For $n = 24$, we calculate all 25 pairs (V, S) with $\deg S = 24$ (notice that in this case V is linear). The asymptotic behavior of the union of zeros of all Van Vleck polynomials whose Stieltjes polynomials have a given degree n when $n \to \infty$ as well as the asymptotics of the zeros of subsequences of Stieltjes polynomials of increasing degrees whose corresponding (monic) Van Vleck polynomials have a limit seems to be an extremely rich and interesting topic, see first steps in Chapters 9–10 and [167, 170].

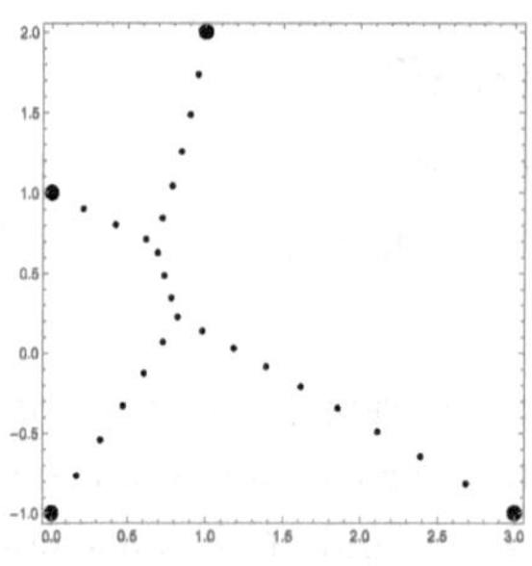

Figure 8.1: Zeros of 25 different linear Van Vleck polynomials whose Stieltjes polynomials are of degree 24. Four average size dots are the zeros of $Q(z) = (z^2+1)(z-3+i)(z-1-2i)$ and large dots are the zeros of different $V(z)$.

8.2 Proofs of generalized Heine's theorems

We start with Theorem 8.4. To settle it, we need a detailed description of the action of a non-degenerate operator T on the linear space Pol_n of all univariate polynomials of degree at most n.

Proof. Substituting $V(z) = v_r z^r + v_{r-1} z^{r-1} + \cdots + v_0$ and $S(z) = s_n z^n + s_{n-1} z^{n-1} + \cdots + s_0$ in (8.5), we get the following system of $(n+r+1)$ equations of a banded shape (i. e., only a fixed and independent of n number of diagonals is non-vanishing in this system):

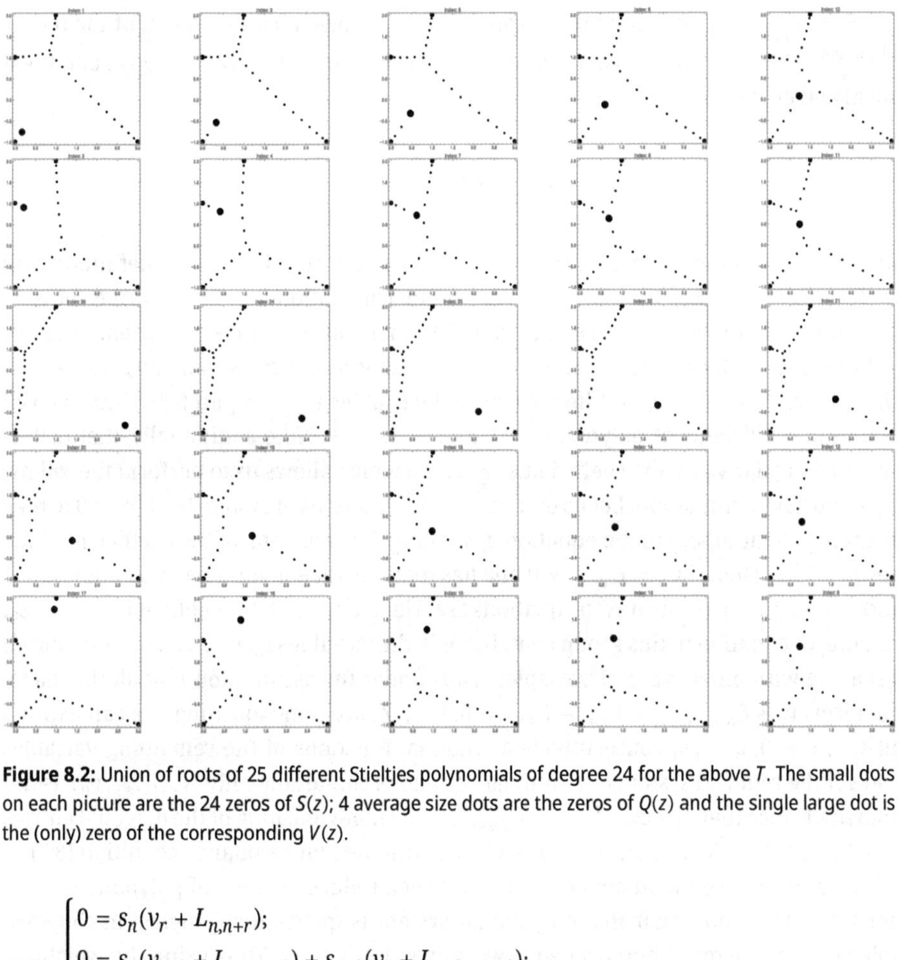

Figure 8.2: Union of roots of 25 different Stieltjes polynomials of degree 24 for the above T. The small dots on each picture are the 24 zeros of $S(z)$; 4 average size dots are the zeros of $Q(z)$ and the single large dot is the (only) zero of the corresponding $V(z)$.

$$\begin{cases}
0 = s_n(v_r + L_{n,n+r}); \\
0 = s_n(v_{r-1} + L_{n,n+r-1}) + s_{n-1}(v_r + L_{n-1,n+r-1}); \\
0 = s_n(v_{r-2} + L_{n,n+r-2}) + s_{n-1}(v_{r-1} + L_{n-1,n+r-2}) + s_{n-2}(v_r + L_{n-2,n+r-2}); \\
\vdots \qquad \vdots \qquad \vdots \qquad \vdots \qquad \vdots \qquad \vdots \\
0 = s_n(v_0 + L_{n,n}) + s_{n-1}(v_1 + L_{n-1,n}) + \cdots + s_{n-r}(v_r + L_{n-r,n}); \\
0 = s_n L_{n,n-1} + s_{n-1}(v_0 + L_{n-1,n-1}) + \cdots + s_{n-r-1}(v_r + L_{n-r-1,n-1}); \\
0 = s_n L_{n,n-2} + s_1 L_{1,r+1} + s_2(v_r + L_{2,r}) + \cdots + s_{n-r-2}(v_r + L_{n-r-2,n-2}); \\
\vdots \qquad \vdots \qquad \vdots \qquad \vdots \qquad \vdots \qquad \vdots \\
0 = s_n L_{n,r} + s_{n-1} L_{n-1,r} + s_{n-2} L_{n-2,r} + \cdots + s_0(v_r + L_{0,r}); \\
0 = s_n L_{n,r-1} + s_{n-1} L_{n-1,r-1} + s_{n-2} L_{n-2,r-1} + \cdots + s_0(v_{r-1} + L_{0,1}); \\
\vdots \qquad \vdots \qquad \vdots \qquad \vdots \qquad \vdots \qquad \vdots \\
0 = s_n L_{n,1} + s_{n-1} L_{n-1,1} + \cdots + s_1(v_0 + L_{1,1}) + s_0(v_1 + L_{0,1}); \\
0 = s_n L_{n,0} + s_{n-1} L_{n-1,0} + \cdots + s_1 L_{0,1} + s_0(v_0 + L_{0,0}).
\end{cases} \tag{8.9}$$

Here $L_{p,q}$ is a polynomial that expresses the coefficient containing s_p of the power z^q in $\sum_{i=1}^k Q_i(z)S^{(i)}$. Obviously, it is linear in the coefficients of $Q_k(z),\dots,Q_1(z)$ and is explicitly given by

$$L_{p,q} = \sum_{r=1}^{k} (p)_r A_{r,q-p+r},$$

where $A_{r,q-p+r}$ is the coefficient of z^{q-p+r} in $Q_r(z)$. In the notation used in definition (8.6) we have $L_{m,m+r} = \mathbb{L}_m$, $m = 0,\dots,n$. We use the convention that $L_{p,q}$ vanishes outside the admissible range of indices and, therefore many of the above coefficients $L_{p,q}$ are in fact equal to 0. (In system (8.9) we assumed that $n \geq r$ for simplicity.) Notice that all equations in (8.9) depend linearly on the variables $v_r,\dots,v_0$ and $s_n,\dots,s_0$ as well as on the coefficients of polynomials $Q_i(z)$, $i = 1,\dots,k$. Note additionally, that (8.9) is lower triangular w. r. t. the coefficients $s_n,\dots,s_0$, which allows us to perform the following important elimination. Let us enumerate the equations of system (8.9) from 0 to $n+r$ assigning the number j to the equation describing the vanishing of the coefficient of the power z^{n+r-j}. Then if $\mathbb{L}_n = L_{n,n+r} \neq 0$ one has that the 0th equation has a solution $s_n = 1$ and $v_r = -L_{n,n+r} \neq 0$. The next n equations are triangular w. r. t. the coefficients $s_n,\dots,s_0$, i. e., the jth equation in this group contains only the variables $s_n, s_{n-1},\dots,s_{n-j}$ (among all s_js) along with other types of variables. Thus under the assumption that all the diagonal terms $v_r + L_{n-i,n+r-i} = \mathbb{L}_{n-i} - \mathbb{L}_n$, $i = 0,1,\dots,n$, are non-vanishing, we can express all s_{n-i}, $i = 0,1,\dots,n$, consecutively as rational functions of the remaining variables and get the reduced system of r rational equations containing only $(v_{r-1},\dots,v_0)$ as unknowns. Notice that in view of $v_r = -L_{n,n+r} \neq 0$ the non-vanishing of the diagonal entries $v_r + L_{n-i,n+r-i}$, $i = 0,1,\dots,n$, coincides exactly with the non-resonance condition (8.7).

Cleaning the common denominators, we get a reduced system of polynomial equations. We show now that this polynomial system is quasi-homogeneous in the variables v_j with the quasi-homogeneous weights $w(v_j) = r - j$. Thus using the weighted-homogeneous version of the Bezout theorem, see, e. g., [61], we get that if the system under consideration defines a complete intersection, i. e., has only isolated solutions, then their number (counted with multiplicities) equals $\binom{n+r}{r}$ in the corresponding weighted projective space. To check the quasi-homogeneity, note that the standard action of $\mathbb{C}^*$ on the set of roots of the polynomial $V(z)$ by simultaneous multiplication assigns the weight $r - j$ to its coefficient v_j. These weights are still valid in the reduced system with the variables $s_n,\dots,s_0$ eliminated. Finally, we have to show that if the coefficients of $Q_k(z),\dots,Q_1(z)$ are algebraically independent, then the eliminated system has exactly $\binom{n+r}{r}$ simple solutions. Indeed, consider the linear space EQ of all systems of r quasi-homogeneous equations in the variables $(v_r,\dots,v_0)$ with the weights $w(v_j) = r - j$. Here the ith equation is weighted-homogeneous of degree $n + i$. We equip this space with the standard monomial basis.

To accomplish the proof of Theorem 8.4, we need two additional standard facts.

Lemma 8.1. *The discriminant Discr ⊂ EQ (i. e., the set of the coefficients of monomials in the equations for which the system has at least one solution of multiplicity greater than 1) is given by an algebraic equation with rational coefficients in the standard monomial basis of EQ.*

Proof. See [73, Chapter 13]. □

Consider some linear parameter space Λ with a chosen basis. Assume that there is a rational map $\Phi : \Lambda \to EQ$ where each coordinate in the standard monomial basis of EQ is given by a rational function with rational coefficients w. r. t. the chosen basis in Λ. The next statement is obvious.

Lemma 8.2. *In the above notation, the pullback of* $\Phi^{-1}(Discr)$ *in* Λ
either
a) *coincides with the whole* Λ*;*
or
b) *is given in the chosen basis by an algebraic equation with rational coefficients.*

It remains to show that there are some values of the coefficients of the polynomials $Q_k(z), \dots, Q_1(z)$ for which (8.5) has exactly $\binom{n+r}{r}$ distinct solutions. Here we are not able to follow the nice inductive argument of [82]. Heine's original proof does not generalize straightforwardly to higher order equations. Instead we can, for example, invoke Theorem 8.7 whose proof is completely independent of the present arguments. It claims, in particular, that for any strict hyperbolicity preserver of the form $T = \sum_{i=m}^{k} Q_i(z)\frac{d^i}{dz^i}$ and any $n \geq m$ there exist exactly $\binom{n+r}{r}$ pairs (V, S). One can additionally choose such a hyperbolicity preserver with $m = 1$ and therefore get a necessary example of an operator with given k and r such that for any $n \geq 1$ it has exactly the maximal possible number of pairs (V, S). □

8.2.1 On eigenvalues for rectangular matrices

To settle Theorem 8.5, let us first reinterpret Problem (8.5) in purely linear algebraic terms. (More discussions can be found in Appendix I, Chapter 13.)

We start with the following natural question.

Problem 8.2. Given an $(l + 1)$-tuple of $(m_1 \times m_2)$-matrices $A, B_1, \dots, B_l$ where $m_1 \leq m_2$, describe the set of all values of parameters $\lambda_1, \dots, \lambda_l$ for which the rank of the linear combination $A + \lambda_1 B_1 + \dots + \lambda_l B_l$ is less than m_1, i. e., such parameters $\lambda_1, \dots, \lambda_l$ for which the linear system $v * (A + \lambda_1 B_1 + \dots + \lambda_l B_l) = 0$ has a non-trivial (left) solution $v \neq 0$. This left solution will be called an *eigenvector of A w. r. t. the linear span of* $B_1, \dots, B_l$.

Let $\mathcal{M}_{m_1,m_2}$ denote the linear space of all $(m_1 \times m_2)$-matrices with complex entries. Below we will consider l-tuples of $(m_1 \times m_2)$-matrices $B_1, \ldots, B_l$, which are linearly independent in $\mathcal{M}_{m_1,m_2}$ and denote their linear span by $\mathcal{L} = \mathcal{L}(B_1, \ldots, B_l)$.

Given a matrix pencil $\mathcal{P} = A + \mathcal{L}$ where $A \in \mathcal{M}_{m_1,m_2}$ denote by $\mathcal{E}_\mathcal{P} \subset \mathcal{P}$ its *eigenvalue locus*, i. e., the set of matrices in $\mathcal{P}$ whose rank is less than the maximal one. Denote by $\mathcal{M}^1 \subset \mathcal{M}_{m_1,m_2}$ the set of all $(m_1 \times m_2)$ matrices with positive corank, i. e., whose rank is less than m_1. Its codimension equals $m_2 - m_1 + 1$, and its degree as an algebraic variety equals $\binom{m_2}{m_1-1}$, see [47, Prop. 2.15].

Consider the natural left-right action of the group $GL_{m_1} \times GL_{m_2}$ on $\mathcal{M}_{m_1,m_2}$, where GL_{m_1} (resp. GL_{m_2}) acts on $(m_1 \times m_2)$-matrices by the left (resp. right) multiplication. This action on $\mathcal{M}_{m_1,m_2}$ has finitely many orbits, each orbit being the set of all matrices of a given (co)rank, see, e. g., [14, Chapter 1, § 2]. Notice that due to the well-known formula of the product of coranks the codimension of the set of matrices of rank $\le r$ equals $(m_1 - r)(m_2 - r)$. Obviously, for any pencil $\mathcal{P}$, one has that the eigenvalue locus coincides with $\mathcal{E}_\mathcal{P} = \mathcal{M}^1 \cap \mathcal{P}$. Thus for a generic pencil $\mathcal{P}$ of dimension l, the eigenvalue locus $\mathcal{E}_\mathcal{P}$ is a subvariety of $\mathcal{P}$ of codimension $m_2 - m_1 + 1$ if $l \ge m_2 - m_1 + 1$, and it is empty otherwise.

The most interesting situation for applications occurs when $l = m_2 - m_1 + 1$ in which case $\mathcal{E}_\mathcal{P}$ is generically a finite set. From now on let us assume that $l = m_2 - m_1 + 1$. Denoting as above the linear span of $B_1, \ldots, B_l$ by $\mathcal{L}$, we say that $\mathcal{L}$ is *transversal to* $\mathcal{M}^1$ if the intersection $\mathcal{L} \cap \mathcal{M}^1$ is finite and *non-transversal to* $\mathcal{M}^1$ otherwise. Notice that due to the homogeneity of $\mathcal{M}^1$ any $(m_2 - m_1 + 1)$-dimensional linear subspace $\mathcal{L}$ transversal to it intersects $\mathcal{M}^1$ only at 0 and that the multiplicity of this intersection at 0 equals $\binom{m_2}{m_1-1}$.

We start with the following simple statement, which will later imply Theorem 8.5.

Lemma 8.3. *If the $(m_2 - m_1 + 1)$-dimensional linear space $\mathcal{L}$ is transversal to $\mathcal{M}^1$, then for any matrix $A \in \mathcal{M}_{m_1,m_2}$, the eigenvalue locus $\mathcal{E}_\mathcal{P}$ of the pencil $\mathcal{P} = A + \mathcal{L}$ consists of exactly $\binom{m_2}{m_1-1}$ points counted with multiplicities.*

Remark 8.10. Notice that since $\mathcal{M}^1 \subset \mathcal{M}(m_1, m_2)$ is an incomplete intersection, then in order to explicitly determine the eigenvalue locus of a given matrix A w. r. t. some $(m_2 - m_1 + 1)$-dimensional linear subspace $\mathcal{L} \subset \mathcal{M}(m_1, m_2)$ one has to solve an overdetermined system of $\binom{m_2}{m_1}$ equations describing the vanishing of all maximal minors of a $(m_1 \times m_2)$-matrix depending on parameters. Fortunately, in our main application, i. e., for the multi-parameter spectral problem (8.5), we encounter only "triangular" rectangular matrices (i. e., with the left-lower corner vanishing) for which the determination of the eigenvalue locus often reduces to a complete intersection, see the proof of Theorem 8.4.

Let us explain how Lemma 8.3 implies Theorem 8.5. Namely, for a non-degenerate operator T, in order to find all its Van Vleck polynomials having (at least one) Stieltjes polynomial of degree at most n, we need to study the action of T on the linear space $\mathbb{C}_n[z]$ of all univariate polynomials of degree at most n. If T has the Fuchs index r, then T maps $\mathbb{C}_n[z]$ to $\mathbb{C}_{n+r}[z]$. Using the standard monomial basis $1, z, z^2, \ldots, z^l$ in $\mathbb{C}_l[z]$, we get that if $n \ge k = \operatorname{ord}(T)$ then the action of T in this basis is represented by a "triangular" banded

$(n+1)\times(n+r+1)$-matrix $A_{T,n}$ with at most $r+k$ non-vanishing diagonals. Here "triangular" means that all entries $a_{i,j}$ of $A_{T,n}$ with $i<j$ vanish. Denote by I_s, $s=0,\dots,r$ the $(n+1)\times(n+r+1)$-matrix whose entries are given by $a_{i,j}=0$ if $i-j\neq s$ and 1 otherwise. Denote by $\mathfrak{L}$ the linear span of $I_0,\dots,I_r$ and notice that $\mathfrak{L}$ is transversal to $\mathcal{M}^1\subset M_{n+1,n+r+1}$ since any matrix belonging to the pencil $\mathfrak{L}$ and different from 0 has full rank.

Notice that adding an arbitrary polynomial $V(x)=v_rz^r+v_{r-1}z^{r-1}+\dots+v_0$ of degree at most r to T corresponds on the matrix level to adding of the linear combination $v_rI_0+v_{r-1}I_1+\dots+v_0I_r$ to the initial matrix $A_{T,n}$. The existence of a non-trivial Stieltjes polynomial of degree at most n corresponds to the fact that the matrix $A_{T,n}+v_rI_0+v_{r-1}I_1+\dots+v_0I_r$ has a non-trivial (left) kernel. Thus, for a given non-degenerate operator T, the problem of finding all the Van Vleck polynomials whose Stieltjes polynomials are of degree at most n is exactly equivalent to the determination of all the eigenvalues of its matrix $A_{T,n}$ w. r. t. the linear space $\mathfrak{L}$ in the above-mentioned sense. Lemma 8.3 has a simple analog for "triangular" rectangular matrices, which is equivalent to Theorem 8.5.

Namely, denote by $TM(m_1,m_2)\subset\mathcal{M}(m_1,m_2)$, $m_1\le m_2$ the set of all "triangular" $m_1\times m_2$-matrices, i. e., with $a_{i,j}=0$ for $i<j$. Let $\mathfrak{L}\subset TM(m_1,m_2)$ be the linear subspace spanned by all (m_2-m_1+1) possible unit matrices $I_1,\dots,I_{m_2-m_1+1}\in TM(m_1,m_2)$. Finally, denote by $TM^1\subset TM(m_1,m_2)$ the set of all "triangular" matrices with positive corank.

Lemma 8.4. *For any matrix $A\in TM(m_1,m_2)$, the eigenvalue locus $\mathcal{E}_{\mathcal{P}}$ of the pencil $\mathcal{P}=A+\mathfrak{L}$ consists of exactly $\binom{m_2}{m_1-1}$ points counted with multiplicities.*

Proof. The same as above. □

The latter lemma settles Theorem 8.5. Let us now prove Proposition 8.1.

Proof. In the above notation, consider the pencil $\mathcal{A}_n(v_r,v_{r-1},\dots,v_0)=A_{T,n}+v_rI_0+v_{r-1}I_1+\dots+v_0I_r$ of $(n+1)\times(n+r+1)$-matrices. One has

$$\mathcal{A}_n(v_r,v_{r-1},\dots,v_0)=\begin{pmatrix}\mathbb{L}_n+v_r & * & * & * & * & \cdots\\ 0 & \mathbb{L}_{n-1}+v_r & * & * & * & \cdots\\ 0 & 0 & \mathbb{L}_{n-2}+v_r & * & * & \cdots\\ 0 & 0 & 0 & \mathbb{L}_{n-3}+v_r & * & \cdots\\ \vdots & \vdots & \vdots & \vdots & \vdots & \vdots\end{pmatrix},$$

where $*$ stands for possibly non-vanishing entries. The obvious necessary condition for such a matrix to have a positive corank is that one of the elements on the main diagonal shown above vanishes, i. e., there exists $i=0,\dots,n$ such that $\mathbb{L}_i+v_r=0$ or, equivalently, $v_r=-\mathbb{L}_i$. Set $v_r=-\mathbb{L}_n$ thus 'killing' the entry in the left-upper corner. Recall that the nth non-resonance condition requires that $\mathbb{L}_n\neq\mathbb{L}_j$, $j=0,\dots,n-1$. Therefore the subtraction of $\mathbb{L}_n$ along the main diagonal will keep all other diagonal entries except for the left-upper corner non-vanishing. The r-dimensional pencil $\mathcal{A}_n(-\mathbb{L}_n,v_{r-1},\dots,v_0)=A_{T,n}-\mathbb{L}_nI_0+v_{r-1}I_1+\dots+v_0I_r$ has its 1-st column of the above matrix presentation vanishing and its main diagonal being the same for all possible values of $v_{r-1},\dots,v_0$. Therefore, by Lemma 8.3 there exist exactly $\binom{n+r}{r}$ eigenvalues $v_{r-1}, v_{r-2}, \ldots, v_0$ counted with multiplicities such that $\mathcal{A}_n(-\mathbb{L}_n, v_{r-1}, \ldots, v_0)$ has a positive corank. Finally, notice that since for any matrix from the pencil $\mathcal{A}_n(-\mathbb{L}_n, v_{r-1}, \ldots, v_0)$ its entries along the main diagonal except for the left-upper corner are non-vanishing, its corank can be at most 1. Moreover, when the corank of such a matrix is 1, then the occurring non-trivial linear combination of the rows that vanish must necessarily include the first row since the second, the third, etc. rows are linearly independent for the above reason. The coefficients of this linear dependence of rows are exactly the coefficients of the corresponding Stieltjes polynomial. The fact that the first row must be included in the linear dependence means that the leading coefficient of this Stieltjes polynomial (which by definition is of degree at most n) must be non-vanishing, i. e., this Stieltjes polynomial is of degree exactly n. Proposition 8.1 follows. □

To prove Theorem 8.6 we need to take a more careful look at the proof of Theorem 8.4. Namely, consider again system (8.9) determining the set of all pairs (V, S) where V is a Van Vleck polynomial and S is the corresponding Stieltjes polynomial of degree at most n. As in the proof of Theorem 8.4 we solve the 0th equation in (8.9) by taking $s_n = 1$ and $v_r = -L_{n,n+r} = -\mathbb{L}_n$. (By Proposition 8.1 finding a solution of (8.9) with $s_n = 1$ and $v_r = -L_{n,n+r} = -\mathbb{L}_n$ leads to a pair (V, S) such that V is of degree exactly equal to r and S is of degree exactly equal to n.) Then we consecutively express the variables $s_{n-1}, s_{n-2}, \ldots, s_0$ from the next n equations of (8.9). The crucial circumstance here is that while doing this we only divide by the differences of the form $\mathbb{L}_n - \mathbb{L}_j, j = n-1, n-2, \ldots, 0$, which are non-vanishing due to the validity of our non-resonance condition. Substituting the obtained expressions for $s_j, j = 0, \ldots, n$ in the remaining r equations in (8.9), we get the required eliminated system of algebraic equations in the variables $v_{r-1}, \ldots, v_0$, which determines a complete intersection. □

To illustrate the above procedure, let us take a concrete example.

Example 8.1. Consider the action of some operator T with the Fuchs index $r = 2$ on the space Pol_1. Its maps Pol_1 to Pol_3 are represented in the monomial bases of Pol_1 and Pol_3 by the matrix

$$\begin{pmatrix} \mathbb{L}_1 & L_{1,2} & L_{1,1} & L_{1,0} \\ 0 & \mathbb{L}_0 & L_{0,1} & L_{0,0} \end{pmatrix}.$$

(Here we used the notation from the proof of Theorem 8.4.) Since $r = 2$, we need to add to T a quadratic Van Vleck polynomial $V(z) = v_2 z^2 + v_1 z + v_0$ with the undetermined coefficients v_2, v_1, v_0, which modifies the above matrix as follows:

$$\begin{pmatrix} \mathbb{L}_1 + v_2 & L_{1,2} + v_1 & L_{1,1} + v_0 & L_{1,0} \\ 0 & \mathbb{L}_0 + v_2 & L_{0,1} + v_1 & L_{0,0} + v_0 \end{pmatrix}.$$

The operator $T + V(z)$ has a linear Stieltjes polynomial $S(z) = s_1 z + s_0$ if and only if the vector (s_1, s_0) is the left kernel of the latter matrix, which leads to the system

$$\begin{cases} 0 = s_1(\mathbb{L}_1 + v_2); \\ 0 = s_1(L_{1,2} + v_1) + s_0(\mathbb{L}_0 + v_2); \\ 0 = s_1(L_{1,1} + v_0) + s_0(L_{0,1} + v_1); \\ 0 = s_1 L_{1,0} + s_0(L_{0,0} + v_0). \end{cases}$$

Setting $s_1 = 1$ and $v_2 = -\mathbb{L}_1$ as was explained earlier, we get $s_0 = \frac{L_{1,2}+v_1}{\mathbb{L}_1-\mathbb{L}_0}$ from the 2nd equation. Substituting the obtained variables in the remaining two equations, we get the system of two equations

$$\begin{cases} (\mathbb{L}_1 - \mathbb{L}_0)(v_0 + L_{1,1}) + (v_1 + L_{1,1})(v_1 + L_{1,2}) = 0; \\ (v_0 + L_{0,0})(v_1 + L_{1,2}) + (\mathbb{L}_1 - \mathbb{L}_0)L_{1,0} = 0. \end{cases}$$

This system determines three (not necessarily distinct) pairs (v_1, v_0), which together with $v_2 = -\mathbb{L}_1$ give us three (not necessarily distinct) quadratic Van Vleck polynomials whose Stieltjes polynomials are of degree exactly 1.

Now we can finally describe the notion of natural multiplicity of a given Van Vleck polynomial of an operator T used in the beginning of the chapter. Let $\widetilde{V}(z) = \tilde{v}_r z^r + \tilde{v}_{r-1} z^{r-1} + \cdots + \tilde{v}_0$ be some fixed Van Vleck polynomial of the Heine–Stieltjes problem (8.5), i. e., there exists a (not necessarily unique) polynomial solution $\widetilde{S}(z)$ of the equation of (8.5) with the chosen $V(z) = \widetilde{V}(z)$. (Below we use notation from the proof of Proposition 8.1.)

Definition 8.2. Given a positive integer n, let us define the *nth multiplicity* $\sharp_n(\widetilde{V})$ of $\widetilde{V}(z)$ as the (usual) local algebraic multiplicity of the intersection of the $(r+1)$-dimensional matrix pencil $\mathcal{A}_n(v_r, v_{r-1}, \ldots, v_0) = A_{T,n} + v_r I_0 + v_{r-1} I_1 + \cdots + v_0 I_r$ consisting of "triangular" $(n+1) \times (n+r+1)$-matrices with the set $TM^1 \subset TM(n+1, n+r+1)$ of positive corank matrices at the matrix $A_{T,n} + \tilde{v}_r I_0 + \tilde{v}_{r-1} I_1 + \cdots + \tilde{v}_0 I_r$. Here (as above) $A_{T,n}$ denotes the matrix of the action of T on the space $\mathbb{C}_n[z]$ taken w. r. t. the standard monomial basis and $A_{T,n} + \tilde{v}_r I_0 + \tilde{v}_{r-1} I_1 + \cdots + \tilde{v}_0 I_r$ is, therefore, the matrix of the action of the operator $T + \widetilde{V}(z)$ on Pol_n. We set $\sharp_n(\widetilde{V}) = 0$ in case, when $A_{T,n} + \tilde{v}_r I_0 + \tilde{v}_{r-1} I_1 + \cdots + \tilde{v}_0 I_r$ does not belong to $TM^1 \subset TM(n+1, n+r+1)$, i. e., the operator $T + \widetilde{V}(z)$ does not annihilate any polynomial of degree at most n.

Remark 8.11. The natural multiplicity of Van Vleck polynomials in Theorems 8.4 and 8.5 while counting those with Stieltjes polynomials of degree at most n is exactly the nth multiplicity from Definition 8.2.

Obviously, for any given Van Vleck polynomial $\widetilde{V}(z)$, the sequence $\{\sharp_n(\widetilde{V})\}$, $n = 0, 1, \ldots$, is a non-decreasing sequence of non-negative integers. Moreover the following stabilization result holds.

Lemma 8.5. *For any non-degenerate operator T, the sequence $\{\sharp_n(\widetilde{V})\}$ of multiplicities of any its Van Vleck polynomial $\widetilde{V}(z)$ stabilizes, i. e., there exists $n_{\widetilde{V}}$ such that for all $n > n_{\widetilde{V}}$ one has $\sharp_n(\widetilde{V}) = \sharp_{n_{\widetilde{V}}}(\widetilde{V})$.*

Proof. Indeed, as was mentioned in, e. g., the proof of Proposition 8.1, the leading coefficient $\tilde{v}_r$ of $\widetilde{V}(z)$ must necessarily coincide with $-\mathbb{L}_m$ for some non-negative m. The sequence $\{|\mathbb{L}_j|\}$ is strictly increasing starting from some j_0, see (8.6). Moreover, by Proposition 8.1, if $\mathbb{L}_n \neq L_j, j = 0, 1, \ldots, n-1$, then the total multiplicity of all Van Vleck polynomials whose leading term equals $-\mathbb{L}_n$ equals $\binom{n+r}{r}$. Therefore, if we take the index value that j_0 such that $|\mathbb{L}_j| > |\mathbb{L}_m|$ for all $j \geq j_0$, then the multiplicities $\sharp_j(\widetilde{V})$ cannot change for $j \geq j_0$ since the total multiplicity increase is obtained on Van Vleck polynomials with a different leading coefficient when j grows. □

8.3 Proofs of generalized Pólya's theorems

Let us now prove Theorem 8.8 by straightforwardly following the recipe of [150], which in its turn is closely related to the proof of the classical Gauss–Lucas theorem.

Proof. Let $(z_1, \ldots, z_n)$ denote the set of all roots of a Stieltjes polynomial $S(z)$ of some degree n satisfying equation (8.8) with α_is being complex and β_js being positive. Then, for each z_i, one has

$$\frac{S^{(k)}(z_i)}{S^{(k-1)}(z_i)} + \sum_{j=1}^{l} \frac{\beta_j}{z_i - \alpha_j} = 0.$$

This equation has the form

$$\sum_{s=1}^{p} \frac{m_s}{z_i - \xi_s} + \sum_{j=1}^{l} \frac{\beta_j}{z_i - \alpha_j} = 0, \tag{8.10}$$

where $(\xi_1, \ldots, \xi_p)$ is the set of all roots of $S^{(k-1)}(z)$ with $p = n - k + 1$ and $(m_1, \ldots, m_p)$ is the set of multiplicities of the roots $(\xi_1, \ldots, \xi_p)$. Notice that by the standard Gauss–Lucas theorem all $(\xi_1, \ldots, \xi_p)$ lie in the convex hull of the set of roots $(z_1, \ldots, z_n)$. Assume now that the convex hull of $(z_1, \ldots, z_n)$ is not contained in the convex hull of $(\alpha_1, \ldots, \alpha_l)$. Then there exists some root z_i and an affine line $L \subset \mathbb{C}$ separating z_i from the rest of z_js together with all α_js and ξ_ms. But then equation (8.10) cannot hold since all the vectors $z_i - \xi_s$ and $z_i - \alpha_j$ lie in the same half-plane. □

Remark 8.12. The above argument works for the roots of Van Vleck polynomials $V(z)$ as well and extends to the case $\beta_j \geq 0$.

To settle a much more delicate Theorem 8.9, we will prove a number of localization results having an independent interest.

8.3.1 Root localization for Van Vleck and Stieltjes polynomials

Definition 8.3. As before the Cauchy transform $\mathcal{C}_\mu(z)$ of a measure μ supported in $\mathbb{C}$ is standardly defined as

$$\mathcal{C}_\mu(z) = \int_{\mathbb{C}} \frac{d\mu(\zeta)}{z-\zeta}. \tag{8.11}$$

Recall that directly from the definition of the root-counting measure $\mu_P(z)$ of a polynomial P one has that for any given polynomial $P(z)$ of degree m, its Cauchy transform is given by $\mathcal{C}_{\mu_P}(z) = \frac{P'(z)}{mP(z)}$.

We start with a rather simple estimate of the absolute value of the Cauchy transform of a probability measure, comp. Lemma 2 in [26].

Lemma 8.6. *Let μ be a probability measure supported in a disk D_0 of radius R_0 and centered at z_0. Then, for any z outside D_0, one has*

$$\frac{1}{|z-z_0-R_0|} \geq |\mathcal{C}_\mu(z)| \geq \frac{1}{2|z-z_0|}. \tag{8.12}$$

Proof. The l. h. s. of the above inequality is quite obvious. From the formula for the Cauchy transform one has that $|\mathcal{C}_\mu(z)|$ will be maximal if one places the whole unit mass of μ at the point which has the least distance to z in the admissible support. In our case such a point p is the intersection of the boundary circle of D_0 with the segment (z, z_0). Its distance to z equals $|z-z_0-R_0|$, which gives the required inequality. To settle the r. h. s., let us assume for simplicity that $z_0 = 0$. Translation invariance of our considerations is obvious. Let us use (8.11) and change the integration variable as follows:

$$\frac{1}{z-\zeta} = \frac{1}{z}\cdot\frac{1}{1-\zeta/z} = \frac{1}{z}\cdot\frac{1}{1-\theta},$$

where $\theta = \frac{\zeta}{z}$. Since z lies outside D_0 and ζ lies inside D_0, one has $|\theta| < 1$, which implies that $\mathrm{Re}(w) \geq \frac{1}{2}$ for $w = \frac{1}{1-\theta}$. Indeed,

$$|w-1| = \frac{|\theta|}{|1-\theta|} = |\theta||w| \leq |w| \Leftrightarrow |w-1| \leq |w| \Leftrightarrow \mathrm{Re}(w) \geq \frac{1}{2}.$$

Therefore,

$$|\mathcal{C}_\mu(z)| = \left|\int_{\mathbb{C}} \frac{d\mu(\zeta)}{z-\zeta}\right| = \frac{1}{|z|}\left|\int_{\mathbb{C}} \frac{d\mu(\zeta)}{1-\theta}\right| = \frac{1}{|z|}\left|\int_{\mathbb{C}} w\,d\mu(\zeta)\right| \geq \frac{1}{|z|}\left|\int_{\mathbb{C}} \mathrm{Re}(w)\,d\mu(\zeta)\right| \geq \frac{1}{2|z|}. \qquad \square$$

Using Lemma 8.6 we can now settle Proposition 8.4.

Proof. Take a pair $(V(z), S(z))$ where $V(z)$ is some Van Vleck polynomial and $S(z)$ is its corresponding Stieltjes polynomial of degree n. Let ξ be the root of either $V(z)$ or $S(z)$, which has the maximal modulus among all roots of the chosen $V(z)$ and $S(z)$. We want to show that as soon as n is large enough there exists a radius $R > 0$ such that $|\xi| \le R$ for any ξ as above. Substituting $V(z)$, $S(z)$, ξ in (8.5) and using (8.4), we get the relation

$$Q_k(\xi)S^{(k)}(\xi) + Q_{k-1}(\xi)S^{(k-1)}(\xi) + \cdots + Q_1(\xi)S'(\xi) = 0,$$

dividing which by its first term we obtain

$$1 + \sum_{j=1}^{k-1} \frac{Q_j(\xi)S^{(i)}(\xi)}{Q_k(\xi)S^{(k)}(\xi)} = 0. \tag{8.13}$$

Notice that the rational function $b_i(z) := \frac{S^{(i+1)}(z)}{(n-i)S^{(i)}(z)}$ is the Cauchy transform of the polynomial $S^{(i)}(z)$. Easy calculation shows that

$$S^{(i)}(z) = \frac{S^{(k)}(z)}{(n-k+1)\cdots(n-i)\prod_{j=i}^{k-1} b_j(z)} \Leftrightarrow \frac{S^{(i)}(z)}{S^{(k)}(z)} = \frac{(n-k)!}{(n-i)!\prod_{j=i}^{k-1} b_j(z)}.$$

Notice additionally that by the usual Gauss–Lucas theorem all roots of any $S^{(i)}(z)$ lie within the convex hull of the set of roots of $S(z)$. In particular, all these roots lie within the disk of radius $|\xi|$. Therefore, using Lemma 8.6 we get

$$\left|\frac{Q_i(\xi)S^{(i)}(\xi)}{Q_k(\xi)S^{(k)}(\xi)}\right| \le \frac{|Q_i(\xi)|}{|Q_k(\xi)|}\frac{(n-k)!}{(n-i)!}2^{k-i}|\xi|^{k-i}. \tag{8.14}$$

Notice that since $Q_k(z)$ is a monic polynomial of degree $k+r$ where r is the Fuchs index of the operator T, one can choose a radius R such that for any z with $|z| > R$ one has $|Q_k(z)| \ge \frac{|z|^{k+r}}{2}$. Now since for any $i = 1, \dots, k-1$ one has $\deg Q_i(z) \le i + r$, we can choose a positive constant K such that $|Q_i(z)| \le K|z|^{i+r}$ for all $i = 1, \dots, k-1$ and $|z| > R$. We want to show that ξ cannot be too large for a sufficiently large n. Using our previous assumptions and assuming additionally that $|\xi| > R$, we get

$$\left|\frac{Q_i(\xi)S^{(i)}(\xi)}{Q_k(\xi)S^{(k)}(\xi)}\right| \le \frac{|Q_i(\xi)|}{|Q_k(\xi)|}\frac{(n-k)!}{(n-i)!}2^{k-i}|\xi|^{k-i} \le \frac{K\cdot 2^{k-i+1}}{(n-i)\cdots(n-k+1)}.$$

Now we can finally choose N_0 so large that for all $n \ge N_0$, all $i = 1, \dots, k-1$, and any $|\xi| > R$, one has that

$$\left|\frac{Q_i(\xi)S^{(i)}(\xi)}{Q_k(\xi)S^{(k)}(\xi)}\right| \le \frac{K\cdot 2^{k-i+1}}{(n-i)\cdots(n-k+1)} < \frac{1}{k-1}.$$

But then obviously relation (8.13) cannot hold for all $n \ge N_0$ and any $|\xi| > R$ since

$$\left|\sum_{j=1}^{k-1} \frac{Q_j(\xi)S^{(i)}(\xi)}{Q_k(\xi)S^{(k)}(\xi)}\right| \le \sum_{j=1}^{k-1}\left|\frac{Q_j(\xi)S^{(i)}(\xi)}{Q_k(\xi)S^{(k)}(\xi)}\right| < \sum_{i=1}^{k-1}\frac{1}{k-1} < 1. \qquad \square$$

Now we will strengthen the arguments in the proof of Proposition 8.4 in order to settle Theorem 8.9. Denote by $\mathcal{R}_{Q_k}$ the maximal distance between the origin and $Conv_{Q_k}$. The following statement holds.

Lemma 8.7. *For any non-degenerate higher Lamé operator T and a given number $\delta > 0$, there exists a positive integer N_δ such that the roots of all Van Vleck polynomials $V(z)$ possessing a Stieltjes polynomial $S(z)$ of degree $\geq N_\delta$ as well as the roots of these Stieltjes polynomials lie in the disk $|z| \leq \mathcal{R}_{Q_k} + \delta$.*

Proof. Notice that once δ is fixed the quotient $\frac{|Q_i(z)|}{|Q_k(z)|}$ is bounded from above for each $i = 1, \ldots, k-1$ if we assume that $|z| \geq \mathcal{R}_{Q_k} + \delta$. Indeed, all the roots of $Q_k(z)$ lie within the disk of radius $\mathcal{R}_{Q_k}$ centered at the origin and each $Q_i(z)$ has a smaller degree than $Q_k(z)$. Consider now again estimate (8.14). Since we now know that ξ lies in some bounded domain for all possible polynomials $V(z)$ and $S(z)$ of sufficiently high degree and that the quotient $\frac{|Q_i(z)|}{|Q_k(z)|}$ is bounded from above outside the disk of radius $\mathcal{R}_{Q_k} + \delta$, we get that the right-hand side of (8.14) goes to 0 when $n \to \infty$ under the assumption that ξ stays outside the latter disk. Looking again at (8.13) we see that by the latter argument it cannot hold for $|\xi| \geq \mathcal{R}_{Q_k} + \delta$ when $n \to \infty$. This contradiction proves the lemma. □

To finish the proof of Theorem 8.9, notice that the choice of the origin is in our hands, i. e., we can make an arbitrary affine shift of the independent variable z and use the same arguments. Since the convex hull $Conv_{Q_k}$ is the intersection of all disks centered at different points and containing $Conv_{Q_k}$, we can for any chosen $\epsilon > 0$ find the intersection $\mathcal{K}$ of finitely many disks in $\mathbb{C}$ such that $\mathcal{K}$ contains $Conv_{Q_k}$ but is contained in $Conv_{Q_k}^{\epsilon}$. (One can choose one such disk for each edge of the boundary of $Conv_{Q_k}$ putting its center sufficiently far away on the line perpendicular to the edge and passing through its middle point.) Since $\mathcal{K}$ is the intersection of finitely many disks, applying Lemma 8.7 we can find such N_ϵ that all roots of all $V(z)$ and $S(z)$ for all $n \geq N_\epsilon$ lie in $\mathcal{K}$. □

8.3.2 Comments on Heine's result and some history

As E. Heine himself mentions in [82] the requirement of algebraic independence of the coefficients of $Q_1(z)$ and $Q_2(z)$ is too strong and restrictive for his purposes. However, he fails to give any other explicit condition guaranteeing the same result (see Theorem 8.1). Heine's original motivation for the consideration (8.2) comes from the classical Lamé equation (8.1) in which case the coefficient of $Q_1(z)$ and $Q_2(z)$ are in fact algebraically dependent, namely, $Q_1(z) = Q_2'(z)/2$. Also in order to prove that the upper bound $\binom{n+l-2}{n}$ is actually achieved for algebraically independent $Q_1(z)$ and $Q_2(z)$, Heine uses an inductive argument, where he forces the coefficient of $Q_1(z)$ and $Q_2(z)$ to become algebraically dependent in a special way. Theorem 8.2 is another clear indication that the algebraic independence is apparently an inappropriate condition for the goal.

Interpretation of Heine's text written in a rather cumbersome 19th century German seemed to create difficulties for mathematicians already in 1870s and up to now, see,

e. g., [135]. The main classical sources, namely, [176] from 1885, [150] from 1912, and [181] from 1939, are not too clear about what Heine actually proved and under which assumptions on $Q_1(z)$ and $Q_2(z)$ one can guarantee that for a given positive n, the number of possible polynomial pairs (V, S) such that the corresponding S has degree exactly n is finite and bounded by $\binom{n+l-2}{n}$. Being aware of a gap in his proof, Heine seems to cover it by a reference to a letter of his friend Leopold Kronecker, who has (under unknown conditions) shown that for a given degree n the eliminant of the system of algebraic equations defining the coefficients of the polynomial $V(z)$ does not vanish identically. This statement is equivalent to the finiteness of the number of these polynomials. Heine mentions also a short note of Kronecker on this topic presented in the January issue of Monatsbericht der Berliner Akademie from 1864. Unfortunately, the track ends here. All one can find in this issue is a phrase about a meeting of the section of physics and mathematics of the Prussian Academy of Science held on November 14, 1864 "Herr Kronecker las über die verschieden Faktoren des Discriminante von Eliminantions-Gleichungen", i. e., "Herr Kronecker gave a lecture on different factors of the discriminant of the elimination equation." In an attempt to overcome this gap, we undertook the task on translating and decoding Heine's arguments. The result of these efforts can be found in the Appendix II, Chapter 13.2.

8.3.3 Related literature

Let us finally mention a few relatively recent references on the (generalized) Lamé equation. Being an object of substantial physical and mathematical importance, it gives, in particular, an example of an equation whose monodromy group can be analyzed in detail, see [30, 120, 121]. It is also closely related to the quasi-exact solvability and integrable models, see [65]. Theory of multi-parameter spectral problems originating from Heine–Stieltjes pioneering studies was developed in sixties, see, e. g., [192] and the references therein. Recently the interest in Heine–Stieltjes polynomials has been stimulated by an unexpected extension of the Bethe ansatz in representation theory, see [156, 138, 160, 161], and a further series of articles by A. Varchenko and his coauthors, e. g., [137]. Starting with [134], substantial progress has been made in the understanding of the asymptotics of the root distributions for these polynomials when either $l \to \infty$ (thermodynamic asymptotics) or $n \to \infty$ (semi-classical asymptotics), see [41], and [42].

8.4 Open problems in Chapter 8

Let us formulate a number of relevant questions and conjectures.

8.1. Is it possible to describe when a linear ordinary differential equation with polynomial coefficients admits at least two linearly independent polynomial solutions?

The prototype result of Varchenko–Scherbak gives a satisfactory answer for equations of the second order, see [161]. An answer to the latter question will allow to detect the appearance of multi-dimensional families of Stieltjes polynomials.

8.2. Under the non-resonance assumption (8.7), is it possible to obtain explicitly the discriminantal surface that shows when a Van Vleck polynomial attains a non-trivial multiplicity?

This question addresses the problem of explicit determination of the discriminantal surface mentioned in Heine's proof. Some discussion of this problem can be found in [165].

8.3. Explain how the number of Van Vleck's polynomials having Stieltjes polynomials for a certain given degree n can drop below $\binom{n+r}{r}$?

8.4. Under what assumptions on T the roots of any its Van Vleck and Stieltjes polynomials lie in the convex hull of its leading coefficient $Q_k(z)$?

The basic examples are provided by Stieltjes' and Pólya's theorems.

8.5. Is it possible to extend the results of this chapter to the case of degenerate higher Lamé operators?

T. Bergkvist [26] has obtained a number of interesting results and conjectures in the case of degenerate ES-operators, see Chapter 3. Motivated by her results, we formulate the following conjecture.

Conjecture 8.1. *For any degenerate Lamé operator and any positive integer N_0, the union of all the roots of polynomials V and S taken over $\deg S \geq N_0$ is always unbounded. Therefore, this property is a key distinction between non-degenerate and degenerate Lamé operators.*

9 Asymptotics of spectral polynomials in the classical case

(The material of this chapter is based on [37].)

9.1 Preliminaries

As we mentioned before, the standard Lamé equation is given by

$$\left\{Q(z)\frac{d^2}{dz^2} + \frac{1}{2}Q'(z)\frac{d}{dz} + V(z)\right\}S(z) = 0, \tag{9.1}$$

where $Q_l(z)$ is a real degree l polynomial with all real and distinct roots and $V(z)$ is a polynomial of degree at most $l-2$ whose choice depends on the type of solution to (9.1) one is looking for. The *generalized Lamé equation* [197] is a second order differential equation of the form

$$\left\{Q(z)\frac{d^2}{dz^2} + P(z)\frac{d}{dz} + V(z)\right\}S(z) = 0, \tag{9.2}$$

where $Q(z)$ is a complex polynomial of degree l and $P(z)$ is a complex polynomial of degree at most $l-1$, see Chapter 1. As it was first shown by Heine [82] for a generic equation of either of the forms (9.1) or (9.2) and an arbitrary positive integer n, there are exactly $\binom{n+l-2}{n}$ polynomials $V(z)$ such that $S(z)$ is a polynomial of degree n. Below we concentrate on the most classical case of a cubic polynomial $Q(z)$ with all real roots as treated in, e. g., [197]. Already Lamé and Liouville knew that in this case the (unique) root of $V(z)$ is real and located between the minimal and the maximal roots of $Q(z)$. The latter property was further generalized by Heine, Van Vleck, and other authors. Note that if $\deg Q(z) = 3$ then $V(z)$ is at most linear and that for a given value of the positive integer n, there are at most $n+1$ such polynomials.

Before formulating our main results, let us briefly review—following mostly [121, § 3] and [153]—some necessary background on the version of the Lamé equation used below and its spectral polynomials. Setting, as one traditionally does, $Q(z) = 4(z-e_1)(z-e_2)(z-e_3)$ with $e_1 > e_2 > e_3$, we can rewrite equation (9.1) as

$$\left\{\frac{d^2}{dz^2} + \frac{1}{2}\sum_{i=1}^{3}\frac{1}{z-e_i}\frac{d}{dz} - \frac{n(n+1)z+E}{4\prod_{i=1}^{3}(z-e_i)}\right\}S(z) = 0. \tag{9.3}$$

(The chosen representation for the linear polynomial $V(z)$ will shortly become clear.) Notice that several equivalent forms of the Lamé equation are classically known. Among those one should mention two algebraic forms, the Jacobian form and the Weierstrassian form, respectively, see [197, § 23.4] and Remark 9.1 below. Equation (9.3)

https://doi.org/10.1515/9783112207741-011

presents the most commonly used real algebraic form of the Lamé equation, which is smooth while the other (real) algebraic form is singular.

A *Lamé solution of the first kind* of equation (9.3) is a solution of the form $S(z) = (z-e_1)^{\kappa_1}(z-e_2)^{\kappa_2}(z-e_3)^{\kappa_3}\tilde{S}(z)$, where each κ_i is either 0 or $\frac{1}{2}$ and $\tilde{S}(z)$ is a polynomial. Lamé solutions that are pure polynomials—i. e., for which $\kappa_1 = \kappa_2 = \kappa_3 = 0$—are said to be of *type 1*, those with a single square root are of *type 2*, those with two square roots are of *type 3* and, finally, those involving three square roots are said to be of *type 4*. ("Types" are sometimes called "species", see, e. g., [197].) One can easily check that (9.3) has a Lamé solution if and only if n is a non-negative integer. Moreover, if n is even, then only solutions of types 1 and 3 exist. Namely, for appropriate choices of the energy constant E, one gets exactly $\frac{n+2}{2}$ distinct independent solutions of type 1 and $\frac{3n}{2}$ independent solutions of type 3. If n is odd, then only solutions of types 2 and 4 exist. In this case, for appropriate choices of the energy constant E, one gets exactly $\frac{3(n+1)}{2}$ distinct independent solutions of type 2 and $\frac{n-1}{2}$ independent solutions of type 4. In both cases the total number of Lamé solutions equals $2n+1$, which coincides with the number of independent spherical harmonics of order n.

Let $R_n(E) = \prod_{j=0}^{2n+1}(E - E_j)$, $n \in \mathbb{Z}_+$, denote the monic polynomial of degree $2n+1$ whose roots are exactly the values of the energy E at which equation (9.3) has a Lamé solution. These polynomials are often referred to as *spectral polynomials* in the literature; their study goes back to Hermite and Halphen. Some relatively recent results in this direction can be found in [21, 121, 184, 185, 193] and [76, 77]. In particular, [121] contains an excellent survey of this topic as well as a comprehensive table with these polynomials (and their modified versions) correcting several mistakes that occurred in previous publications. Article [77] is apparently the first attempt to give a (somewhat) closed formula for $R_n(E)$; for this the authors use yet another family of polynomials, which they call *elliptic Bernoulli polynomials*. Since explicit formulas for $R_n(E)$ seem to be rather complicated and difficult to handle, below we study the asymptotics of the root distribution of appropriately scaled versions of $R_n(E)$ (Corollary 9.1). In spite of a more than 150 years long history of the Lamé equation, the only source discussing questions similar to ours that we were able to locate is [41]. We should also mention that the results below are actually much more precise and therefore supersede those of *loc. cit.*

Let us now introduce eight spectral polynomials $R_n^{\kappa_1,\kappa_2,\kappa_3}(E)$ related to the eight types of Lamé solutions mentioned above, namely

$$R_n(E) = \begin{cases} R_n^{0,0,0}(E)R_n^{\frac{1}{2},\frac{1}{2},0}(E)R_n^{\frac{1}{2},0,\frac{1}{2}}(E)R_n^{0,\frac{1}{2},\frac{1}{2}}(E) & \text{when } n \text{ is even,} \\ R_n^{\frac{1}{2},0,0}(E)R_n^{0,\frac{1}{2},0}(E)R_n^{0,0,\frac{1}{2}}(E)R_n^{\frac{1}{2},\frac{1}{2},\frac{1}{2}}(E) & \text{when } n \text{ is odd.} \end{cases} \tag{9.4}$$

Note for example that if n is even, then $R_n^{0,0,0}(E)$ is the (unique) monic polynomial of degree $\frac{n+2}{2}$ whose roots are precisely the values of E for which equation (9.3) has a pure polynomial solution.

The results of the present paper actually hold not only for equation (9.3) but also for generalized Lamé equations (cf. (9.2)) of the form

$$\left\{\frac{d^2}{dz^2}+\sum_{i=1}^{3}\frac{\alpha_i}{z-e_i}\frac{d}{dz}+\frac{V(z)}{\prod_{i=1}^{3}(z-e_i)}\right\}S(z)=0, \tag{9.5}$$

where $\alpha_i > 0$, $i = 1, 2, 3$, and $V(z)$ is an undetermined linear polynomial. This case was thoroughly treated by Stieltjes in [176]. In particular, he proved that for any positive integer n, there exist exactly $n+1$ polynomials $V_{n,j}(z)$, $1 \le j \le n+1$, such that (9.5) has a polynomial solution $S(z)$ of degree n. Moreover, the unique root $t_{n,j}$ of $V_{n,j}(z)$ lies in the interval (e_3, e_1), and these $n+1$ roots are pairwise distinct. Consider now the polynomial

$$Sp_n(\lambda)=\prod_{j=1}^{n+1}(\lambda-t_{n,j}),$$

which is the scaled version of the spectral polynomial $R_n^{0,0,0}(E)$. More exactly, for any even n, one has that $Sp_n(\lambda)=\frac{R_n^{0,0,0}(n(n+1)\lambda)}{(n(n+1))^{\frac{n+2}{2}}}$. Assume further that the set $\{t_{n,j} : 1 \le j \le n+1\}$ is ordered so that $t_{n,1} < t_{n,2} < \cdots < t_{n,n+1}$ and associate to it the finite measure

$$\nu_n=\frac{1}{n+1}\sum_{j=1}^{n+1}\delta(z-t_{n,j}),$$

where $\delta(z-a)$ is the Dirac measure supported at a. The measure ν_n thus obtained is clearly a real probability measure that one usually refers to as the root-counting measure of $Sp_n(\lambda)$.

Below we shall make use of some well-known notions from the theory of special functions that can be found in, e. g., [2]. In fact several of our formulas rely on various integral representations and transformation properties of the Gauss hypergeometric series

$$F(a,b,c;z)={}_2F_1(a,b,c;z)=\frac{\Gamma(c)}{\Gamma(a)\Gamma(b)}\sum_{m=0}^{\infty}\frac{\Gamma(a+m)\Gamma(b+m)}{\Gamma(c+m)}\frac{z^m}{m!},$$

where Γ denotes Euler's Gamma function, and the complete elliptic integral of the first kind

$$\mathbb{K}(\zeta)=\int_0^1\frac{dt}{\sqrt{(1-t^2)(1-\zeta^2t^2)}}. \tag{9.6}$$

Recall from [2, § 15.3.1] that if $\Re(c) > \Re(b) > 0$ then a convenient way of rewriting $F(a,b,c;z)$ is given by

$$F(a,b,c;z)=\frac{\Gamma(c)}{\Gamma(b)\Gamma(c-b)}\int_0^1 t^{b-1}(1-t)^{c-b-1}(1-tz)^{-a}\,dt. \tag{9.7}$$

The above integral represents a univalent analytic function in the z-plane cut along the real axis from 1 to ∞, and thus it gives the analytic continuation of $F(a,b,c;z)$. To simplify our formulas, let us also define a real-valued function of three real variables $(x_1,x_2,x_3) \mapsto \omega(x_1,x_2,x_3)$ by setting

$$\omega(x_1,x_2,x_3) = \frac{(|x_1|+|x_2|)|x_3|}{|x_1||x_2-x_3|+|x_2||x_1-x_3|}. \tag{9.8}$$

Note that $\omega(x_1,x_2,x_3)$ is symmetric in its first two arguments and that

$$0 \le \omega(x_1,x_2,x_3) \le 1 \quad \text{whenever } x_1 < 0,\ x_2 > 0 \text{ and } x_1 \le x_3 \le x_2. \tag{9.9}$$

We should emphasize that the latter conditions on x_1, x_2, x_3—hence also the bounds for $\omega(x_1,x_2,x_3)$ given in (9.9)—always hold in the present context.

We are now ready to state our main results.

Theorem 9.1. *For any choice of real numbers $e_1 > e_2 > e_3$ and positive numbers α_1, α_2, α_3, the sequence of measures $\{\nu_n\}_{n\in\mathbb{Z}_+}$ strongly converges to the probability measure ν_Q supported on the interval $[e_3,e_1]$ with density ρ_Q given by any of the following equivalent expressions:*

(i)

$$\rho_Q(s) = \frac{1}{\pi}\sqrt{\frac{1+\sqrt{1-\omega(e_1-e_2,e_3-e_2,s-e_2)^2}}{(e_1-e_3)|s-e_2|\omega(e_1-e_2,e_3-e_2,s-e_2)}} \times \mathbb{K}\left(i\sqrt{\frac{2\sqrt{1-\omega(e_1-e_2,e_3-e_2,s-e_2)^2}}{1-\sqrt{1-\omega(e_1-e_2,e_3-e_2,s-e_2)^2}}}\right),$$

where ω is the function defined in (9.8).

(ii)

$$\rho_Q(s) = \frac{1}{2}\sqrt{\frac{1+\sqrt{1-\omega(e_1-e_2,e_3-e_2,s-e_2)^2}}{(e_1-e_3)|s-e_2|\omega(e_1-e_2,e_3-e_2,s-e_2)}} \times F\left(\frac{1}{2},\frac{1}{2},1;-\frac{2\sqrt{1-\omega(e_1-e_2,e_3-e_2,s-e_2)^2}}{1-\sqrt{1-\omega(e_1-e_2,e_3-e_2,s-e_2)^2}}\right).$$

(iii)

$$\rho_Q(s) = \sqrt{\frac{\omega(e_1-e_2,e_3-e_2,s-e_2)}{2(e_1-e_3)|s-e_2|(1+\omega(e_1-e_2,e_3-e_2,s-e_2))}} \times F\left(\frac{1}{2},\frac{1}{2},1;\frac{1-\omega(e_1-e_2,e_3-e_2,s-e_2)}{1+\omega(e_1-e_2,e_3-e_2,s-e_2)}\right).$$

(iv)

$$\rho_Q(s) = \frac{1}{\sqrt{2[(e_1+e_3-2e_2)(s-e_2)+2(e_1-e_2)(e_2-e_3)+(e_1-e_3)|s-e_2|]}}$$
$$\times F\left(\frac{1}{2},\frac{1}{2},1;\frac{(e_1+e_3-2e_2)(s-e_2)+2(e_1-e_2)(e_2-e_3)-(e_1-e_3)|s-e_2|}{(e_1+e_3-2e_2)(s-e_2)+2(e_1-e_2)(e_2-e_3)+(e_1-e_3)|s-e_2|}\right),$$

that is,

$$\rho_Q(s) = \frac{1}{2\pi}\int_{e_2}^{e_1}\frac{dx}{\sqrt{(e_1-x)(x-e_2)(x-e_3)(x-s)}} \quad \text{when } e_3 < s < e_2,$$

$$\rho_Q(s) = \frac{1}{2\pi}\int_{e_3}^{e_2}\frac{dx}{\sqrt{(e_1-x)(e_2-x)(x-e_3)(s-x)}} \quad \text{when } e_2 < s < e_1.$$

Theorem 9.2. *The density function ρ_Q defined in Theorem 9.1 satisfies the following Heun differential equation:*

$$8Q(s)\rho_Q''(s) + 8Q'(s)\rho_Q'(s) + Q''(s)\rho_Q(s) = 0, \tag{9.10}$$

where $Q(s) = (s-e_1)(s-e_2)(s-e_3)$. Both indices of this equation at the finite regular singularities e_1, e_2, e_3 vanish while its indices at ∞ equal $\frac{1}{2}$ and $\frac{3}{2}$.

Corollary 9.1. *The root-counting measures for each of the eight normalized spectral polynomials $R_n^{\kappa_1,\kappa_2,\kappa_3}(n(n+1)t)$ as well as that of the normalized spectral polynomial $R_n(n(n+1)t)$ converge to the measure ν_Q defined in Theorem 9.1, see Figure 9.1 below.*

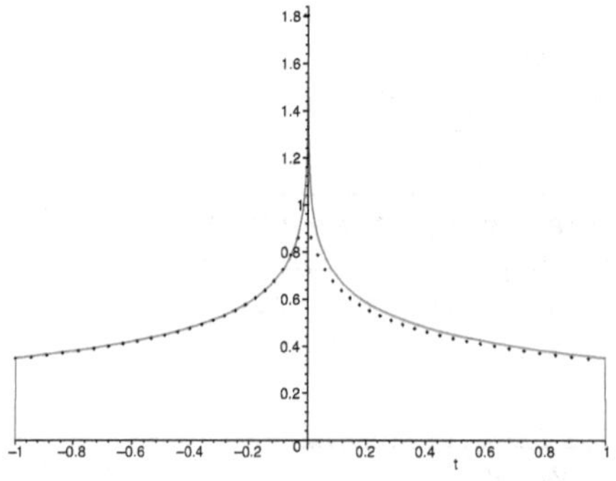

Figure 9.1: Comparison of the theoretical density with the numerical density of the measure ν_Q for $Q(z) = z^3 - z$.

Remark 9.1. Note that equation (9.3) is usually lifted to the elliptic curve Ψ defined by $y^2 = Q(z)$ and that on Ψ this equation takes the Weierstrassian form

$$\left\{\frac{d^2}{du^2} - [n(n+1)\wp(u) + E]\right\} S(u) = 0,$$

where u is the canonical coordinate on the universal covering of Ψ given by

$$u = \int_z^\infty \frac{dw}{\sqrt{Q(w)}}.$$

If $Q(z)$ has all real roots, then $n(n+1)\wp(u)$ becomes a real periodic finite gap potential on $\mathbb{R}$, see [93]. Depending on the parity on n, four of the eight spectral polynomials $R_n^{\kappa_1,\kappa_2,\kappa_3}(E)$ determine the ends of bands of the spectrum of the corresponding Schrödinger operator on $\mathbb{R}$. Hopefully our results will find applications to the spectral theory of finite gap potentials.

To end this introduction, let us mention that in the present context one can actually "guess" the last two formulas in Theorem 9.1 (iv) by WKB-type considerations. However, such arguments fail to provide an accurate mathematical proof. By contrast, our methods use solely rigorous results involving orthogonal polynomials, elliptic integrals, and hypergeometric functions, and we are currently unaware of any possible "shortcuts" in this set-up. Finally, we should also note that so far neither our methods nor WKB-type "guesses" could provide an answer to these questions in more general situations, see § 9.3.

9.2 Proofs

Proof of Theorem 9.1. After some amount of work, it essentially follows from the main result of [112]. First we express the polynomial $Sp_n(t)$ as the characteristic polynomial of a certain matrix. In order to make this matrix tridiagonal so as to simplify the calculations, we assume wlog that

$$Q(z) = (z - e_3)z(z - e_1) = z^3 + vz^2 + wz$$

with $e_3 < e_2 = 0 < e_1$. Set

$$T = (z^3 + vz^2 + wz)\frac{d^2}{dz^2} + (\alpha z^2 + \beta z + \gamma)\frac{d}{dz} - \theta_n(z - t),$$

where v, w, α, β, γ are fixed constants and θ_n, t are variables. Assuming that $S(z) = a_0 z^n + a_1 z^{n-1} + \cdots + a_n$ with undetermined coefficients a_i, $0 \le i \le n$, we are looking for the values of θ_n, t and a_i, $0 \le i \le n$, such that $T(S(z)) = 0$. Note that $T(S(z))$ is in general a polynomial of degree $n+1$ whose leading coefficient equals $a_0[n(n-1) + \alpha n - \theta_n]$. To get a non-trivial solution, we therefore set

$$\theta_n = n(n - 1 + \alpha).$$

Straightforward computations show that the coefficients of the successive powers $z^n, z^{n-1}, \dots, z^0$ in $T(S(z))$ can be expressed in the form of a matrix product $M_n A$, where $A = (a_0, a_1, \dots, a_n)^T$ and M_n is the following tridiagonal $(n+1) \times (n+1)$ matrix:

$$M_n := \begin{pmatrix} \lambda - \xi_{n,1} & \alpha_{n,2} & 0 & 0 & \cdots & 0 & 0 \\ \gamma_{n,2} & \lambda - \xi_{n,2} & \alpha_{n,3} & 0 & \cdots & 0 & 0 \\ 0 & \gamma_{n,3} & \lambda - \xi_{n,3} & \alpha_{n,4} & \cdots & 0 & 0 \\ \vdots & \vdots & \ddots & \ddots & \ddots & \vdots & \vdots \\ 0 & 0 & 0 & \ddots & \ddots & \alpha_{n,n} & 0 \\ 0 & 0 & 0 & \cdots & \gamma_{n,n} & \lambda - \xi_{n,n} & \alpha_{n,n+1} \\ 0 & 0 & 0 & \cdots & 0 & \gamma_{n,n+1} & \lambda - \xi_{n,n+1} \end{pmatrix}$$

with

$$\begin{aligned} \xi_{n,i} &= -\frac{v(n-i)(n-i+1) + \beta(n-i+1)}{\theta_n}, \quad i \in \{1, \dots, n+1\}, \\ \alpha_{n,i} &= \frac{(n-i)(n-i+1) + a(n-i+1)}{\theta_n} - 1, \quad i \in \{2, \dots, n+1\}, \\ \gamma_{n,i} &= \frac{w(n-i+1)(n-i+2) + \gamma(n-i+2)}{\theta_n}, \quad i \in \{2, \dots, n+1\}. \end{aligned} \tag{9.11}$$

A similar matrix can be found in [82] and also in [188]. The matrix M_n depends linearly on the indeterminate λ, which appears only on its main diagonal. If the linear system $M_n A = 0$ has a non-trivial solution $A = (a_0, a_1, \dots, a_n)^T$, then the determinant of M_n has to vanish. This gives the polynomial equation

$$Sp_n(\lambda) = \det(M_n) = 0.$$

The sequence of polynomials $\{Sp_n(\lambda)\}_{n \in \mathbb{Z}_+}$ does not seem to satisfy any finite linear recurrence relation. In order to overcome this difficulty and to be able to use the technique of 3-term recurrence relations with variable coefficients (which is applicable since M_n is tridiagonal), we extend the above polynomial sequence by introducing an additional index. Namely, define

$$Sp_{n,i}(\lambda) = \det M_{n,i}, \quad i \in \{1, \dots, n+1\},$$

where $M_{n,i}$ is the principal upper $(i \times i)$-submatrix of M_n. One can easily check (see, e. g., [15, p. 20]) that the following 3-term relation holds:

$$Sp_{n,i}(\lambda) = (\lambda - \xi_{n,i}) Sp_{n,i-1}(\lambda) - \psi_{n,i} Sp_{n,i-2}(\lambda), \quad i \in \{1, \dots, n+1\}, \tag{9.12}$$

where $\xi_{n,i}$ is as in (9.11) and

$$\psi_{n,i} = \alpha_{n,i} \gamma_{n,i}, \quad i \in \{2, \dots, n+1\}. \tag{9.13}$$

Here we used the (standard) initial conditions $Sp_{n,0}(\lambda) = 1$, $Sp_{n,-1}(\lambda) = 0$. It is well known that if all $\xi_{n,i}$s are real and all $\psi_{n,i}$s are positive, then the polynomials $Sp_{n,i}(\lambda)$, $i \in \{0, \dots, n+1\}$, form a (finite) sequence of orthogonal polynomials. In particular, all their roots are real. Under the assumptions of Theorem 9.1 the reality of $\xi_{n,i}$ is obvious. Assuming that the positivity of $\psi_{n,i}$ is also settled (see Lemma 9.1 below), let us complete the proof of Theorem 9.1. For this we invoke [112, Theorem 1.4] which, translated in our notation, claims that if there exist two continuous functions $\xi(\tau)$ and $\psi(\tau)$, $\tau \in [0,1]$ such that

$$\lim_{i/(n+1)\to\tau} \xi_{i,n} = \xi(\tau), \qquad \lim_{i/(n+1)\to\tau} \psi_{i,n} = \psi(\tau), \quad \psi(\tau) \geq 0 \quad \forall \tau \in [0,1],$$

then the density of the asymptotic root-counting measure of the polynomial sequence $\{Sp_n(\lambda)\}_{n\in\mathbb{Z}_+} = \{Sp_{n,n+1}(\lambda)\}_{n\in\mathbb{Z}_+}$ is given by

$$\int_0^1 \omega_{[\xi(\tau)-2\sqrt{\psi(\tau)},\xi(\tau)+2\sqrt{\psi(\tau)}]}\,d\tau,$$

where for any $x < y$ one has

$$\omega_{[x,y]}(s) = \begin{cases} \frac{1}{\pi\sqrt{(y-s)(s-x)}} & \text{if } s \in [x,y], \\ 0 & \text{otherwise.} \end{cases}$$

From the explicit formulas for $\xi_{n,i}$ and $\psi_{n,i}$ (see (9.11) and (9.13)), one easily gets

$$\xi(\tau) = \lim_{i/(n+1)\to\tau} \xi_{i,n} = -v(1-\tau)^2,$$
$$\psi(\tau) = \lim_{i/(n+1)\to\tau} \psi_{i,n} = -w(1-(1-\tau)^2)(1-\tau)^2.$$

Notice that the above limits are independent of the coefficients α, β, γ and that by the assumption on $Q(z)$ made at the beginning of this section one has $w = e_3e_1 < 0$, which in its turn implies that $\psi(\tau) \geq 0$ for $\tau \in [0,1]$. The required density $\rho_Q(s)$ is therefore given by

$$\rho_Q(s) = \int_0^1 \frac{d\tau}{\pi\sqrt{(\xi(\tau)+2\sqrt{\psi(\tau)}-s)(s-\xi(\tau)+2\sqrt{\psi(\tau)})}^{\,+}}$$
$$= \int_0^1 \frac{d\tau}{\pi\sqrt{-4w(1-(1-\tau)^2)(1-\tau)^2-(v(1-\tau)^2+s)^2}^{\,+}},$$

where $\sqrt{\cdot}^{\,+}$ is meant to remind that the integrand vanishes whenever the expression under the square root becomes negative. Introducing $\sqrt{\theta} = 1-\tau$, we get

$$\rho_Q(s) = \frac{1}{2\pi}\int_0^1 \frac{d\theta}{\sqrt{\theta[(4w-v^2)\theta^2-(4w+2vs)\theta-s^2]}^{\,+}}.$$

In order to get rid of $\sqrt{\cdot}^{+}$, we rewrite

$$\rho_Q(s) = \frac{1}{2\pi\sqrt{v^2 - 4w}} \int_{\theta_{\min}(s)}^{\theta_{\max}(s)} \frac{d\theta}{\sqrt{\theta(\theta - \theta_{\min}(s))(\theta_{\max}(s) - \theta)}}, \tag{9.14}$$

where $\theta_{\min}(s)$ and $\theta_{\max}(s)$ are the minimal and maximal roots of the equation

$$(4w - v^2)\theta^2 - (4w + 2vs)\theta - s^2 = 0$$

with respect to θ, that is,

$$\begin{cases} \theta_{\min}(s) = -(v^2 - 4w)^{-1}[2w + vs + 2\sqrt{w(w + vs + s^2)}], \\ \theta_{\max}(s) = -(v^2 - 4w)^{-1}[2w + vs - 2\sqrt{w(w + vs + s^2)}]. \end{cases}$$

A few remarks are in order at this stage. First, by the real-rootedness of $Q(z)$ one has $v^2 - 4w = (e_1 - e_3)^2 > 0$. Second, we claim that

$$[\xi(\tau) - 2\sqrt{\psi(\tau)}, \xi(\tau) + 2\sqrt{\psi(\tau)}] \subseteq [e_3, e_1]$$

for $\tau \in [0, 1]$. Indeed,

$$\begin{aligned} e_1 - \xi(\tau) - 2\sqrt{\psi(\tau)} &= e_1 + v(1-\tau)^2 - 2\sqrt{-w(1-(1-\tau)^2)(1-\tau)^2} \\ &= e_1(1-(1-\tau)^2) - 2\sqrt{e_1(1-(1-\tau)^2)}\sqrt{-e_3(1-\tau)^2} - e_3(1-\tau)^2 \\ &= \big[\sqrt{e_1(1-(1-\tau)^2)} - \sqrt{-e_3(1-\tau)^2}\big]^2 \geq 0, \end{aligned}$$

and similarly

$$\begin{aligned} \xi(\tau) - 2\sqrt{\psi(\tau)} - e_3 &= -v(1-\tau)^2 - 2\sqrt{-w(1-(1-\tau)^2)(1-\tau)^2} - e_3 \\ &= e_1(1-\tau)^2 - 2\sqrt{e_1(1-\tau)^2}\sqrt{-e_3(1-(1-\tau)^2)} - e_3(1-(1-\tau)^2) \\ &= \big[\sqrt{e_1(1-\tau)^2} - \sqrt{-e_3(1-(1-\tau)^2)}\big]^2 \geq 0. \end{aligned}$$

It follows that whenever s is such that the integrand in the above formulas is non-vanishing, one has $e_3 \leq s \leq e_1$. Hence

$$w(w + vs + s^2) = -e_3e_1(s - e_3)(e_1 - s) \geq 0 \quad \text{and} \quad 2w + vs = e_1(e_3 - s) + e_3(e_1 - s) < 0$$

for all s as above. Therefore

$$\theta_{\min}(s) + \theta_{\max}(s) = -2(2w + vs)(v^2 - 4w) > 0 \quad \text{and} \quad \theta_{\min}(s)\theta_{\max}(s) = s^2(v^2 - 4w) \geq 0$$

from which we conclude that $\theta_{\max}(s) > \theta_{\min}(s) \geq 0$ for $e_3 < s < e_1$.

Recall the definition and properties of the function ω given by (9.8)–(9.9) and note that

$$\omega(e_1, e_3, s) = \frac{(e_1 - e_3)|s|}{e_1(s - e_3) - e_3(e_1 - s)} = -\frac{|s|\sqrt{v^2 - 4w}}{2w + vs} \tag{9.15}$$

so that $\theta_{\min}(s)$ and $\theta_{\max}(s)$ may actually be rewritten as follows:

$$\theta_{\min}(s) = \frac{|s|(1 - \sqrt{1 - \omega(e_1, e_3, s)^2})}{(e_1 - e_3)\omega(e_1, e_3, s)}, \quad \theta_{\max}(s) = \frac{|s|(1 + \sqrt{1 - \omega(e_1, e_3, s)^2})}{(e_1 - e_3)\omega(e_1, e_3, s)}.$$

Using these expressions combined with the fact that for any $x < y$ one has

$$\int_x^y \frac{d\theta}{\sqrt{\theta(\theta - x)(y - \theta)}} = \frac{2}{\sqrt{x}}\mathbb{K}\left(\sqrt{\frac{x - y}{x}}\right)$$

(which readily follows from (9.6)), we deduce from (9.14) that

$$\rho_Q(s) = \frac{1}{\pi}\sqrt{\frac{1 + \sqrt{1 - \omega(e_1, e_3, s)^2}}{(e_1 - e_3)|s|\omega(e_1, e_3, s)}}\,\mathbb{K}\left(i\sqrt{\frac{2\sqrt{1 - \omega(e_1, e_3, s)^2}}{1 - \sqrt{1 - \omega(e_1, e_3, s)^2}}}\right). \tag{9.16}$$

Now from (9.6)–(9.7) and the well-known identities $\Gamma(1) = 1$, $\Gamma(\frac{1}{2}) = \sqrt{\pi}$, one easily gets

$$\mathbb{K}(\zeta) = \frac{\pi}{2}F\left(\frac{1}{2}, \frac{1}{2}, 1; \zeta^2\right),$$

hence

$$\rho_Q(s) = \frac{1}{2}\sqrt{\frac{1 + \sqrt{1 - \omega(e_1, e_3, s)^2}}{(e_1 - e_3)|s|\omega(e_1, e_3, s)}}\,F\left(\frac{1}{2}, \frac{1}{2}, 1; -\frac{2\sqrt{1 - \omega(e_1, e_3, s)^2}}{1 - \sqrt{1 - \omega(e_1, e_3, s)^2}}\right) \tag{9.17}$$

by (9.16). It is a remarkable fact due to Kummer and Goursat that for special choices of the numbers a, b, c the hypergeometric series $F(a, b, c; z)$ obeys certain quadratic transformation laws. To complete the proof, we need precisely such transformation properties, namely formulas 15.3.19–20 in [2] with $a = \frac{1}{4}$, $b = \frac{1}{2}$ and $c = 1$:

$$\frac{1}{\sqrt{1 - \sqrt{z}}}F\left(\frac{1}{2}, \frac{1}{2}, 1; -\frac{2\sqrt{z}}{1 - \sqrt{z}}\right) = \sqrt{\frac{2}{1 + \sqrt{1 - z}}}\,F\left(\frac{1}{2}, \frac{1}{2}, 1; \frac{1 - \sqrt{1 - z}}{1 + \sqrt{1 - z}}\right).$$

The above identity for $z = 1 - \omega(e_1, e_3, s)^2$ together with (9.17) then yields

$$\rho_Q(s) = \sqrt{\frac{\omega(e_1, e_3, s)}{2(e_1 - e_3)|s|(1 + \omega(e_1, e_3, s))}} F\left(\frac{1}{2}, \frac{1}{2}, 1; \frac{1 - \omega(e_1, e_3, s)}{1 + \omega(e_1, e_3, s)}\right) \tag{9.18}$$

which by (9.15) amounts to

$$\rho_Q(s) = \frac{1}{\sqrt{2[(e_1 + e_3)s - 2e_1e_3 + (e_1 - e_3)|s|]}} \times F\left(\frac{1}{2}, \frac{1}{2}, 1; \frac{(e_1 + e_3)s - 2e_1e_3 - (e_1 - e_3)|s|}{(e_1 + e_3)s - 2e_1e_3 + (e_1 - e_3)|s|}\right). \tag{9.19}$$

To prove the last two formulas of Theorem 9.1 (iv), we use (9.19) and (9.7) with $a = b = \frac{1}{2}$ and $c = 1$ in order to get an integral representation of $\rho_Q(s)$ in which one makes the following variable substitutions:

$$t = -\frac{e_3(e_1 - x)}{e_1(x - e_3)} \quad \text{if } e_3 < s < 0, \qquad t = -\frac{e_1(x - e_3)}{e_3(e_1 - x)} \quad \text{if } 0 < s < e_1.$$

The desired expressions are then obtained by straightforward computations.

We thus established all formulas stated in Theorem 9.1 in the special case when $e_3 < e_2 = 0 < e_1$. The general case reduces to this one by noticing that if $e_3 < e_2 < e_1$ and $Q(z) = (z-e_3)(z-e_2)(z-e_1)$ then all the above arguments may be used for the polynomial $Q(z + e_2)$. Hence the expressions for $\rho_Q(s)$ given in Theorem 9.1 in the general case are obtained from (9.16)–(9.19) simply by replacing e_1, e_3, and s with $e_1 - e_2$, $e_3 - e_2$, and $s - e_2$, respectively. □

Lemma 9.1. *Let $Q(z) = z^3 + vz^2 + wz$ and $P(z) = \alpha z^2 + \beta z + \gamma$ be two polynomials such that $Q(z)$ has three real distinct roots $e_1 > e_2 = 0 > e_3$ and*

$$\frac{P(z)}{Q(z)} = \frac{\alpha_1}{z - e_1} + \frac{\alpha_2}{z - e_2} + \frac{\alpha_3}{z - e_3}$$

with $\alpha_1, \alpha_2, \alpha_3 > 0$. Then the coefficients $\psi_{n,i}$ defined in (9.13) *are all positive.*

Proof. One immediately gets $\alpha > 0$, which implies that $\theta_n > 0$ for any positive integer n. Recall from (9.11) and (9.13) that $\psi_{n,i} = \alpha_{n,i}\gamma_{n,i}$. Now

$$\alpha_{n,i} = \theta_n^{-1}\{[(n - i)(n - i + 1) - n(n - 1)] + \alpha[(n - i + 1) - n]\}$$

and since both α and θ_n are positive, it follows that $\alpha_{n,i} < 0$. Let us show that under our assumptions one also has $\gamma_{n,i} < 0$. For this, it is clearly enough to show that both w and γ are negative. Obviously, $w = e_3e_1 < 0$ and since $\alpha_1, \alpha_2, \alpha_3 > 0$, the quadratic polynomial $P(z)$ has two real roots interlacing with e_3, e_2, e_1. Therefore $P(z)$ has one positive and one negative root and positive leading coefficient hence $\gamma = P(0) < 0$. □

Proof of Theorem 9.2. In order to deduce the differential equation satisfied by $\rho_Q(s)$, we note first that the restrictions of $\rho_Q(s)$ to (e_3, e_2) and (e_2, e_1), respectively, are two branches of the same multi-valued analytic function (this can be seen, e. g., from the last two expressions in Theorem 9.1). Thus it suffices to derive the linear differential equation satisfied by $\rho_Q(s)$ restricted to, say, (e_2, e_1). Specializing formula (iv) of Theorem 9.1 to this case, we get

$$\rho_Q(s) = \frac{I_Q(s)}{2\pi\sqrt{(e_1-e_2)(s-e_3)}},$$

where

$$I_Q(s) = \int_0^1 \frac{dw}{\sqrt{w(1-w)(1-\frac{(e_2-e_3)(e_1-s)}{(e_1-e_2)(s-e_3)}w)}}$$

and $s \in (e_2, e_1)$. By (9.7) we see that up to a constant factor $I_Q(s)$ is the same as the hypergeometric series $F(\frac{1}{2}, \frac{1}{2}, 1; \frac{(e_2-e_3)(e_1-s)}{(e_1-e_2)(s-e_3)})$, which is known to satisfy the following Riemann differential equation (see, e. g., [2, § 15.6]):

$$I_Q''(s) + \left(\frac{1}{s-e_1} + \frac{1}{s-e_2}\right)I_Q'(s) + \frac{(e_3-e_2)(e_3-e_1)}{4(s-e_3)Q(s)}I_Q(s) = 0.$$

Substituting $I_Q(s) = K\sqrt{s-e_3}\rho_Q(s)$, where $K := 2\pi\sqrt{e_1-e_2}$ is a constant, in the latter equation we get after some straightforward algebraic manipulations

$$\rho_Q''(s) + \left(\sum_{i=1}^{3}\frac{1}{s-e_i}\right)\rho_Q'(s) + \frac{3s-\sum_{i=1}^{3}e_i}{4\prod_{i=1}^{3}(s-e_i)}\rho_Q(s) = 0, \tag{9.20}$$

which after multiplication by $8Q(s)$ coincides with the required equation (9.10). To calculate the indices, recall from, e. g., [94] that for a second order linear differential equation $\rho''(s)+a_1(s)\rho'(s)+a_2(s)\rho(s) = 0$ its indicial equation at a finite regular or regular singular point $\tilde{s}$ has the form

$$\zeta(\zeta-1) + \alpha_1\zeta + \alpha_2 = 0,$$

where $\alpha_1 = \lim_{s\to\tilde{s}}(s-\tilde{s})a_1(s)$ and $\alpha_2 = \lim_{s\to\tilde{s}}(s-\tilde{s})^2a_2(s)$. Thus for (9.20) the indicial equation at each e_i, $i = 1, 2, 3$, has the form $\zeta^2 = 0$, which implies that both corresponding indices vanish and that a solution to (9.20) might have a logarithmic singularity at any of these points, see Remark 9.2 below. The indicial equation of $\rho''(s)+a_1(s)\rho'(s)+a_2(s)\rho(s) = 0$ at ∞ has the form

$$\zeta(\zeta+1) - \tilde{\alpha}_1\zeta + \tilde{\alpha}_2 = 0,$$

where $\tilde{\alpha}_1 = \lim_{s\to\infty} s a_1(s)$ and $\tilde{\alpha}_2 = \lim_{s\to\infty} s^2 a_2(s)$. Therefore the indicial equation at ∞ for (9.20) is $\zeta(\zeta+1) - 3\zeta + \frac{3}{4} = 0$ whose roots are $\frac{1}{2}$ and $\frac{3}{2}$. □

Remark 9.2. It is not difficult to show that

$$\rho_Q(s) \approx \frac{1}{2\pi} \frac{\log\left(\frac{16(e_1-e_2)(e_2-e_3)}{(e_1-e_3)|s-e_2|}\right)}{\sqrt{(e_1-e_2)(e_2-e_3)}} \quad \text{as } s \to e_2,$$

so that $\rho_Q(s)$ always has a logarithmic singularity at e_2.

Proof of Corollary 9.1. The main idea of the proof of Corollary 9.1 is that the polynomial part of any Lamé solution to (9.3) itself satisfies a very similar differential equation. Let us illustrate this in the case of Lamé solutions of type 2. (The other cases can be dealt with in the same way.) Assume that $\tilde{S}(z)$ is a polynomial of degree n such that $S(z) := (z-e_1)^{\frac{1}{2}}(z-e_2)^{\frac{1}{2}}\tilde{S}(z)$ solves the equation

$$4Q(z)\left[S''(z) + \frac{1}{2}\left(\sum_{i=1}^{3} \frac{1}{z-e_i}\right)S'(z)\right] + V(z)S(z) = 0$$

for some linear polynomial $V(z)$. Recall that the (unique) root of each such $V(z)$ is also a root of the scaled spectral polynomial $R_n^{\frac{1}{2},\frac{1}{2},0}(n(n+1)\lambda)$. Substituting $S(z) = (z-e_1)^{\frac{1}{2}}(z-e_2)^{\frac{1}{2}}\tilde{S}(z)$, one gets after some straightforward calculations that $\tilde{S}(z)$ satisfies the equation

$$4Q(z)\left[\tilde{S}''(z) + \frac{1}{2}\left(\frac{3}{z-e_1} + \frac{3}{z-e_2} + \frac{1}{z-e_3}\right)\tilde{S}'(z)\right] + \tilde{V}(z)\tilde{S}(z) = 0, \tag{9.21}$$

where $\tilde{V}(z) = V(z)+(z-e_1)-(z-e_2)+4(z-e_3)$. Notice that in the coefficient in front of $\tilde{S}'(z)$ the numerators of the first two simple fractions are increased by 1 while the remaining one is unchanged. Thus we are practically in the situation covered by Theorem 9.1. The only difference is that we are not considering the asymptotic distribution of the roots of polynomials $\tilde{V}(z)$ but that of $V(z)$. However, since $\tilde{V}(z) = V(z)+(z-e_1)-(z-e_2)+4(z-e_3)$ and the leading coefficient θ_n of $V(z)$ tends to ∞ as $n \to \infty$, it follows that both families of polynomials actually have the same asymptotic root distribution. Finally, note that by (9.4) the scaled spectral polynomial $R_n(n(n+1)\lambda)$ equals the product of four corresponding polynomials $R_n^{\kappa_1,\kappa_2,\kappa_3}(n(n+1)\lambda)$ and that all eight polynomial families $\{R_n^{\kappa_1,\kappa_2,\kappa_3}(n(n+1)\lambda)\}_{n\in\mathbb{Z}_+}$ have the same asymptotic root distribution when $n \to \infty$. This implies that the family $\{R_n(n(n+1)\lambda)\}_{n\in\mathbb{Z}_+}$ itself has the same limiting root distribution. □

9.3 Open problems in Chapter 9

9.1. More difficulties occur when dealing with the more general Lamé equation (9.2) since in this case Van Vleck polynomials and their roots cannot be found by means of a determinantal equation. They are in fact related to a more complicated situation when the rank of a certain non-square matrix is less than the maximal one.

Problem 9.1. Describe the asymptotic distribution of the roots of all Van Vleck polynomials for equation (9.2) when $n \to \infty$.

An illustration of this asymptotic distribution is given in Figure 9.2.

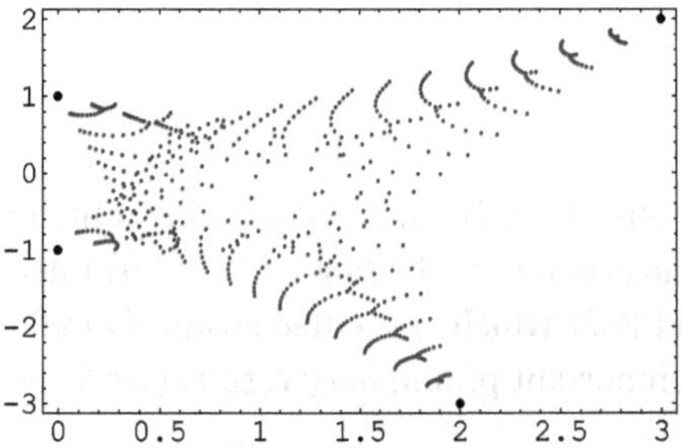

Figure 9.2: The union of the roots of 861 quadratic Van Vleck polynomials corresponding to Stieltjes polynomials of degree 40 for the classical Lamé equation $Q(z)S''(z) + \frac{Q'(z)}{2}S'(z) + V(z)S(z) = 0$ with $Q(z) = (z^2+1)(z-3i-2)(z+2i-3)$.

10 Asymptotics of spectral polynomials, complex case

(The material of this chapter is based on [167] and [168].)

10.1 Preliminaries

In this chapter we study of spectral polynomials of the Heun equation with complex coefficients. It is divided in two interrelated parts where in the first part we formulate two main statements (see Propositions 10.1 and 10.2), which are settled positively in the second part of this chapter by using several important preliminary results (see Theorems 10.2 and 10.3) proven in the first part.[1]

As above a *generalized Lamé equation* is a second order differential equation of the form

$$\left\{Q(z)\frac{d^2}{dz^2}+P(z)\frac{d}{dz}+V(z)\right\}S(z)=0, \tag{10.1}$$

where $Q(z)$ is a complex polynomial of degree l and $P(z)$ is a complex polynomial of degree at most $l-1$, see [197].

Below we concentrate on the classical case $l = \deg Q(z) = 3$, which is better known under the name *the Heun differential equation*, see, e. g., [157] and study the union of all roots of polynomials $V(z)$ whose Stieltjes polynomials have degree n as $n \to \infty$. Note that if $l = \deg Q(z) = 3$ then $V(z)$ is at most linear and that for a given value of the positive integer n there are at most $n+1$ such polynomials.

Before we move further, let us formulate appropriate versions of two main results of [162] generalizing the statements of Heine and Pólya, see Chapter 8.

Theorem 10.1. *For any polynomial $Q(z)$ of degree l and any polynomial $P(z)$ of degree at most $l-1$,*

- *There exists N such that for any $n \geq N$, there exist exactly $\binom{n+l-2}{n}$ polynomials $V(z), \deg V(z) = l-2$ counted with appropriate multiplicity such that* (10.1) *has a polynomial solution $S(z)$ of degree exactly n.*
- *For any $\epsilon > 0$, there exists N_ϵ such that for any $n \geq N_\epsilon$ any root of any above $V(z)$ and $S(z)$ lie in the ϵ-neighborhood of $Conv_Q$.*

Applying the latter result to the situation $l = 3$, i. e., to the Heun equation, we can introduce the set $\mathcal{V}_n$ consisting of polynomials $V(z)$ giving a polynomial solution $S(z)$

1 The reason for this division is that these two theorems are to a very large extent independent of the results of the second part, which are technically more complicated. An interested reader might understand the whole picture from the first part not going into too many technical details.

https://doi.org/10.1515/9783112207741-012

of (10.1) of degree n; each such $V(z)$ appearing the number of times equal to its multiplicity. Then by the above results the set $\mathcal{V}_n$ will contain exactly $n+1$ linear polynomials for all sufficiently large n. As in Chapter 9 it will be convenient to introduce a sequence $\{Sp_n(\lambda)\}$ of *spectral polynomials* where the nth spectral polynomial is defined by

$$Sp_n(\lambda) = \prod_{j=1}^{n+1}(\lambda - t_{n,j}),$$

where $t_{n,j}$ is the unique root of the jth polynomial in $\mathcal{V}_n$ in any fixed ordering. ($Sp_n(\lambda)$ are well defined for all sufficiently large n.)

Associate to $Sp_n(\lambda)$ the finite measure

$$\nu_n = \frac{1}{n+1}\sum_{j=1}^{n+1}\delta(z - t_{n,j}),$$

where $\delta(z-a)$ is the Dirac measure supported at a. The measure ν_n obtained in this way is clearly a real probability measure which one usually refers to as the *root-counting measure* of the polynomial $Sp_n(\lambda)$.

The starting point of this project was numerical results for the distribution of roots of $Sp_n(\lambda)$ illustrated in Figure 10.1.

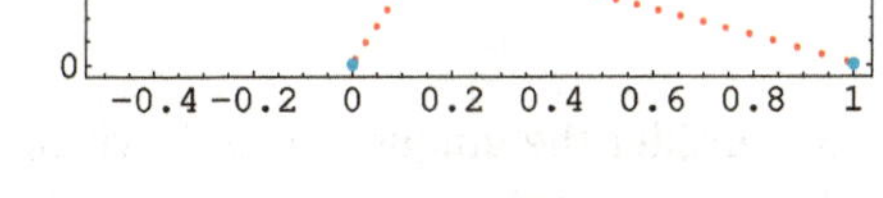

Figure 10.1: The roots of the spectral polynomial $Sp_{50}(\lambda)$ for the classical Lamé equation $\{Q(z)\frac{d^2}{dz^2} + \frac{1}{2}Q'(z)\frac{d}{dz}+V(z)\}S(z) = 0$, with $Q(z) = z(z-1)(z+\frac{1}{2}-i)$.

These numerical experiments strongly suggested that the following statement holds.

Proposition 10.1. *For any equation* (10.1) *the sequence* $\{\nu_n\}$ *of the root-counting measures of its spectral polynomials converges to a probability measure* ν *supported on the union of three curved segments located inside* $Conv_Q$ *and connecting the three roots of* $Q(z)$ *with a certain interior point, see Figure* 10.1. *Moreover, the limiting measure* ν *depends only on* $Q(z)$, *i. e., is independent of* $P(z)$.

An elegant explicit description of the support of ν was suggested by Professor K. Takemura.

Denote the three roots of $Q(z)$ by a_1, a_2, a_3. For $i \in \{1,2,3\}$ consider the curve γ_i given as the set of all b satisfying the relation

$$\int_{a_j}^{a_k} \sqrt{\frac{b-t}{(t-a_1)(t-a_2)(t-a_3)}}\,dt \in \mathbb{R}, \tag{10.2}$$

here j and k are the remaining two indices in $\{1,2,3\}$ in any order and the integration is taken over the straight interval connecting a_j and a_k. One can see that the roots a_i belong to γ_i and that these three curves connect the corresponding a_i with a common point within $Conv_Q$. Take a segment of γ_i connecting a_i to the common intersection point of all γs. Let us denote the union of these three segments by Γ_Q.

Proposition 10.2. *The support of the limiting root-counting measure ν coincides with the above Γ_Q.*

In the second part of this chapter we completely settle the above Propositions 10.1 and 10.2 using some methods and results presented here. In particular, Theorems 10.2 and 10.3 below will be highly instrumental in proving the existence and uniqueness of the above measure ν.

Generalizing the technique of Chapter 9 (see [112]), we mainly study another probability measure which is easily described and from which the measure ν is obtained by the inverse balayage. In other words, the support of ν will be contained in the support of the measure that we construct and they have the same logarithmic potential outside the support of the latter one. This measure will be uniquely determined by the choice of a root of $Q(z)$, and thus we are in fact constructing three different measures having the same measure ν as their inverse balayage.

10.1.1 Constructing the measure

Choosing one of the three vertices a_i, $i = \{1,2,3\}$ consider the unique ellipse E_i which: a) passes through a_i and b) has a_j, a_k as its foci. The constructed probability measure M_i is supported on the domain $\tilde{E}_i$ bounded by the ellipse E_i. We need the following notion.

Given two distinct points $a_1 \neq a_2$ on $\mathbb{C}$, define the *arcsine measure* $\omega_{[a_1,a_2]}$ of the interval $[a_1,a_2]$ as the measure supported on $[a_1,a_2]$ and whose density at a point $t \in [a_1,a_2]$ equals $\frac{1}{\pi\sqrt{|(t-a_1)(t-a_2)|}}$.

To describe the measure M_i, consider the family of straight lines parallel to the tangent line to the ellipse E_i at a_i. Take now the family Φ_i of intervals obtained by intersection of the latter straight lines with the elliptic domain $\tilde{E}_i$. Denote by $-v_i$ the vector connecting a_i with its opposite point on E_i, i. e., draw the straight line through a_i and the center of E_i till it hits E_i again and take the difference of the latter and the former points. (One can easily check that if we introduce a new variable $z_i = z - a_i$ and express $Q(z) = z_i^3 + v_i z_i^2 + w_i z_i$, then the above vector will be exactly $-v_i$ in the expression for

$Q(z)$, which explains our notation.) Now parameterize the above family Φ_i of the intervals by their middle points using the formula $-v_i\theta^2$, $\theta \in [0, 1]$. Consider the family ν_θ of arcsine measures of these intervals. Finally the required measure M_i is obtained by the averaging of ν_θ w. r. t. parameter θ, i. e., $M_i = \int_0^1 \nu_\theta d\theta$, see Figure 10.2b).

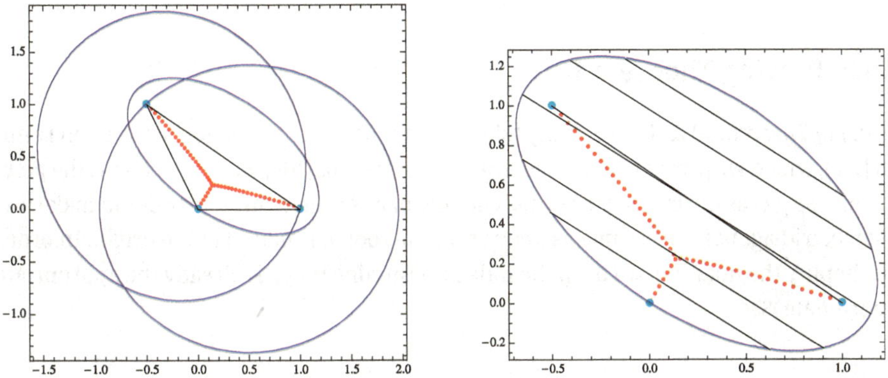

Figure 10.2: a) The measure ν and the three ellipses E_1, E_2, E_3 for $Q(z) = z(z-1)(z + \frac{1}{2} - i)$. b) The measure ν, ellipse E_1, and several straight segments belonging to the family Φ_1.

As above, the Cauchy transform $\mathcal{C}_\nu(z)$ and the logarithmic potential $u_\nu(z)$ of a (complex-valued) measure ν supported in $\mathbb{C}$ are given by

$$\mathcal{C}_\nu(z) = \int_{\mathbb{C}} \frac{d\nu(\xi)}{z - \xi} \quad \text{and} \quad u_\nu(z) = \int_{\mathbb{C}} \log|z - \xi| d\nu(\xi).$$

Definition 10.1. Given two positive measures μ_1 and μ_2 in the complex plane such that *supp* $\mu_2 \subset$ *supp* μ_1, we say μ_2 is *obtained by the inverse balayage* from μ_1 if their logarithmic potentials coincide in $\mathbb{C} \setminus$ *supp* μ_1.

Now we can finally formulate the main results of the first part of this chapter.

Theorem 10.2. *Each of the measures M_i, $i \in \{1, 2, 3\}$ have ν as its inverse balayage, i. e., ν and M_i have the same logarithmic potential (and the same Cauchy transform) outside the ellipse E_i and the support of ν is contained inside the support of M_i.*

Remark 10.1. Notice that the rigorous proof of the existence of ν is postponed until the second part. In the proofs of the first part of this chapter we assume the validity of this existence.

Denote by $\mathcal{C}_{Q_i}(z)$ the Cauchy transform of the measure M_i, $i = 1, 2, 3$. The following result demonstrates that each Cauchy transform $\mathcal{C}_{Q_i}(z)$ satisfies outside the elliptic domain $\tilde{E}_i$ the following peculiar linear non-homogeneous second order differential equation (similar to the one obtained earlier in Chapter 9, see [162]).

Theorem 10.3. *The Cauchy transforms $C_{Q_i}(z)$ of the measures M_i, $i = 1, 2, 3$ defined in Theorem 10.2 satisfy outside the ellipses E_i the same linear non-homogeneous differential equation given by*

$$Q(z)C''_{Q_i}(z) + Q'(z)C'_{Q_i}(z) + \frac{Q''(z)}{8}C_{Q_i}(z) + \frac{Q'''(z)}{24} = 0. \tag{10.3}$$

10.2 Proving Theorem 10.2

Proof of Theorem 10.2. It essentially follows from the stronger version of the main result of [112], which we present below, see, e. g., Appendix II of [169]. First we express the polynomial $Sp_n(\lambda)$ as the characteristic polynomial of a certain matrix. In order to make this matrix tridiagonal, we assume as above that the root a_i is placed at the origin. In order to simplify the notation, we drop the index i, assuming that z is already the appropriate coordinate. Set

$$Q(z) = z^3 + vz^2 + wz.$$

Consider the operator

$$T = (z^3 + vz^2 + wz)\frac{d^2}{dz^2} + (\alpha z^2 + \beta z + \gamma)\frac{d}{dz} - \theta_n(z - \lambda),$$

where $v, w, \alpha, \beta, \gamma$ are fixed coefficients of $Q(z)$ and $P(z)$ respectively and θ_n, λ are variables. Assuming that $S(z) = u_0 z^n + u_1 z^{n-1} + \cdots + u_n$ with undetermined coefficients u_i, $0 \le i \le n$, and we will be looking for the values of θ_n, λ, and u_i, $0 \le i \le n$, such that $T(S(z)) = 0$. Note that $T(S(z))$ is in general a polynomial of degree $n + 1$ whose leading coefficient equals $u_0[n(n-1) + \alpha n - \theta_n]$. To get a non-trivial solution, we therefore set

$$\theta_n = n(n - 1 + \alpha).$$

Straightforward computations show that the coefficients of the successive powers $z^n, z^{n-1}, \ldots, z^0$ in $T(S(z))$ can be expressed in the form of a matrix product $M_n U$, where $U = (u_0, u_1, \ldots, u_n)^T$ and M_n is the following tridiagonal $(n+1) \times (n+1)$ matrix

$$M_n := \begin{pmatrix} \lambda - \xi_{n,1} & \alpha_{n,2} & 0 & 0 & \cdots & 0 & 0 \\ \gamma_{n,2} & \lambda - \xi_{n,2} & \alpha_{n,3} & 0 & \cdots & 0 & 0 \\ 0 & \gamma_{n,3} & \lambda - \xi_{n,3} & \alpha_{n,4} & \cdots & 0 & 0 \\ \vdots & \vdots & \ddots & \ddots & \ddots & \vdots & \vdots \\ 0 & 0 & 0 & \ddots & \ddots & \alpha_{n,n} & 0 \\ 0 & 0 & 0 & \cdots & \gamma_{n,n} & \lambda - \xi_{n,n} & \alpha_{n,n+1} \\ 0 & 0 & 0 & \cdots & 0 & \gamma_{n,n+1} & \lambda - \xi_{n,n+1} \end{pmatrix}$$

with

$$\begin{aligned}
\xi_{n,i} &= -\frac{v(n-i)(n-i+1)+\beta(n-i+1)}{\theta_n}, \quad i \in \{1,\dots,n+1\},\\
\alpha_{n,i} &= \frac{(n-i)(n-i+1)+\alpha(n-i+1)}{\theta_n} - 1, \quad i \in \{2,\dots,n+1\}, \qquad (10.4)\\
\gamma_{n,i} &= \frac{w(n-i+1)(n-i+2)+\gamma(n-i+2)}{\theta_n}, \quad i \in \{2,\dots,n+1\}.
\end{aligned}$$

A similar matrix can be found in Chapter 9, see [82] and also in [187]. The matrix M_n depends linearly on the indeterminate λ, which appears only on its main diagonal. Obviously if the linear homogeneous system $M_n U = 0$ is to have a non-trivial solution $U = (u_0, u_1, \dots, u_n)^T$ the determinant of M_n has to vanish. This gives the required polynomial equation

$$Sp_n(\lambda) = \det(M_n) = 0.$$

It seems that the sequence of polynomials $\{Sp_n(\lambda)\}_{n\in\mathbb{Z}_+}$ does not satisfy any finite linear recurrence relation. In order to overcome this difficulty and to be able to use the technique of 3-term recurrence relations with variable coefficients (which is applicable since M_n is tridiagonal), we extend the above polynomial sequence by introducing an additional parameter as in the previous chapter. Namely, define

$$Sp_{n,i}(\lambda) = \det M_{n,i}, \quad i \in \{1,\dots,n+1\},$$

where $M_{n,i}$ is the principal upper $(i \times i)$-submatrix of M_n. One can easily check (see, e. g., [15, p. 20]) that the following 3-term relation holds:

$$Sp_{n,i}(\lambda) = (\lambda - \xi_{n,i})Sp_{n,i-1}(\lambda) - \psi_{n,i}Sp_{n,i-2}(\lambda), \quad i \in \{1,\dots,n+1\}, \qquad (10.5)$$

where $\xi_{n,i}$ is as in (10.4) and

$$\psi_{n,i} = \alpha_{n,i}\gamma_{n,i}, \quad i \in \{2,\dots,n+1\}. \qquad (10.6)$$

Here we use the (standard) initial conditions $Sp_{n,0}(\lambda) = 1$, $Sp_{n,-1}(\lambda) = 0$. It is well known that if all $\xi_{n,i}$s are real and all $\psi_{n,i}$s are positive, then the polynomials $Sp_{n,i}(\lambda)$, $i \in \{0,\dots,n+1\}$, form a finite sequence of orthogonal polynomials. In particular, all their roots are real. In our case, however, these coefficients are complex. To complete the proof of Theorem 10.2, we state the following generalization of [112, Theorem 1.4], which translated in our notation claims the following.

Theorem 10.4 (A. Kuijlaars–W. Van Assche). *If there exist two continuous functions $\xi(\tau)$ and $\psi(\tau)$, $\tau \in [0,1]$, such that*

$$\lim_{i/(n+1)\to\tau} \xi_{i,n} = \xi(\tau), \qquad \lim_{i/(n+1)\to\tau} \psi_{i,n} = \psi(\tau), \quad \forall \tau \in [0,1],$$

then the asymptotic root-counting measure ν (if it exists) of the polynomial sequence $\{Sp_n(\lambda)\}_{n\in\mathbb{Z}_+} = \{Sp_{n,n+1}(\lambda)\}_{n\in\mathbb{Z}_+}$ *and the average M of the arcsine measures given by*

$$M = \int_0^1 \omega_{[\xi(\tau)-2\sqrt{\psi(\tau)},\xi(\tau)+2\sqrt{\psi(\tau)}]} d\tau$$

have the same logarithmic potential outside the union of their supports.

Recall that for a pair of distinct complex number $\alpha_1 \neq \alpha_2$, the arcsine measure $\omega_{[\alpha_1,\alpha_2]}$ is the measure supported on $[\alpha_1, \alpha_2]$ and whose density at a point $t \in [\alpha_1, \alpha_2]$ equals $\frac{1}{\pi\sqrt{|(t-\alpha_1)(t-\alpha_2)|}}$.

Remark 10.2. Theorem 10.4 is not explicitly stated in [112], but it is very similar and its proof is parallel to that of Theorem 1.4 from this paper. As we mentioned above, complete proof of Theorem 10.4 can be found in Appendix II of [169]. More general result can be found in an earlier article [98].

From the explicit formulas for $\xi_{n,i}$ and $\psi_{n,i}$ (see (10.4) and (10.6)), one easily gets

$$\xi(\tau) = \lim_{i/(n+1)\to\tau} \xi_{i,n} = -v(1-\tau)^2,$$
$$\psi(\tau) = \lim_{i/(n+1)\to\tau} \psi_{i,n} = -w(1-(1-\tau)^2)(1-\tau)^2.$$

Notice that the above limits are independent of the coefficients α, β, γ of the polynomial $P(z)$. □

Lemma 10.1. *The parametric curve Γ given in the above notation by the formula* $\xi(\tau) \pm 2\sqrt{\psi(\tau)}$, $\tau \in [0,1]$ *is the ellipse passing through the origin and given in coordinates* $x = \mathrm{Re}(z)$, $y = \mathrm{Im}(z)$ *by the equation*

$$a_{11}x^2 + 2a_{12}xy + a_{22}y^2 + 2a_{13}x + 2a_{23}y = 0 \tag{10.7}$$

where

$$a_{11} = C^2 + 4D^2, \quad a_{12} = -(AC + 4BD), \quad a_{22} = A^2 + 4B^2,$$
$$a_{13} = 2D(BC - AD), \quad a_{23} = -2B(BC - AD)$$

and $A = -\mathrm{Re}(v)$, $B = -\mathrm{Im}(u)$, $C = -\mathrm{Im}(v)$, $D = \mathrm{Re}(u)$.

Proof. We express the functions ξ and ψ as

$$\begin{cases} \xi(\tau) = -v(1-\tau)^2 = -v\theta^2 = -v\sin^2\varphi, \\ \psi(\tau) = -w(1-(1-\tau)^2)(1-\tau^2) = -w(1-\theta^2)\theta^2 = -w\sin^2\varphi\cos^2\varphi, \end{cases} \tag{10.8}$$

where $\tau \in [0,1]$, $\theta := 1-\tau \in [0,1]$, and $\sin\varphi := \theta$, $\varphi \in [0,\pi/2]$. Then

$$\xi(\tau) \pm 2\sqrt{\psi(\tau)} = -v\sin^2\varphi \pm \sqrt{-w}\sin 2\varphi.$$

Thus the curve $\Gamma \subset \mathbb{C}$ is given by the parametrization $\Gamma(\varphi) = -v\sin^2\varphi \pm \sqrt{-w}\sin 2\varphi$, where $v, w, z \in \mathbb{C}$ and $\varphi \in [0, \pi/2]$. Set $w = u^2$, so that $\sqrt{-w} = iu$. Then Γ has the form

$$\Gamma(\varphi) = -v\sin^2\varphi \pm iu\sin 2\varphi = (-\operatorname{Re}(v) - \mathrm{i}\operatorname{Im}(v))\sin^2 \pm \mathrm{i}(\operatorname{Re}(u) + \mathrm{i}\operatorname{Im}(u))\sin 2\varphi.$$

We, therefore, get the following system for its real and imaginary parts:

$$\begin{cases} x(\varphi) = A\sin^2\varphi + B\sin 2\varphi \\ y(\varphi) = C\sin^2\varphi + D\sin 2\varphi. \end{cases} \tag{10.9}$$

Here $A = -\operatorname{Re}(v)$, $B = -\operatorname{Im}(u)$, $C = -\operatorname{Im}(v)$, $D = \operatorname{Re}(u)$, and $\varphi \in [-\pi/2, \pi/2]$ since Γ is π-periodic.

To show that Γ is an ellipse passing through the origin and satisfying (10.7), substitute (10.9) into the expression $a_{11}x^2(\varphi) + 2a_{12}x(\varphi)y(\varphi) + a_{22}y^2(\varphi) + 2a_{13}x(\varphi) + 2a_{23}y(\varphi)$, where the coefficients $a_{i,j}$ are defined in the statement of Lemma 10.1. Simple calculations then show that the latter expression vanishes identically, i. e., for all values of φ.

To prove that (10.7) describes a real ellipse (and not some other real affine quadric), consider the determinant

$$\Delta := \begin{vmatrix} a_{11} & a_{12} & a_{13} \\ a_{12} & a_{22} & a_{23} \\ a_{13} & a_{23} & 0 \end{vmatrix} = -4(BC - AD)^4.$$

It is well known that if Δ is negative then we have a real ellipse ($\Delta > 0$ corresponds to an imaginary ellipse, i. e., an empty set of solutions). Thus unless $BC - AD = 0$ (which describes the situation with all three roots of $Q(z)$ being collinear) then Γ is a real ellipse. To find its semiaxes a and b, we calculate the quantities

$$\delta := \begin{vmatrix} a_{11} & a_{12} \\ a_{12} & a_{22} \end{vmatrix} = 4(BC - AD)^2; \quad \iota := a_{11} + a_{22} = A^2 + C^2 + 4(B^2 + C^2).$$

It is known that the roots $\lambda_{1,2}$ of the characteristic equation $\lambda^2 - \iota\lambda + \delta = 0$ are equal to $2a^2$ and $2b^2$ (in particular, both need to be positive), where a, b are the semiaxes of the ellipse under consideration. We arrive therefore at

$$\begin{cases} a = \frac{1}{2}\sqrt{\iota + \sqrt{\iota^2 - 4\delta}} \\ b = \frac{1}{2}\sqrt{\iota - \sqrt{\iota^2 - 4\delta}} \end{cases}$$

and $\sqrt{\iota^2 - 4\delta} = \sqrt{((A - 2D)^2 + (C + 2B)^2)((A + 2D)^2 + (C - 2B)^2)}$. For the sake of completeness, the eccentricity c of our ellipse can be expressed as

$$c = \sqrt{\frac{-\Delta}{\delta^2}\sqrt{\iota^2 - 4\delta}} = \frac{1}{2}\sqrt[4]{((A - 2D)^2 + (C + 2B)^2)((A + 2D)^2 + (C - 2B)^2)}. \qquad \square$$

Lemma 10.2. *The foci of the ellipse coincide with the two roots of the polynomial $Q(z)$ different from the origin.*

Proof. The coordinates of the center $\mathfrak{c} = (x_\mathfrak{c}, y_\mathfrak{c})$ of our ellipse satisfy

$$\left.\begin{aligned} a_{11}x_\mathfrak{c} + a_{12}y_\mathfrak{c} + a_{13} &= 0 \\ a_{12}x_\mathfrak{c} + a_{22}y_\mathfrak{c} + a_{23} &= 0 \end{aligned}\right\} \quad \Rightarrow \quad x_\mathfrak{c} = \frac{A}{2}, \quad y_\mathfrak{c} = \frac{C}{2}.$$

Recalling that $Q(z) = z(z^2+vz+w) = z(z^2+vz+u^2)$, we need to show that the coordinates (x_f, y_f) of the foci f of Γ satisfy the equation

$$f = x_f + \mathrm{i}y_f = \frac{-v \pm \sqrt{v^2 - 4u^2}}{2}.$$

To do this we express them through A, B, C, D. First, we see that $\operatorname{Re}(v^2 - 4u^2) = A^2 + 4B^2 - C^2 - 4D^2$. Using the relation

$$\sqrt{\xi + \mathrm{i}\eta} = \sqrt{\frac{r+\xi}{2}} + \mathrm{i}\sqrt{\frac{r-\xi}{2}},$$

where $r = \sqrt{\xi^2 + \eta^2}$, we get

$$\begin{aligned} r &= \sqrt{(\operatorname{Re}(v^2 - 4u^2))^2 + (\operatorname{Im}(v^2 - 4u^2))^2} \\ &= 4c^2 = \sqrt{((A-2D)^2 + (C+2B)^2)((A+2D)^2 + (C-2B)^2)} \end{aligned}$$

and

$$\begin{aligned} x_f &= \frac{A}{2} \pm \frac{1}{2\sqrt{2}}\sqrt{4c^2 + (A^2 + 4B^2 - C^2 - 4D^2)} \\ y_f &= \frac{C}{2} \pm \frac{1}{2\sqrt{2}}\sqrt{4c^2 - (A^2 + 4B^2 - C^2 - 4D^2)}. \end{aligned}$$

Straightforward calculation shows that the center $\mathfrak{c}$ and the foci f_1 and f_2 lie on the same line given by the equation

$$y = \frac{4c^2 - (A^2 + 4B^2 - C^2 - 4D^2)}{2(AC + 4BD)}\left(x - \frac{A}{2}\right) + \frac{C}{2}. \tag{10.10}$$

Finally, we check that the spacing between the center and either focus equals to the eccentricity c, which settles the lemma. This follows, for example, from the expression for the coordinates of the intersection points between (10.10) and the circle $(x - x_\mathfrak{c})^2 + (y - y_\mathfrak{c})^2 = c^2$. ☐

10.3 Proof of Theorem 10.3

We start with the following integral representation of the required Cauchy transform.

Lemma 10.3. *The Cauchy transform $\mathcal{C}_0(z)$ of the measure M_0 associated with the root of the polynomial $Q(z) = z(z^2 + vz + w)$ at the origin is given by*

$$\mathcal{C}_0(z) = \int_0^1 \frac{d\theta}{\sqrt{(v^2 - 4w)\theta^4 + (2vz + 4w)\theta^2 + z^2}}. \tag{10.11}$$

Proof. Indeed, recall that the Cauchy transform $\mathcal{C}_{[a_1,a_2]}$ of the arcsine measure $\omega_{[a_1,a_2]}$ of the interval $[a_1, a_2]$ equals

$$\mathcal{C}_{[a_1,a_2]} = \frac{1}{\sqrt{z - a_1)(z - a_2)}}.$$

The measure M_0 is obtained by the averaging of the family of arcsine measures, namely

$$M_0 = \int_0^1 \omega_{[\xi(\tau)-2\sqrt{\psi(\tau)},\xi(\tau)+2\sqrt{\psi(\tau)}]}\,d\tau,$$

where $\xi(\tau) = -v(1-\tau)^2 = -v\theta^2$, $\psi(\tau) = -w(1-(1-\tau)^2)(1-\tau)^2 = -w(1-\theta^2)\theta^2$, and $\theta = 1-\tau$. Since the Cauchy transform of the average of a family of measures equals the average of the family of their Cauchy transforms, one gets (after certain simplifications)

$$\begin{aligned}\mathcal{C}_0(z) &= \int_0^1 \frac{d\tau}{(z - (\xi(\tau) - 2\sqrt{\psi(\tau)})(z - (\xi(\tau) + 2\sqrt{\psi(\tau)})} \\ &= \int_0^1 \frac{d\theta}{\sqrt{(v^2 - 4w)\theta^4 + (2vz + 4w)\theta^2 + z^2}}.\end{aligned}$$

□

10.3.1 Special case

We first provide the proof of Theorem 10.3 for a specific case $Q(z) = z(4z^2 - 1)$ where the calculations are somewhat simpler and then address the general case. By Lemma 10.3 the Cauchy transform $\mathcal{C}_0(z)$ of the measure M_0 associated with the root of $Q(z)$ at the origin is then given by the integral

$$\mathcal{C}_0(z) := \int_0^1 \frac{d\theta}{\sqrt{\theta^4 - \theta^2 + z^2}}.$$

We want to find a differential equation satisfied by $C_0(z)$ w. r. t. the variable z. Unfortunately, we do not know how to do it directly, and our proof requires a number of intricate variable changes and manipulations. We first change $t = 2\theta^2 - 1$ and consider

$$C_0(z) = I_0(s) = \frac{1}{\sqrt{2}} \int_{-1}^{1} \frac{dt}{\sqrt{t+1}\sqrt{t^2+s}}, \tag{10.12}$$

where $s := 4z^2 - 1$. Introduce now a family of functions $I_\nu(s)$ indexed by $\nu \geq 0$ and defined by

$$I_\nu(s) := \frac{1}{\sqrt{2}} \int_{-1}^{1} \frac{t^\nu dt}{\sqrt{t+1}\sqrt{t^2+s}}.$$

Lemma 10.4. *For $\nu \geq 0$, the following three relations are satisfied:*

$$\frac{\partial I_{\nu+2}}{\partial s} = -\frac{1}{2} I_\nu - s\frac{\partial I_\nu}{\partial s}, \tag{10.13}$$

$$\frac{\partial}{\partial s}(I_2 + I_1) = -\frac{1}{4} I_0 + \frac{1}{2\sqrt{1+s}}, \tag{10.14}$$

$$\frac{\partial}{\partial s}(I_3 - I_1) = -\frac{3}{4} I_1 - \frac{1}{4} I_0. \tag{10.15}$$

Proof. Relation (10.13) can be proved directly

$$\begin{aligned}
\frac{\partial I_{\nu+2}}{\partial s} &= -\frac{1}{2\sqrt{2}} \int_{-1}^{1} \frac{t^{\nu+2} dt}{\sqrt{t+1}(t^2+s)^{3/2}} \\
&= -\frac{1}{2\sqrt{2}} \int_{-1}^{1} \frac{(t^{\nu+2} + st^\nu) dt}{\sqrt{t+1}(t^2+s)^{3/2}} + \frac{1}{2\sqrt{2}} \int_{-1}^{1} \frac{st^\nu dt}{\sqrt{t+1}(t^2+s)^{3/2}} \\
&= -\frac{1}{2} I_\nu - s\frac{\partial I_\nu}{\partial s}.
\end{aligned}$$

Relation (10.14) is easy to verify by integration by parts. Indeed,

$$\frac{\partial}{\partial s}(I_2 + I_1) = -\frac{1}{2\sqrt{2}} \int_{-1}^{1} \frac{t+1}{\sqrt{t+1}} \frac{tdt}{(t^2+s)^{3/2}} = \frac{1}{2\sqrt{1+s}} - \frac{1}{4} I_0.$$

Similarly, by integration by parts one gets

$$\frac{\partial}{\partial s}(I_3 - I_1) = -\frac{1}{2\sqrt{2}} \int_{-1}^{1} \frac{t+1}{\sqrt{t^2-1}} \frac{tdt}{(t^2+s)^{3/2}} = -\frac{3}{4} I_1 - \frac{1}{4} I_0. \qquad \square$$

Now, we express $\partial I_2/\partial s$ from (10.13), substitute it in (10.14), and single out $\partial I_1/\partial s$:

$$\frac{\partial I_1}{\partial s} = s\frac{\partial I_0}{\partial s} + \frac{1}{4}I_0 + \frac{1}{2\sqrt{1+s}}. \tag{10.16}$$

Adding (10.14) with (10.15) and reducing $\partial I_3/\partial s$, $\partial I_2/\partial s$ with the help of (10.13), we obtain

$$4(1+s)\frac{\partial I_1}{\partial s} = I_0 + I_1.$$

Through (10.16) we get

$$I_1 = 4s(1+s)\frac{\partial I_0}{\partial s} + sI_0 + 2\sqrt{s+1}.$$

Differentiating both sides of the latter relation w. r. t. s and using (10.16) again, we obtain the required linear non-homogeneous differential equation satisfied by $I_0(s)$

$$16s(1+s)\frac{\partial^2 I_0}{\partial s^2} + 16(1+2s)\frac{\partial I_0}{\partial s} + 3I_0 = -\frac{2}{\sqrt{1+s}}. \tag{10.17}$$

In order to recover the required equation (10.3) for $C_0(z)$, we have to change s back to z. Using straightforward relations

$$\frac{\partial I_0}{\partial s} = \frac{1}{8z}\frac{\partial C_0}{\partial z} \quad \text{and} \quad \frac{\partial^2 I_0}{\partial s^2} = \frac{1}{64z^3}\left(z\frac{\partial^2 C_0}{\partial z^2} - \frac{\partial C_0}{\partial z}\right)$$

we obtain after some obvious simplifications the equation

$$z(4z^2-1)\frac{\partial^2 C_0}{\partial z^2} + (12z^2-1)\frac{\partial C_0}{\partial z} + 3zC_0(z) + 1 = 0$$

which coincides with (10.3) for $Q(z) = z(4z^2-1)$. Thus our special case of Theorem 10.3 is settled.

Notice also that (10.17) can be solved explicitly. The general solution of the corresponding linear homogeneous equation is an arbitrary linear combination of a complete elliptic integral of the first kind $y_1(s)$ and of an associated Legendre function of the second kind $y_2(s)$ given by

$$\begin{cases} y_1(s) = \frac{2}{\pi\sqrt[4]{1+s}}\mathbb{K}\left(\frac{\sqrt{1+s}-1}{2\sqrt{1+s}}\right) \\ y_2(s) = \mathbb{Q}_{-1/4}(1+2s), \end{cases}$$

here $\mathbb{K}(x)$ and $\mathbb{Q}(x)$ are the complete elliptic integral and the associated Legendre function of the second kind respectively.

The general solution to (10.17) depends on two arbitrary constants C_1, C_2 and is given by

$$I_0 = C_1 y_2 + C_2 y_2 + y_2 \int y_1 \frac{g}{f_2} \frac{ds}{W} - y_1 \int y_2 \frac{g}{f_2} \frac{ds}{W},$$

where $g(s) = -\frac{2}{\sqrt{1+s}}$, $f_2(s) = 16s(1+s)$, $W(s) = y_1(s)y_2'(s) - y_2(s)y_1'(s)$. However, we need its particular solution and thus have to determine the corresponding particular values of C_1, C_2. (To find them we evaluated the integral (10.12) for two different values of s. Moreover, analyzing the polynomial $(t+1)(t^2+s) = t^3 + t^2 + st + s$, we observed that it is positive on $[-1,1]$ for $s > 0$ and that (10.12) is divergent when $s = 0$.)

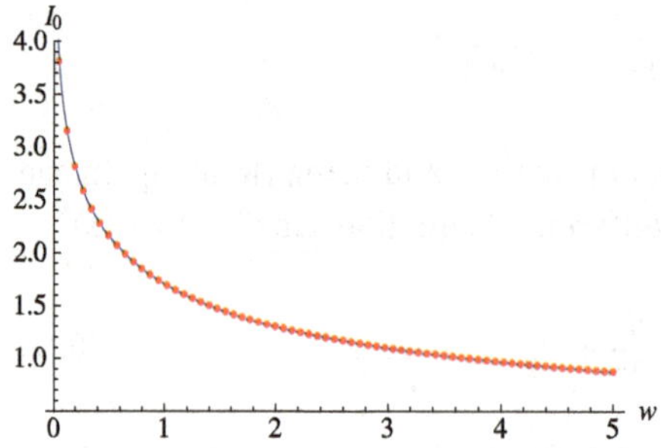

Figure 10.3: The graph of $I_0(s)$ obtained from (10.17) and its numerical values obtained from (10.12).

Figure 10.3 compares the appropriate solution of (10.17) giving $I_0(s)$ with the values of $I_0(s)$ calculated numerically using integral (10.12) for a number of values of s (which are shown by red dots).

10.3.2 General case

The scheme of the proof is exactly the same as in the above special case, but calculations are somewhat messier. Assuming that $v^2 - 4w \neq 0$, we need to find a differential equation satisfied by integral (10.40). We change variables as follows:

$$s = -16w\frac{z^2 + vz + w}{(v^2 - 4w)^2}, \qquad u = v\frac{v + 2z}{v^2 - 4w}, \qquad a = v^2 - 4w$$

and denote $I_0(s, u, a) = C_0(z)$. (Here as above we assume that v and w are some fixed complex numbers.) It also helps to change the variable θ in (10.40) by using $2\theta^2 = t + 1$, and then we finally get

$$C_0(z) = I_0(s) = \frac{1}{\sqrt{2a}} \int_{-1}^{1} \frac{dt}{\sqrt{t+1}\sqrt{(t+u)^2 + s}}.$$

As above we introduce a family of functions $I_v(s)$, $s \geq 0$ given by the formula

$$I_\nu(s) := \frac{1}{\sqrt{2a}} \int_{-1}^{1} \frac{(t+u)^\nu dt}{\sqrt{t+1}\sqrt{(t+u)^2+s}}.$$

Analogously to Lemma 10.4, one can prove the next statement.

Lemma 10.5. *The following relations are valid for $I_\nu(s)$, $s \geq 0$:*

$$\frac{\partial I_{\nu+2}}{\partial s} = -\frac{1}{2}I_\nu - s\frac{\partial I_\nu}{\partial s}, \tag{10.18}$$

$$\frac{\partial}{\partial s}(I_2 + I_1) = u\frac{\partial I_1}{\partial s} - \frac{1}{4}I_0 + \frac{1}{2\sqrt{2}\sqrt{(u+1)^2+s}}, \tag{10.19}$$

$$\frac{\partial}{\partial s}(I_3 - I_1) = (u^2 - 2u)\frac{\partial I_1}{\partial s} - \frac{3}{4}I_1 + \frac{u-1}{4}I_0 + \frac{u}{\sqrt{2}\sqrt{(u+1)^2+s}}. \tag{10.20}$$

Now, we use (10.18) for expressing $\partial I_2/\partial s$ and then we single out $\partial I_1/\partial s$ from (10.19) given by

$$\frac{\partial I_1}{\partial s} = \frac{s}{1-u}\frac{\partial I_0}{\partial s} + \frac{I_0}{4(1-u)} + \frac{1}{2\sqrt{a}(1-u)\sqrt{(u+1)^2+s}}. \tag{10.21}$$

Adding (10.19) and (10.20), employing (10.18) again, and using (10.21), we get the relation

$$(u-1)I_1 = -4s(s+(u-1)^2)\frac{\partial I_0}{\partial s} - sI_0 - \frac{2(s-u^2+1)}{\sqrt{a}\sqrt{(u+1)^2+s}}. \tag{10.22}$$

Eventually, taking the derivative of the both sides of the latter equation w. r. t. s and using (10.21) again, we finally get a linear differential equation in the variable s satisfied by $I_0(s)$

$$16s(s+(u-1)^2)\frac{\partial^2 I_0}{\partial s^2} + 16(2s+(u-1)^2)\frac{\partial I_0}{\partial s} + 3I_0 + \frac{2}{\sqrt{2}}\frac{s+(u+1)(5u+1)}{\sqrt{(u+1)^2+s}} = 0. \tag{10.23}$$

In order to get an equation for $\mathcal{C}_0(z)$ w. r. t. the variable z, we use

$$\frac{\partial \mathcal{C}_0}{\partial z} = \frac{\partial s}{\partial z}\frac{\partial I_0}{\partial s} + \frac{\partial u}{\partial z}\frac{\partial I_0}{\partial u}$$

and

$$\frac{\partial^2 \mathcal{C}_0}{\partial z^2} = \frac{\partial^2 s}{\partial z^2}\frac{\partial I_0}{\partial s} + \left(\frac{\partial s}{\partial z}\right)^2\frac{\partial^2 I_0}{\partial s^2} + 2\frac{\partial s}{\partial z}\frac{\partial u}{\partial z}\frac{\partial^2 I_0}{\partial s \partial u} + \left(\frac{\partial u}{\partial z}\right)^2\frac{\partial^2 I_0}{\partial u^2}.$$

With the help of $\frac{\partial I_0}{\partial u} = 2I_1$ and (10.22), we obtain

$$\frac{\partial C_0}{\partial z} = \left(\frac{\partial s}{\partial z} + 2\frac{\partial u}{\partial z}\frac{s}{1-u}\right)\frac{\partial I_0}{\partial s} + 2\frac{\partial u}{\partial z}\frac{I_0}{4(1-u)} + \frac{\partial u}{\partial z}\frac{1}{\sqrt{a}(1-u)\sqrt{(u+1)^2+s}}.$$

Now, we get

$$\frac{\partial I_0}{\partial s} = \frac{(v^2-4w)(vz+2w)}{16wz}\frac{\partial C_0}{\partial z} + \frac{v(v^2-4w)}{32wz}C_0 + \frac{v(v^2-4w)}{32wz(v+z)}. \tag{10.24}$$

Further, we use

$$\frac{\partial^2 I_0}{\partial u^2} = -4\frac{\partial I_0}{\partial s} - 4s\frac{\partial^2 I_0}{\partial s^2}$$

and

$$\frac{\partial^2 I_0}{\partial s \partial u} = 2(u-1)\frac{\partial^2 I_0}{\partial s^2} + \frac{3s+4(u-1)^2}{2s(u-1)}\frac{\partial I_0}{\partial s} + \frac{3I_0}{8s(u-1)} + \frac{3s+5u^2+6u+1}{4\sqrt{a}(u-1)s(s+(u-1)^2)^{3/2}}.$$

We can now express $\partial^2 C_0/\partial z^2$ through $\partial^2 I_0/\partial s^2$, $\partial I_0/\partial s$, and I_0 as follows:

$$\begin{aligned}\frac{\partial^2 C_0}{\partial z^2} = {} & \left(\left(\frac{\partial s}{\partial z}\right)^2 + 4(u-1)\left(\frac{\partial s}{\partial z}\right)\left(\frac{\partial u}{\partial z}\right) - 4s\left(\frac{\partial u}{\partial z}\right)^2\right)\frac{\partial^2 I_0}{\partial s^2}\\ & + \left(\frac{\partial^2 s}{\partial z^2} + \left(\frac{\partial s}{\partial z}\right)\left(\frac{\partial u}{\partial z}\right)\frac{3s+4(u-1)^2}{s(u-1)} - 4\left(\frac{\partial u}{\partial z}\right)^2\right)\frac{\partial I_0}{\partial s} +\\ & + \left(\frac{\partial s}{\partial z}\right)\left(\frac{\partial u}{\partial z}\right)\frac{3I_0}{4s(u-1)} + \left(\frac{\partial s}{\partial z}\right)\left(\frac{\partial u}{\partial z}\right)\frac{3s+5u^2+6u+1}{4\sqrt{a}(u-1)s(s+(u-1)^2)^{3/2}}.\end{aligned}$$

This leads to

$$\begin{aligned}\frac{\partial^2 C_0}{\partial z^2} = {} & -\frac{256wz^2}{(v^2-4w)^3}\frac{\partial^2 I_0}{\partial s^2}\\ & - 16\frac{4w^2(w+z^2)+4vwz(w+2z^2)+v^2w(w+5z^2)+v^3(2wz-z^3)}{(v^2-4w)^2(2w+vz)(w+z(v+z))}\frac{\partial I_0}{\partial s}\\ & + \frac{3v(v+2z)}{4(2w+vz)(w+z(v+z))}I_0\\ & + \frac{v(v+2z)(3v^4+8v^3z-24vwz-4w(2w+3z^2)+v^2(5z^2-8w))}{4(v^2-4w)(v+z)^3(2w+vz)(w+z(v+z))}.\end{aligned}$$

From the latter equation and (10.24), we finally get

$$\begin{aligned}\frac{\partial^2 I_0}{\partial s^2} = {} & -\frac{(v^2-4w)^3}{256wz^2}\frac{\partial^2 C_0}{\partial z^2} - \frac{(v^2-4w)^2 c_1}{256w^2z^3(w+z(v+z))}\frac{\partial C_0}{\partial z}\\ & - \frac{v(v^2-4w)^2 c_2}{1024wz^2(w+z(v+z))}C_0 - \frac{v(v^2-4w)^2 c_3}{1024w^2z^3(v+z)^3(w+z(v+z))},\end{aligned}$$

where

$$\begin{aligned} c_1 &= 4w^2(w+z^2) + 4vwz(w+2z^2) + v^2w(w+5z^2) + v^3(2wz - z^3), \\ c_2 &= 8vwz + v^2(w-2z^2) + 4w(w+4z^2)), \\ c_3 &= v^4(w-2z^2) + 12vwz(w+3z^2) + 4wz^2(3w+4z^2) + v^3(8wz-4z^3) \\ &\quad + v^2(4w^2 + 27wz^2 - 2z^4). \end{aligned}$$

Plugging these formulae into (10.23), we arrive at

$$4z(z^2+vz+w)\mathcal{C}_0''(z) + 4(3z^2+2vz+w)\mathcal{C}_0'(z) + (3z+v)\mathcal{C}_0(z) + 1 = 0,$$

which can be equivalently expressed as

$$Q(z)\mathcal{C}_0''(z) + Q'(z)\mathcal{C}_0'(z) + Q''(z)\mathcal{C}_0(z)/8 + Q'''(z)/24 = 0,$$

with $Q(z) = 4z(z^2+vz+w)$. (Notice that multiplication of $Q(z)$ by a non-vanishing constant is irrelevant in our considerations.) □

10.4 Proving the existence of the measure

In the above notation, our main question in the second part of this chapter is as follows.

Problem 10.1. Does the sequence $\{v_n\}$ converge (in the weak sense) to some limiting measure v? If the convergence takes place, describe the limiting measure v?

Below we answer both parts of this question, see Theorem 10.5. Denote the three roots of $Q(z)$ by a_1, a_2, a_3. For $i \in \{1,2,3\}$ consider the curve γ_i given as the set of all $b \in \mathbb{C}$ satisfying the relation

$$\gamma_i : \quad \int_{a_j}^{a_k} \sqrt{\frac{b-t}{(t-a_1)(t-a_2)(t-a_3)}}\,dt \in \mathbb{R}, \tag{10.25}$$

here j and k are the remaining two indices in $\{1,2,3\}$ in any order and the integration is taken over the straight interval connecting a_j and a_k. One can see that a_i belongs to γ_i and show that these three curves connect all a_is with the unique common point b_0 lying within Δ_Q, see Lemma 10.11. Take the segment of γ_i connecting a_i with the common intersection point b_0 and denote this segment by Γ_i, see Figure 10.4. Finally, denote the union of these three segments by Γ_Q.

Notation 10.1. Let b_0 denote the point where $\gamma_i, \gamma_j, \gamma_k$ meet. Recall that we denote the segment of γ_i connecting a_i and b_0 by Γ_i. Let D_i be the domain bounded by Γ_j, Γ_k and by the side $\overline{a_j a_k}$.

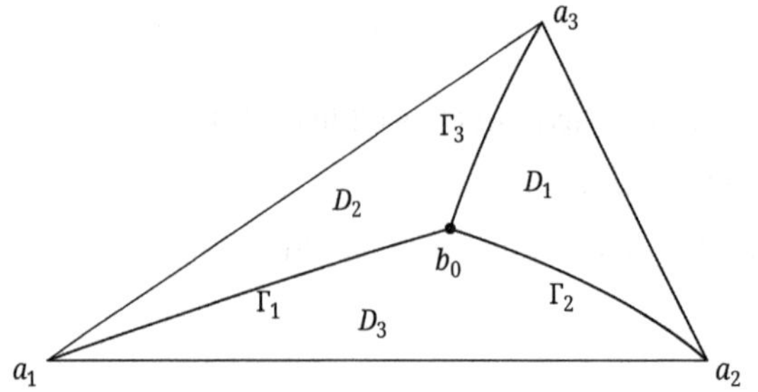

Figure 10.4: Used notation.

For $R(z) = \frac{(b-z)}{(z-a_1)(z-a_2)(z-a_3)}$, introduce the quadratic differential $\Psi = R(z)dz^2$. Take its (horizontal) trajectory, i. e., a level curve

$$\mathfrak{I} \int_{z_0}^{z} \sqrt{R(t)}dt = \text{const}, \tag{10.26}$$

where z_0 is some fixed point. Assume that b is located inside Δ_Q where as above $Q(z) = (z-a_1)(z-a_2)(z-a_3)$. If $R(z)$ is non-vanishing and regular at some $z = z^*$, then the curve $\mathfrak{H} : \mathfrak{I} \int_{z_0}^{z} \sqrt{R(t)}dt = \text{const}$ passing through z^* is analytic in a neighborhood of z^*, and the tangential direction to $\mathfrak{H}$ at z^* is given by $(\mathfrak{R}\sqrt{R(z^*)}, -\mathfrak{I}\sqrt{R(z^*)})$. To see this, note that locally near z^* one has

$$\int_{z^*}^{z} \sqrt{R(t)}dt \sim \sqrt{R(z^*)}(z-z^*) + O((z-z^*)^2). \tag{10.27}$$

Analogously, the vertical trajectory $\mathfrak{V}$ of Ψ (which is given by $\mathfrak{R} \int_{z_0}^{z} \sqrt{R(t)}dt = \text{const}$) passing through z^* is also analytic in a neighborhood of z^*, and its tangential direction at z^* is given by $(\mathfrak{I}\sqrt{R(z^*)}, \mathfrak{R}\sqrt{R(z^*)})$. Note that the orientation of $\mathfrak{H}$ and $\mathfrak{V}$ depends on the choice of a branch of $\sqrt{R(z)}$.

Notation 10.2. For a fixed $z^* \in \mathbb{C}\setminus\{a_1, a_2, a_3, b\}$, denote by $\theta_1, \theta_2, \theta_3$, and ϕ the arguments of the complex numbers $a_1 - z^*, a_2 - z^*, a_3 - z^*$, and $b - z^*$ resp. Let $\theta_{j'j}, \phi_j$ be the arguments of $a_{j'} - a_j$, $b - a_j$. Finally, let $\tilde{\phi}_i$ be the argument of $a_i - b$.

Our first result is as follows.

Theorem 10.5.

(i) *For any equation* (10.3), *the sequence* $\{\nu_n\}$ *of the root-counting measures of its spectral polynomials converges to a probability measure* ν *depending only on the leading coefficient* $Q(z)$.

(ii) *The support of the limiting root-counting measure* ν *coincides with* Γ_Q.

Remark 10.3. Knowing the support of ν, it is also possible to define its density along the support using the linear differential equation satisfied by its Cauchy transform, see the first part of this chapter and Theorem 4 of [167]. In case when $Q(z)$ has all real zeros, the density is explicitly given in the previous chapter, see [44].

An essential role in the proof of Theorem 10.5 is played by the description of the behavior of the Stokes lines of (10.3). Important contribution also comes from a generalization of the technique of [112]. In particular, in [167] using the latter technique we were able to find an additional probability measure which is easily described and from which the measure ν is obtained by the inverse balayage, see the beginning of the chapter. In other words, the support of ν will be contained in the support of the measure which we construct and they will have the same logarithmic potential outside the support of the latter one. This measure is uniquely determined by the choice of a root of $Q(z)$ and thus we, in fact, construct three different measures having the same measure ν as their inverse balayage.

Our second result describes the asymptotic behavior of Stieltjes polynomials of increasing degrees when the sequence of their (normalized) Van Vleck polynomials has a limit. This result is a special case of a more general statement of [170] but we explain below in much more detail the interaction of the limiting measure with the appropriate rational Strebel differential.

Namely, take a sequence $\{S_{n,i_n}(z)\}$, $\deg S_{n,i_n}(z) = n$ of its Stieltjes polynomials for a Heun equation (10.3) such that the sequence of normalized Van Vleck polynomials $\{\widetilde{V}_{n,i_n}(z)\}$ converges to some monic linear polynomial $\widetilde{V}(z)$. (Here by normalization we mean the division by the leading coefficient, i. e., each $\widetilde{V}_{n,i_n}(z)$ is the monic polynomial proportional to $V_{n,i_n}(z)$.)

Notice that since each $\widetilde{V}_{n,i_n}(z)$ is linear for all sufficiently large n, then the existence of the limiting polynomial $\widetilde{V}(z)$ is the same as the existence of the limit of the sequence of (unique) roots $\{b_{n,i_n}\}$ of $\{V_{n,i_n}(z)\}$. Part 2 of Theorem 10.1 guarantees the existence of plenty of such converging sequences, and Theorem 10.5 claims that the limit $\tilde{b}$ of these roots must necessarily belong to Γ_Q.

Finally, denote by μ_{n,i_n} the root-counting measure of the corresponding Stieltjes polynomial $S_{n,j_n}(z)$.

Theorem 10.6. *In the above notation the sequence $\{\mu_{n,i_n}\}$ of the root-counting measures of the corresponding Stieltjes polynomials $\{S_{n,j_n}(z)\}$ weakly converges to the unique probability measure $\mu_{\widetilde{V}}$ whose Cauchy transform $\mathcal{C}_{\widetilde{V}}(z)$ satisfies almost everywhere in $\mathbb{C}$ the equation*

$$\mathcal{C}_{\widetilde{V}}^2(z) = \widetilde{V}(z)/Q(z).$$

To formulate our further results, we assume that the reader is familiar with the results of Chapter 7 on quadratic differentials, see also [179].

10.5 Proving Theorem 10.6

Our scheme follows roughly the scheme suggested in [27]. We need to prove that under the assumptions of Theorem 10.6 the sequence $\{\mu_{n,i_n}\}$ of root-counting measures of the sequence of Stieltjes polynomials $\{S_{n,i_n}(z)\}$ converges weakly to a probability measure $\mu_{\widetilde{V}}$ whose Cauchy transform $C_{\widetilde{V}}(z)$ satisfies almost everywhere in $\mathbb{C}$ the equation

$$C^2_{\widetilde{V}}(z) = \widetilde{V}(z)/Q(z). \tag{10.28}$$

Since such a measure is positive, it is unique by Proposition 7.2.

To simplify the notation, we denote by $\{\bar{S}_n(z)\}$ the chosen sequence $\{S_{n,i_n}(z)\}$ of Stieltjes polynomials whose sequence of normalized Van Vleck polynomials $\{\widetilde{V}_{n,i_n}(z)\}$ converge to $\widetilde{V}(z)$, and let $\{\bar{\mu}_n\}$ denote the sequence of its root-counting measures. Also let $\bar{\mu}_n^{(i)}$ be the root-counting measure of the ith derivative $\bar{S}_n^{(i)}(z)$. Assume now that NN is a subsequence of natural numbers such that

$$\bar{\nu}^{(j)} = \lim_{n\to\infty, n\in NN} \bar{\mu}_n^{(j)}$$

exists for $j = 0, 1, 2$. The next lemma shows that the Cauchy transform of $\bar{\nu} = \bar{\nu}^{(0)}$ satisfies the required algebraic equation.

Lemma 10.6. *The measures $\bar{\nu}^0$, $\bar{\nu}^1$, $\bar{\nu}^2$ are all equal and the Cauchy transform $C_{\bar{\nu}}(z)$ of their common limit satisfies equation* (10.28) *for almost every z.*

Proof. We have

$$\frac{\bar{S}_n^{(j+1)}(z)}{(n-j)\bar{S}_n^{(j)}(z)} \to \int \frac{d\bar{\nu}^{(j)}(\zeta)}{z-\zeta}$$

with convergence in L^1_{loc}, and by passing to a subsequence again we can assume that we have pointwise convergence almost everywhere. From the relation

$$Q(z)\bar{S}_n''(z) + P(z)\bar{S}_n'(z) + V_n(z)\bar{S}_n(z) = 0$$

it follows that

$$\frac{Q(z)\bar{S}_n''(z)}{n(n-1)\bar{S}_n(z)} + \frac{V_n(z)}{n(n-1)} = -\frac{P(z)\bar{S}_n'(z)}{n(n-1)\bar{S}_n(z)}.$$

One can immediately check that $-V_n(z)/n(n-1) \to \widetilde{V}(z)$, while the expression in the right-hand side converges pointwise to 0 almost everywhere in $\mathbb{C}$ due to the presence of the factor $n(n-1)$ in the denominator. Thus, for almost all $z \in \mathbb{C}$, one has

$$\frac{\bar{S}_n''(z)}{n(n-1)\bar{S}_n(z)} \to \frac{\widetilde{V}(z)}{Q(z)}$$

when $n \to \infty$ and $n \in NN$. If $u^{(j)}$ denotes the logarithmic potential of $\bar{\nu}^{(j)}$, then one has

$$u^{(2)} - u^{(0)} = \lim_{n\to\infty} \frac{1}{n} \log\left|\frac{\bar{S}_n''(z)}{n(n-1)\bar{S}_n(z)}\right| = \lim_{n\to\infty} \frac{1}{n}(\log|\widetilde{V}(z)| - \log|Q(z)|) = 0.$$

On the other hand, we have that $u^{(0)} \geq u^{(1)} \geq u^{(2)}$, see Lemma 10.7 below. Hence all the potentials $u^{(j)}$ are equal, and all $\mu_j = \Delta u^{(j)}/2\pi$ are equal as well. Finally we get

$$\mathcal{C}^2_{\tilde{\nu}}(z) = \lim_{n\to\infty} \frac{\bar{S}_n'(z)}{n\bar{S}_n(z)} \cdot \frac{\bar{S}_n''(z)}{(n-1)\bar{S}_n'(z)} = \lim_{n\to\infty} \frac{\bar{S}_n''(z)}{n(n-1)\bar{S}_n(z)} = \widetilde{V}(z)/Q(z)$$

for almost all z.

Lemma 10.7 (See Chapter 2 and Lemma 8 of [27]). *Let $\{p_m\}$ be a sequence of polynomials such that $n_m := \deg p_m \to \infty$ and there exists a compact set K containing the zeros of all p_m. Finally, let μ_m and μ_m' be the root-counting measures of p_m and p_m' resp. If $\mu_m \to \nu$ and $\mu_m' \to \nu'$ with compact support and u and u' are the logarithmic potentials of ν and ν', then $u' \leq u$ in the whole $\mathbb{C}$. Moreover, $u = u'$ in the unbounded component of $\mathbb{C} \setminus \operatorname{supp} \nu$.*

Proof. Assume wlog that p_m are monic. Let K be a compact set containing the zeros of all p_m. We have

$$u(z) = \lim_{m\to\infty} \frac{1}{n_m} \log|p_m(z)|$$

and

$$u'(z) = \lim_{m\to\infty} \frac{1}{n_m - 1} \log\left|\frac{p_m'(z)}{n_m}\right| = \lim_{m\to\infty} \frac{1}{n_m} \log\left|\frac{p_m'(z)}{n_m}\right|$$

with convergence in L^1_{loc}. Hence

$$u'(z) - u(z) = \lim_{m\to\infty} \frac{1}{n_m} \log\left|\frac{p_m'(z)}{n_m p_m(z)}\right| = \lim_{m\to\infty} \frac{1}{n_m} \log\left|\int \frac{d\mu_m(\zeta)}{z-\zeta}\right|.$$

Now, if ϕ is a positive test function, it follows that

$$\begin{aligned}\int \phi(z)(u'(z) - u(z))d\lambda(z) &= \lim_{m\to\infty} \int \phi(z) \log\left|\int \frac{d\mu_m(\zeta)}{z-\zeta}\right| d\lambda(z) \\ &\leq \lim_{m\to\infty} \int \phi(z) \int \frac{d\mu_m(\zeta)}{|z-\zeta|} d\lambda(z) \leq \lim_{m\to\infty} \iint \frac{\phi(z)d\lambda(z)}{|z-\zeta|} d\mu_m(\zeta),\end{aligned}$$

where λ denotes Lebesgue measure in the complex plane. Since $\frac{1}{|z|}$ is locally integrable, the function $\int \phi(z)|z-\zeta|^{-1}d\lambda(z)$ is continuous, and hence bounded by a constant M for all z in K. Since $\operatorname{supp} \mu_m \subseteq K$, the last expression in the above inequality is bounded by M/n_m, hence the limit when $m \to \infty$ equals to 0. This proves $u' \leq u$.

In the complement to $supp\,\nu$, u is harmonic and u' is subharmonic, hence $u' - u$ is a negative subharmonic function. Moreover, in the complement of K, $p_n'/(n_m p_m)$ converges uniformly on compact sets to the Cauchy transform $\mathcal{C}_\nu(z)$ of ν. Since $\mathcal{C}_\nu(z)$ is a nonconstant holomorphic function in the unbounded component of $\mathbb{C} \setminus K$, then by the above $u' - u$ identically vanishes there. By the maximum principle for subharmonic functions, it follows that $u' - u$ vanishes in the unbounded component of $\mathbb{C} \setminus supp\,\nu$ as well. □

To complete the proof of Theorem 10.6, we need to show that the whole sequence converges and not just some subsequence. Assume now that the sequence $\bar{\mu}_n$ does not converge to $\bar{\nu}$. Then we can find a subsequence NN' such that $\bar{\mu}_n$ stay away from some fixed neighborhood of $\bar{\nu}$ in the weak topology for all $n \in NN'$. Again by compactness, we can find a subsequence NN^* of NN' such that all the limits for root-counting measures for derivatives exist for $j = 0, \dots, k$. But then $\bar{\nu}^{(0)}$ must coincide with $\bar{\nu}$ by the uniqueness and the latter lemma. We get a contradiction to the assumption that μ_n stay away from $\bar{\nu}$ for all $n \in NN'$ and hence all $n \in NN$. □

10.6 On Strebel differential $\Psi = \frac{(b-z)dz^2}{(z-a_1)(z-a_2)(z-a_3)}$

The main result of this section is as follows.

Theorem 10.7. *For a given Strebel differential Ψ as in the title, the union of its singular trajectories starting at a_1, a_2, a_3 is contained in the convex hull Δ_Q of these roots if and only if $b \in \Gamma_Q$ where $Q(z) = (z - a_1)(z - a_2)(z - a_3)$. (For the definition of Γ_Q see § 10.1.)*

Assume that the points a_1, a_2, a_3 are not collinear in the complex plane. Let i, j, k be a permutation of 1, 2, 3. Recall that we defined the curve γ_i, $i = 1, 2, 3$ by the condition

$$\Im f_{j,k}(b) = 0 \quad \text{where } f_{jk}(b) = \int_{a_j}^{a_k} \sqrt{\frac{b-t}{(t-a_1)(t-a_2)(t-a_3)}}\,dt. \tag{10.29}$$

Lemma 10.8. *Each γ_i is smooth and they can only intersect transversally.*

Proof. Indeed, we have

$$f_{jk}'(b) = \int_{a_j}^{a_k} \sqrt{\frac{-1}{(t-a_1)(t-a_2)(t-a_3)(t-b)}}\,dt. \tag{10.30}$$

Since the right-hand side in (10.30) is a complete elliptic integral, it represents a period of an elliptic curve, which implies that the right-hand side is non-vanishing, which in its turn implies the smoothness of γ_i. To show that γ_i and γ_j for $i \neq j$ can only intersect transversally, notice the following. If they are tangent at some point b^*, then $f_{jk}'(b^*)/f_{ik}'(b^*) \in \mathbb{R}$, but this can never happen since the ratio of periods of a (nonsingular) elliptic curve cannot be real. □

Lemma 10.9. *The following three relations hold:*

$$\int_{a_j}^{a_k} \sqrt{\frac{a_i - t}{(t-a_1)(t-a_2)(t-a_3)}}\,dt = \int_{a_j}^{a_k} \sqrt{\frac{1}{(a_j - t)(t-a_k)}}\,dt = \pi, \tag{10.31}$$

$$\int_{a_j}^{a_k} \sqrt{\frac{b - t}{(t-a_1)(t-a_2)(t-a_3)}}\,dt + \int_{b}^{a_i} \sqrt{\frac{b - t}{(t-a_1)(t-a_2)(t-a_3)}}\,dt = \pi, \tag{10.32}$$

$$\left(\int_{a_j}^{a_k} + \int_{a_k}^{a_i} + \int_{a_i}^{a_j}\right) \sqrt{\frac{b - t}{(t-a_1)(t-a_2)(t-a_3)}}\,dt = 2\pi. \tag{10.33}$$

Proof. To prove the first relation, make an affine change of variable $\tilde{z} = cz + d$ where $c \neq 0$. Set $\tilde{a}_i = ca_i + d$ and $\tilde{b} = cb + d$. If a_1, a_2, a_3, b correspond to $\tilde{a}_1, \tilde{a}_2, \tilde{a}_3, \tilde{b}$ resp., then we have

$$\int_{z'}^{z} \sqrt{\frac{b - t}{(t-a_1)(t-a_2)(t-a_3)}}\,dt = \int_{\tilde{z}'}^{\tilde{z}} \sqrt{\frac{\tilde{b} - \tilde{t}}{(\tilde{t}-\tilde{a}_1)(\tilde{t}-\tilde{a}_2)(\tilde{t}-\tilde{a}_3)}}\,d\tilde{t}, \tag{10.34}$$

where $\tilde{t} = ct + d$ and $\tilde{z}' = cz' + d$. Hence we can always place two points a_j, a_k on the real axis and the third point a_i in the upper half plane. The second relation follows from the fact that the l. h. s. of the second relation equals $\frac{1}{2}\int_C \sqrt{(b-t)/(t-a_1)(t-a_2)(t-a_3)}dt$ where C is any circle bounding a disk containing the triangle Δ_Q. To show that this integral equals π consider its limit when the radius of C tends to infinity. Similar considerations settle the third relation. □

Lemma 10.10. *Let b be a point in the triangle Δ_Q and let l_b be the straight line passing through b and parallel to the side $\overline{a_j a_k}$. Then there exists a unique point b' on l_b such that*

$$\int_{a_j}^{a_k} \sqrt{\frac{b' - t}{(t-a_1)(t-a_2)(t-a_3)}}\,dt \in \mathbb{R}. \tag{10.35}$$

Moreover, $b' \in \Delta_Q \cap l_b$.

Proof. By (10.34) we can wlog assume that the points a_j, a_k lie on the real axis, $a_j < a_k$, and the point a_i lies in the upper half plane. Let us show that the imaginary part of $f_{j,k}(b)$ is a monotone decreasing function when b runs from left to right along the line l_b. Take $c_1, c_2 \in l_b$ such that $\Re c_1 < \Re c_2$, let t be a real number such that $a_j < t < a_k$. Decomposing them into the real and imaginary parts $c_1 = c_1^r + \sqrt{-1}c_1^i$, $c_2 = c_2^r + \sqrt{-1}c_2^i$, $a_i = d^r + \sqrt{-1}d^i$, we get $c_1^r - t < c_2^r - t$, $d^i > 0$, and $c_1^i = c_2^i$. Since $1/((a_k - t)(t - a_j)) > 0$ and

$$\frac{c_l - t}{a_i - t} = \frac{(c_l^r - t)(d^r - t) + c_l^i d^i + \sqrt{-1}(c_l^i(d^r - t) - (c_l^r - t)d^i)}{(d^r - t)^2 + (d^i)^2} \quad (l = 1, 2), \tag{10.36}$$

we get

$$\Im\left[\frac{c_1 - t}{(a_i - t)(a_k - t)(t - a_j)}\right] > \Im\left[\frac{c_2 - t}{(a_i - t)(a_k - t)(t - a_j)}\right].$$

Thus

$$\Im\sqrt{\frac{c_1 - t)}{(a_i - t)(a_k - t)(t - a_j)}} > \Im\sqrt{\frac{c_2 - t}{(a_i - t)(a_k - t)(t - a_j)}}$$

and

$$\Im\int_{a_j}^{a_k}\sqrt{\frac{c_1 - t}{(t - a_1)(t - a_2)(t - a_3)}}dt > \Im\int_{a_j}^{a_k}\sqrt{\frac{c_2 - t}{(t - a_1)(t - a_2)(t - a_3)}}dt, \tag{10.37}$$

proving the required monotonicity. Notice that for any $a_j < t < a_k$ the imaginary part of $(c - t)/(a_i - t)$ is always positive if $c \in l_b$ is to the left of Δ_Q, and negative if $c \in l_b$ is to the right of Δ_Q. Hence condition (10.35) cannot hold if $b' \notin \Delta_Q$. The result follows by the mean value theorem. □

Remark 10.4. Thus the three curves $\gamma_1, \gamma_2, \gamma_3$ determined by (10.25) have to intersect the triangle Δ_Q. Relation (10.33) implies that if two of these curves meet at a certain point, then the third curve also passes through the same point. By Lemma 10.10 any two of these curves meet at (at least) one point.

Lemma 10.11. *The curves $\gamma_1, \gamma_2, \gamma_3$ determined by (10.25) meet at exactly one point which lies inside Δ_Q.*

Proof. If Δ_Q is an equilateral triangle, then γ_i is the straight line that passes through a_i and is perpendicular to the side $\overline{a_j a_k}$. Assume that for some Δ_Q, two curves γ_i and γ_j meet at more than one point. Deform this Δ_Q into the equilateral triangle. During this deformation these two curves experience a deformation during which they should touch each other tangentially. But this contradicts Lemma 10.8. □

The above formula for the tangent direction implies the following statement.

Lemma 10.12. *In the above notation,*
1. *One singular horizontal trajectory emanates from each simple pole $z = a_j$ of $R(z)$; its tangent direction is given by $\theta_{ij} + \theta_{kj} - \phi_j$ and points inside Δ_Q.*
2. *One singular vertical trajectory emanates from each simple pole $z = a_i$ of $R(z)$; its tangent direction is given by $\theta_{ij} + \theta_{kj} - \phi_j + \pi$ and points outside Δ_Q.*
3. *Three singular horizontal trajectories emanate from a simple zero $z = b$ and their tangent directions are given by $(\tilde{\phi}_1 + \tilde{\phi}_2 + \tilde{\phi}_3 + \pi(1 + 2m))/3$ $(m = 0, 1, 2)$.*

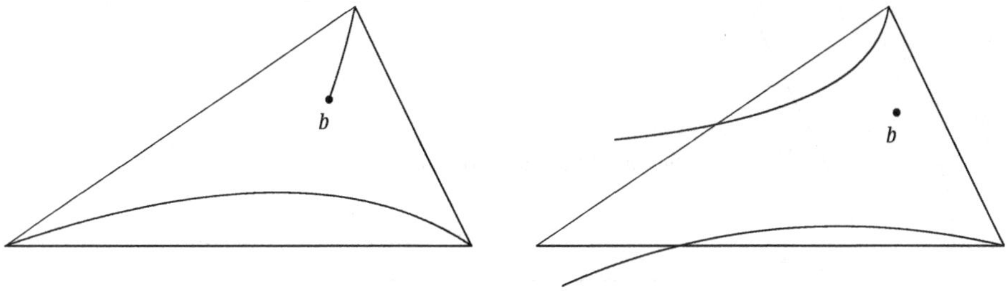

Case $b \in \Gamma_3$ Case $b \in D_1$

Figure 10.5: Illustration of the two cases in Proposition 10.3.

Proof. The only thing we need to check is that if b is in Δ_Q, then $\min(\theta_{ij}, \theta_{kj}) \le \phi_j \le \max(\theta_{ij}, \theta_{kj})$ and $\min(\theta_{ij}, \theta_{kj}) \le \theta_{ij} + \theta_{kj} - \phi_j \le \max(\theta_{ij}, \theta_{kj})$. Hence the tangent vector to the horizontal trajectory emanating from any pole points inside Δ_Q. □

Proposition 10.3.

(i) *If $b \notin \Delta_Q$ and a quadratic differential $\Psi = \frac{(b-z)dz^2}{(z-a_1)(z-a_2)(z-a_3)}$ is Strebel, then at least one of its singular horizontal trajectories emanating from its poles a_1, a_2, a_3 leaves Δ_Q.*

(ii) *If $b \in \Gamma_i \setminus \{b_0\}$ $(i = 1, 2, 3)$, then a_j and a_k are connected by a singular horizontal trajectory γ and it does not contain the point $z = b$, see Figure* 10.5.

(iii) *If $b \in D_j$, then the singular horizontal trajectory γ', which starts at a_k (resp. a_i), goes inside the triangle Δ_Q, crosses the side $\overline{a_k a_i}$ (resp. $\overline{a_k a_j}$), and leaves the triangle Δ_Q.*

Proof. Item (i) is completely obvious. Indeed, since K_Ψ is compact, there should be a singular trajectory connecting one of the poles to the zero b. Since b is located outside Δ_Q, the result follows.

To prove (ii) note that if b coincides with a_i, then the horizontal trajectory emanating from a_j is the straight segment $\overline{a_j a_k}$. Now take $b \in \Gamma_i$ sufficiently close to a_i. Then the horizontal trajectory emanating from a_k passes close to a_j because the direction of the horizontal trajectory changes continuously with b unless the horizontal trajectory hits a singular point. Assume that the horizontal trajectory emanating from a_k does not pass through the point a_j. Let $\theta_{j'j}, \phi_j, \theta$ be the arguments of $a_{j'} - a_j$, $b - a_j$, $z - a_j$ resp. If z is sufficiently close to a_j, then the direction of the horizontal trajectory is approximately given by $(\theta+\theta_{ij}+\theta_{kj}-\phi_j)/2$. It follows from elementary affine geometry that the horizontal trajectories are approximately parabolas whose focus is a_j and the angle of the axis of the symmetry is $\theta_{ij} + \theta_{kj} - \phi_j$.

Hence the horizontal trajectory emanating from a_k goes around the point a_j and intersects the vertical trajectory emanating from a_j, see Figure 10.6. Denote their intersection point by d and consider the integral

$$\left(\int_{a_j}^{d} + \int_{d}^{a_k}\right)\sqrt{\frac{b-t}{(t-a_1)(t-a_2)(t-a_3)}}\,dt, \tag{10.38}$$

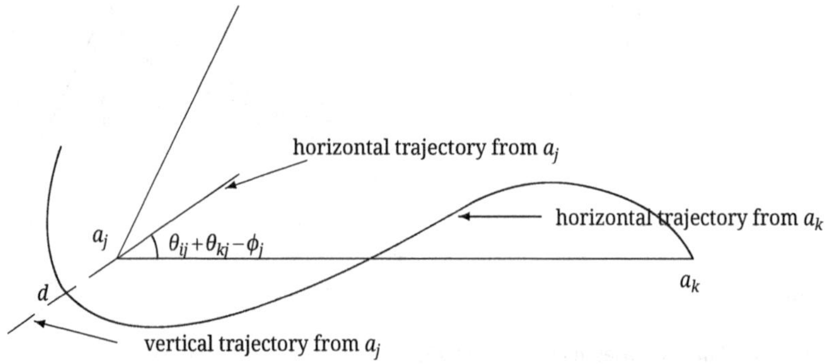

Figure 10.6: Behavior of horizontal and vertical trajectories near a_j.

where the path from a_j to d is taken along the vertical trajectory from a_j and the path from d to a_k is taken along the horizontal trajectory from a_k. By definition of horizontal and vertical trajectories, the value of the integral from d to a_k is real and the one from a_j to d is pure imaginary. Since the integration path does not hit a singular point, the imaginary part (resp. the real part) varies monotonely as the integration variable passes the vertical (resp. horizontal) trajectories. Hence the imaginary part of the value of (10.38) is not zero, but this contradicts the definition of Γ_i.

Therefore we obtain that if $b \in \Gamma_i$ and b is sufficiently close to a_i, then a_j and a_k are connected by a smooth horizontal trajectory γ, which does not hit the singular point $z = b$. Let us move the point b along Γ_i away from a_i. Then a_j and a_k are still connected by a smooth horizontal trajectory as long as the connecting horizontal trajectory does not hit the singular point $z = b$. On the other hand, if the horizontal trajectory passes through $z = b$, then a_j, a_k, and b will be connected by a horizontal trajectory and the integrals

$$\int_{a_j}^{b} \sqrt{\frac{b-t}{(t-a_1)(t-a_2)(t-a_3)}}\,dt, \quad \int_{a_i}^{a_k} \sqrt{\frac{b-t}{(t-a_1)(t-a_2)(t-a_3)}}\,dt \tag{10.39}$$

attain real values. Hence the point b is also contained in Γ_j, and we conclude that $b = b_0$. Therefore, we have shown that a_j and a_k are connected by a smooth horizontal trajectory in case $b \in \Gamma_i$.

To prove (iii) choose $\tilde{b} \in \Gamma_i \setminus \{b_0\}$, and let $l_{\tilde{b}}$ be the straight line that passes through $\tilde{b}$ and is parallel to the side $\overline{a_j a_k}$. Equation (10.34) implies that we can wlog assume that the points a_j, a_k lie on the real axis, $a_j < a_k$ and the point a_i lies in the upper half plane. Let $b \in l_{\tilde{b}} \cap \Delta_Q$ such that $\Re\tilde{b} < \Re b$. Then $b \in D_j$ by Lemma 10.10.

Let z be a point such that $\Im z < \Im b$. By comparing the arguments of $b - z$ and $\tilde{b} - z$, one can easily conclude that the horizontal trajectory of the quadratic differential $\Psi = (b - z)dz^2/Q(z)$ emanating from a_k is located under the similar horizontal trajectory of $\widetilde{\Psi} = (\tilde{b} - z)dz^2/Q(z)$, see Figure 10.7. Since the horizontal trajectory of $\widetilde{\Psi}$ emanating

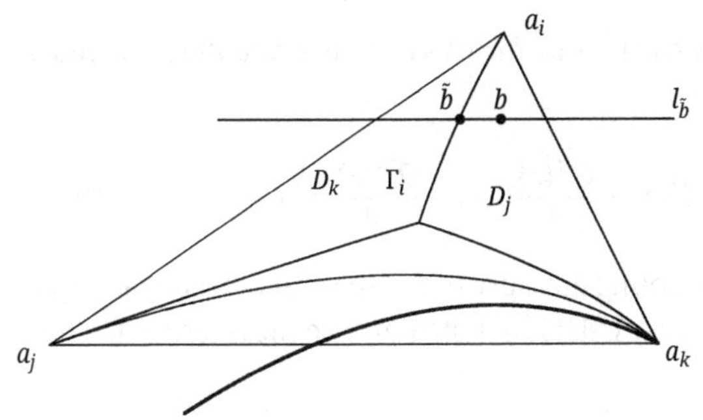

Figure 10.7: Why singular trajectories emanating from poles leave Δ_Q.

from a_k hits the point a_j, one has that the horizontal trajectory of Ψ emanating from a_k must intersect the side $\overline{a_j a_k}$ and leave the triangle Δ_Q. The intersection point of this trajectory with the side $\overline{a_j a_k}$ cannot coincide with a_j since in that case the integral (10.25) will be real, which contradicts Lemma 10.10. Let b vary in D_j. The horizontal trajectory intersects the side $\overline{a_j a_k}$ as long as the intersection point does not coincide with a_j or a_k, or the horizontal trajectory hits the singular point b. Hence (iii) is settled for any $b \in D_i$. □

Remark 1. One can show that item (i) of the latter proposition holds independently of whether Ψ is Strebel or not, but we do not need this fact.

10.7 Proving Theorem 10.5

For this proof we need Theorem 10.8 below whose formulation uses the definition of the following measures. Given $Q(z) = (z - a_1)(z - a_2)(z - a_3)$ choose one of three roots a_i and shift the variable $z = z - a_i$. (Abusing our notation, we use the same letter for the shifted variable.) Then in this new coordinate one has

$$Q(z) = z^3 + v_i z^2 + w_i, \quad i = 1, 2, 3.$$

Define the functions $\xi_i(\tau) = -v_i(1-\tau)^2$, $\tau \in [0,1]$ and $\psi_i(\tau) = -w_i(1-(1-\tau)^2)(1-\tau)^2$, $\tau \in [0,1]$. Let $\omega_i(\tau)$, $i = 1, 2, 3.$ be the arcsine measure supported on the interval $[\xi_i(\tau) - 2\sqrt{\psi_i(\tau)}, \xi_i(\tau) + 2\sqrt{\psi_i(\tau)}]$ in the complex plane. Finally, define the measure M_i, $i = 1, 2, 3$ by averaging

$$M_i = \int_0^1 \omega_i(\tau) d\tau.$$

Results of the beginning of this chapter, see [167], claim that each M_i is supported on an ellipse that is uniquely determined by the triple of roots a_1, a_2, a_3 with the root a_i playing a special role. Moreover all these three measures have the property that outside their respective supports their Cauchy transforms satisfy one and the same linear inhomogeneous differential equation

$$Q(z)\mathcal{C}''(z) + Q'(z)\mathcal{C}'(z) + \frac{Q''(z)}{8}\mathcal{C}(z) + \frac{Q'''(z)}{24} = 0. \tag{10.40}$$

Recall that ν_n denotes the root-counting measure of the spectral polynomial $Sp_n(\lambda)$. The weak limit of the sequence $\{\nu_n\}$ (if it exists) is denoted by ν. In these terms the main result of [167] is as follows.

Theorem 10.8. *If in the above notation the measure ν exists, then each of the measures M_i, $i \in \{1,2,3\}$ has ν as its inverse balayage, i. e., ν and M_i have the same logarithmic potential near infinity and the support of ν is contained inside the support of M_i.*

In fact the proof of Theorem 10.8 presented in [167] (as well as of the original Theorem 1.4 of [112]) extends verbatim to converging **subsequences** of the original sequence $\{\nu_n\}$.

Thus any two converging subsequences of measures from $\{\nu_n\}$ have the same limiting logarithmic potential near infinity. Notice additionally that the support of these measure must necessarily belong to Γ_Q, which is the main result of § 10.4 above. Thus the limiting measures have the same logarithmic potential in the complement $\mathbb{C} \setminus \Gamma_Q$. But then they should coincide since both of them are $\bar{z}$-derivative of the same function.

Let us now prove Theorem 10.5. We show first that the whole sequence $\{\nu_n\}$ of root-counting measures for the whole sequence of $\{Sp_n(\lambda)\}$ converges. This argument resembles that at the very end of § 10.3. Indeed, by item (2) of Theorem 10.1, for any $\epsilon > 0$ $\exists N_\epsilon$ such that for all $n \geq N_\epsilon$ all roots of all $Sp_n(\lambda)$ lie in the ϵ-neighborhood of Γ_Q. Therefore, by compactness the sequence $\{\nu_n\}$ contains a lot of (weakly) converging subsequences. Theorem 10.8 and the argument following it show that any two of such converging subsequences have the same (weak) limiting measure, which we denote by ν. Let us show that then the whole sequence $\{\nu_n\}$ is converging to the same ν. Indeed, assume that $\{\nu_n\}$ is not converging to ν. Then we can find a subsequence N' of the natural numbers such that ν_n stays away from a fixed neighborhood of ν in the weak topology for all $n \in N'$. Again by compactness we can find a subsequence N of N' such that the limit of $\{\nu_n\}$ over N exists and is equal to ν by the above argument. But this contradicts to the assumptions that ν_n stays away from ν for all $n \in N'$.

To show that the whole Γ_Q must be the support of the limiting measure ν, notice that the Cauchy transform of ν satisfies (10.40) whose only singularities are a_1, a_2, a_3, and ∞. One can check that the unique solution of (10.40) with the asymptotics $\frac{1}{z}$ near infinity has a non-trivial monodromy at each singularity a_1, a_2, a_3. Notice that the Cauchy transform of ν coincides with this solution extended from infinity to the whole $\mathbb{C} \setminus \Gamma_Q$. But then the density of ν, which is the $\bar{z}$-derivative of this solution restricted to $\mathbb{C} \setminus \Gamma_Q$, cannot vanish at any generic point of Γ_Q, i. e., outside of b_0. □

10.8 Open problems in Chapter 10

10.1. It is very tempting to extend the methods and results of the present chapter to the case of the 'generalized' Heun equations, which are of the form

$$\left\{Q_{k+1}(z)\frac{d^k}{dz^k} + Q_k(z)\frac{d^{k-1}}{dz^{k-1}} + \cdots + Q_2(z)\frac{d}{dz} + V(z)\right\}S(z) = 0,$$

where $\deg Q_{k+1}(z) = k+1$ and $\deg Q_i(z) \le i$ for $i = 2, 3, \ldots, k$. As in the introduction, for each positive (and sufficiently large) integer n there exist $n+1$ polynomials $V(z)$ counted with appropriate multiplicities such that for each of these $V(z)$ the above equation has a polynomial solution $S(z)$ of degree n. Thus one can define the corresponding spectral polynomials and study the asymptotics of their root-counting measures. Large scale numerical experiments support the following.

Conjecture 10.1. *For any 'generalized' Heun equation, the sequence $\{\nu_n\}$ of the root-counting measures of its spectral polynomials converges to a probability measure ν supported on a curvilinear planar tree located inside $\mathrm{Conv}_{Q_{k+1}}$ and whose leaves (i. e., vertices of valency 1) is the set of all roots of $Q_{k+1}(z)$, see Figure 10.8. Moreover, the limiting measure ν depends only on $Q_{k+1}(z)$, i. e., is independent of the other coefficient of the equation.*

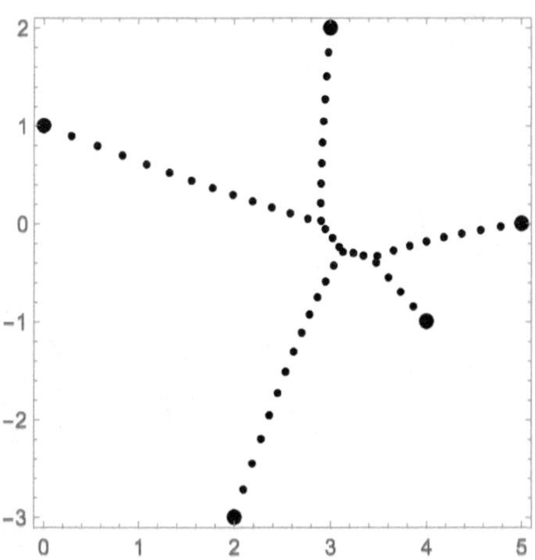

Figure 10.8: The measure ν for the operator $Q(z)\frac{d^4}{dz^4}$ with $Q(z) = (z-5)(z-)(z-4+i)(z-2+3i)(z-3-2i)$. (Here i is the imaginary unit.)

10.2. Here is our guess about the ODE satisfied by the relevant Cauchy transform.

Problem 10.2. Under the assumption that the latter conjecture holds (which is very likely), is it true that the Cauchy transform $\mathcal{C}_\nu$ of the limiting root-counting measure ν satisfies a linear ode of the form

$$Q_{k+1}(z)\mathcal{C}_\nu^{(k)}(z) + a_1 Q'_{k+1}(z)\mathcal{C}_\nu^{(k-1)}(z) + \cdots + a_k Q_{k+1}^{(k)}\mathcal{C}_\nu(z) + a_{k+1}Q_{k+1}^{(k+1)}(z) = 0,$$

where $a_1, \dots, a_{k+1}$ are some universal constants, i. e., independent of $Q_{k+1}(z)$ (but maybe dependent on the order k of the operator)?

10.3. A similar set-up was developed in [162] for linear differential operators of order exceeding 2 of the form $\mathfrak{q} = \sum_{i=1}^{k} Q_i(z)\frac{d^i}{dz^i}$ where $\deg Q_i(z) \le i$ and $\deg Q_k(z) = k$. This topic was continued in [170] where it is shown that a very natural analog of the main object of the present chapter, i. e., the quadratic differentials $\Psi = -\tilde{V}(z)dz^2/Q(z)$, appears for operators of higher order as well. It also has almost all closed trajectories that are continuous but, in general, only piecewise smooth. One needs to develop a notion of a Strebel differential for order > 2, which (to the best of our knowledge) is a completely open problem at the moment. So we pose our question in minimal possible generality.

Problem 10.3. How to define a notion of a rational cubic Strebel differential

$$\Psi = U_1(z)dz^3/U_2(z),$$

where $\deg U_1(z) = 1$ and $\deg U_2(z) = 4$.

Some related information can be found in Chapter 12.

11 Asymptotics of Stieltjes polynomials

(The material of this chapter is based on [84].)

11.1 Preliminaries

The following two main results of [164] were formulated in the previous chapter.

Proposition 11.1. *For any non-degenerate higher Lamé operator T there exist finitely many (up to a scalar multiple) Stieltjes polynomials $S(z)$ of any sufficiently large degree. Their corresponding Van Vleck polynomials $V(z)$ all have degree exactly r. Here r is the Fuchs index of T.*

The next localization result is of a special importance to us.

Theorem 11.1. *For any non-degenerate higher Lamé operator T and any $\epsilon > 0$, there exists a positive integer N_ϵ such that the zeros of all its Van Vleck polynomials $V(z)$ possessing a Heine–Stieltjes polynomial $S(z)$ of degree $n \geq N_\epsilon$ as well as all zeros of these Heine–Stieltjes polynomials belong to $Conv^\epsilon_{Q_k}$. Here $Conv_{Q_k}$ is the convex hull of all zeros of the leading coefficient $Q_k(z)$ and $Conv^\epsilon_{Q_k}$ is its ϵ-neighborhood in the usual Euclidean distance on $\mathbb{C}$.*

Remark 11.1. Proposition 11.1 together with Theorem 11.1 show that the zeros of all Van Vleck polynomials having Stieltjes polynomials of degree greater or equal to a sufficiently large positive integer n are confined in some fixed ϵ-neighborhood $Conv^\epsilon_{Q_k}$ of $Conv_{Q_k}$. Therefore, there exist plenty of converging subsequences $\{\widetilde{V}_{n,i_n}(z)\}$ of normalized Van Vleck polynomials. Here $\widetilde{V}_{n,i_n}(z)$ is a monic polynomial proportional to $V_{n,i_n}(z)$ and $V_{n,i_n}(z)$ is some Van Vleck polynomial having Stieltjes polynomial of degree n.

It seems natural to pose the following two questions.

Problem 11.1. What happens with the set $\{\widetilde{V}_{n,i}(z)\}$ of all normalized Van Vleck polynomials having Stieltjes polynomials of degree exactly n when $n \to \infty$?

Problem 11.2. What happens with the subsequence $\{S_{n,i_n}(z)\}$ of Stieltjes polynomials whose corresponding sequence $\{\widetilde{V}_{n,i_n}(z)\}$ of normalized Van Vleck polynomials has a limit?

At the present moment we do not have even a conjectural answer to Problem 11.1. Some initial steps in this direction can be found in [168] and Chapters 9 and 10. Problem 11.2, however, has a satisfactory answer reported in this chapter.

Assume now that a subsequence $\{\widetilde{V}_{n,i_n}(z)\}$ of normalized Van Vleck polynomials of an operator T converges when $n \to \infty$ to some monic polynomial $\widetilde{V}(z)$ of degree r. (As above r is the Fuchs index of T.)

https://doi.org/10.1515/9783112207741-013

Our first result is a far reaching generalization of the main theorem of [27] together with the main result of [134] describing the asymptotic root distribution of Heine–Stieltjes polynomials under the assumptions that one picks a sequence with the portion of roots in each of the intervals (a_i, a_{i+1}) converging to some number in $[0,1]$. The latter process corresponds to the choice of a sequence of Van Vleck polynomials converging (up to a scalar factor) to some limiting polynomial $\widetilde{V}(z)$.

Theorem 11.2. *For any non-degenerate higher Lamé operator T of order k, take any subsequence $\{\widetilde{V}_{n,i_n}(z)\}$ of its normalized Van Vleck polynomials converging to some monic polynomial $\widetilde{V}(z)$. Then the sequence $\{\mu_{n,i_n}\}$ of the root-counting measures of the corresponding Stieltjes polynomials $\{S_{n,j_n}(z)\}$ weakly converges to a probability measure $\mu_{\eth,\widetilde{V}}$ whose Cauchy transform $C_{T,\widetilde{V}}(z)$ satisfies almost everywhere in $\mathbb{C}$ the equation*

$$C^k_{T,\widetilde{V}}(z) = \frac{\widetilde{V}(z)}{Q_k(z)}. \tag{11.1}$$

Figures 8.1 and 8.2 illustrate the above Theorem 11.2.

Our next result describes some properties of a probability measure in case when it satisfies the assumptions of Theorem 11.2.

Theorem 11.3. *Consider a rational function $R(z) = \frac{V(z)}{Q(z)}$ where $V(z)$ and $Q(z)$ are monic polynomials with $\deg Q(z) - \deg V(z) = k \geq 2$. Assume that there exists a compactly supported probability measure μ on $\mathbb{C}$ whose Cauchy transform $C_\mu(z)$ satisfies almost everywhere (with respect to the standard Lebesgue measure on $\mathbb{C}$) the equation*

$$C^k_\mu(z) = \frac{V(z)}{Q(z)}. \tag{11.2}$$

Then, for a given function $R(z)$, such a measure μ is unique and its support is a curvilinear forest with leaves at the roots of $V(z)$ and/or $Q(z)$. This support is straightened out in the local canonical coordinate $\Psi(z) = \int_{z_0}^{z} \sqrt[k]{\frac{V(t)}{Q(t)}}\,dt$.

Remark 11.2. In fact we will first prove Theorem 11.3 and then use it to settle Theorem 11.2.

11.2 On probability measure with Cauchy transform whose power is a rational function

This section is devoted to the proof of Theorem 11.3. Below we discuss a number of properties of a compactly supported probability measure μ whose Cauchy transform $C_\mu(z)$ satisfies almost everywhere in $\mathbb{C}$ the equation

$$C^k_\mu(z) = \frac{V(z)}{Q(z)}, \tag{11.3}$$

where $V(z)$ and $Q(z)$ are monic polynomials without common factors.

It is clear that $C_\mu(z)$ can have a pole of order at most 1, which means that $Q(z)$ can only have roots of multiplicity $\leq k$ if (11.3) is satisfied. Suppose now that $Q(z)$ has roots of multiplicity exactly k, say $Q(z) = (z - a_1)^k \cdot \dots \cdot (z - a_m)^k \tilde{Q}(z)$, where $\tilde{Q}(z)$ only has roots of multiplicity $< k$. Taking the $\frac{\partial}{\partial \bar{z}}$-derivative in the distributional sense yields

$$\begin{aligned}\pi\mu &= \frac{\partial}{\partial \bar{z}}\left[\frac{1}{(z-a_1)\cdot \dots \cdot (z-a_m)}\left(\frac{V(z)}{\tilde{Q}(z)}\right)^{1/k}\right]\\ &= \sum_{j=1}^{m} \delta_{a_j}\left(\frac{V(a_j)}{\tilde{Q}(a_j)}\right)^{1/k} + \frac{1}{(z-a_1)\cdot \dots \cdot (z-a_m)}\frac{\partial}{\partial \bar{z}}\left(\frac{V(z)}{\tilde{Q}(z)}\right)^{1/k},\end{aligned}$$

where δ_{a_j} is a point mass located at $z = a_j$. Proposition 11.2 below states that the positive measure $\tilde{\mu} := \frac{\partial}{\partial \bar{z}}(\frac{V(z)}{\tilde{Q}(z)})^{1/k}$ must be supported on a finite union of analytic curves. If C_μ^+ and C_μ^- denote the one-sided limits of the Cauchy transform as we approach the support of μ, then the Sokhotsky–Plemelj formula tells us that $\tilde{\mu} = \rho(s)ds$ where $\rho(s) = \frac{(C_\mu^+(s) - C_\mu^-(s))}{2}\mathbf{n}$, with $\mathbf{n}$ being a properly oriented normal to the support, and ds is the usual arc length. Clearly ρ vanishes only at the roots of $V(z)$. From this it is immediate that the point masses δ_{a_j} cannot lie in the closure of the support of $\tilde{\mu}$ since the total mass of μ is assumed to be finite.

The goal of this section is to prove that the support of a positive measure μ whose Cauchy transform satisfies (11.3) is a (curvilinear) forest. By the above remarks we can restrict our attention to measures whose Cauchy transform satisfies (11.3) with $Q(z)$ only having roots of multiplicity $< k$. The reader can easily check the validity of all the arguments given below if we add to the measure μ a finite number of isolated point masses.

Proposition 11.2. *If μ is a probability measure satisfying (11.3) almost everywhere, then the support supp(μ) of μ is the union of finitely many smooth curve segments, and each of these curves is mapped to a straight line by the (locally defined) mapping*

$$\Psi(z) = \int_{z_0}^{z} \sqrt[k]{\frac{V(t)}{Q(t)}}\,dt.$$

Remark 11.3. Proof of this proposition repeats more or less literally that of Lemma 4 in [27] and is included for the sake of completeness, see also Chapter 2. We need the following technical statement.

Lemma 11.1 (Corollary 2 of [27]). *For a finite set $A \subset \mathbb{C}$, a convex domain U, and a measurable function $\chi : U \to A$, the claim that $\frac{\partial \chi}{\partial \bar{z}} \geq 0$ (as a distribution) is equivalent to the existence of real numbers c_a, $a \in A$ such that $\chi(z) = a$ almost everywhere in G_a where*

$$G_a = \{z \in U; c_a + \Re(az) \geq c_b + \Re(bz); \forall b \in A\}.$$

In other words, any (local) domain where χ attains a constant value is (locally) an angle given by linear inequalities.

Remark 11.4. Proof of Lemma 11.1 is based on Corollary 1 of [27] claiming that for a convex domain $U \subset \mathbb{C}$, a finite set $A \subset \mathbb{C}$, and a subharmonic function v defined in U such that $2\frac{\partial v}{\partial z} \in A$ almost everywhere one has that v coincides with the maximum of a number of linear (non-homogeneous) functions and is, therefore, convex.

Proof of Proposition 11.2. Let us first prove it in a neighborhood of $z_0 \in supp(\mu)$, which is neither a zero nor a pole of $\frac{V(z)}{Q(z)}$. Choose a branch $B(z)$ of $\sqrt[k]{\frac{V(z)}{Q(z)}}$ in some simply connected neighborhood Ω of z_0 and define Ψ as some concrete primitive function of $B(z)$ in Ω. Let U be some small convex neighborhood of $\Psi(z_0)$ such that Ψ maps some neighborhood of z_0 bijectively on U. By (11.3) we can write

$$\mathcal{C}_\mu(z) = \chi(\Psi(z))B(z)$$

for $z \in \Psi^{-1}(U)$, where χ takes values in the set of kth roots of unity. Using the variable $w = \Psi(z)$, one gets

$$\pi\mu = \frac{\partial \mathcal{C}_\mu}{\partial \bar{z}} = \frac{\partial \chi(\Psi(z))}{\partial \bar{z}} B(z) = \Psi^*\left(\frac{\partial \chi}{\partial \bar{w}}\right) \cdot \frac{\overline{\partial \Psi}}{\partial z} \cdot B(z) = \Psi^*\left(\frac{\partial \chi}{\partial \bar{w}}\right) \cdot |B(z)|^2,$$

where Ψ^* denotes the pullback of distributions defined in U by the map Ψ. Since the measure μ is positive, one gets the relation $\frac{\partial \chi}{\partial \bar{w}} \geq 0$, which should be interpreted in the distributional sense. By the above mentioned Corollary 2, U is the union of sets G_a whose boundaries are finite unions of line segments such that χ is constant in each G_a. Therefore, $\mathfrak{S}_\mu \cap \Psi^{-1}(U) = \Psi^{-1}(supp\ \frac{\partial \chi}{\partial \bar{z}})$ is the union of finitely many curve segments that are mapped to straight lines by Ψ.

If z_0 is a zero or a pole of $\frac{V(z)}{Q(z)}$, we take a disk D centered at z_0 and not containing any other zeros or poles of $\frac{V(z)}{Q(z)}$. If γ is any ray emanating from z_0, we can define single-valued branches of $\sqrt[k]{\frac{V(z)}{Q(z)}}$ and Ψ in $D \setminus \gamma$. Notice that Ψ is continuous up to z_0. Let U be any half disc centered at $\Psi(z_0)$ and contained in $\Psi(D \setminus \gamma)$. Then by the first part of the proof $supp(\mu)$ has the required properties in $\Psi^{-1}(D)$. Varying γ and U, we see that the same holds in a full neighborhood of z_0. □

Lemma 11.2. *Suppose that the Cauchy transform $\mathcal{C}_\mu(z)$ of μ satisfies (*11.3*) and $u(z)$ is the logarithmic potential of μ. If Ψ^{-1} is a (locally defined) inverse of Ψ, which is a primitive to $\sqrt[k]{\frac{V}{Q}}$, then the function $u \circ \Psi^{-1}$ coincides with the maximum of some number of linear functions and, in particular, is convex (in its domain).*

Proof. Since u is subharmonic, we need to check that its derivative belongs to a finite set. We prove our lemma in a neighborhood of any z_0, which is neither a zero of $V(z)$ nor a zero of $Q(z)$. Choose a branch $B(z)$ of $\sqrt[k]{\frac{V(z)}{Q(z)}}$ in some simply connected neighborhood Ω of z_0 and define $\Psi(z)$ as some concrete primitive function of $B(z)$ in Ω. Let U be some small convex neighborhood of $\Psi(z_0)$ such that $\Psi(z)$ maps some neighborhood of z_0 bijectively on U. We want to show that $u \circ \Psi^{-1}$ is convex in U. By (11.3) we can write

$$\mathcal{C}_\mu(z) = \chi(\Psi(z))B(z)$$

for $z \in \Psi^{-1}(U)$, where χ takes values in the set of kth roots of unity. Since $\mathcal{C}_\mu(z) = 2\frac{\partial u}{\partial z}$, we have using the variable $w = \Psi(z)$

$$2\frac{\partial}{\partial w}u(\Psi^{-1}(w)) = 2\frac{\partial u}{\partial z}(\Psi^{-1}(w))(B(\Psi^{-1}(w)))^{-1}$$
$$= \mathcal{C}_\mu(\Psi^{-1}(w))(B(\Psi^{-1}(w)))^{-1} = \chi(w).$$

Therefore the locally defined function $u \circ \Psi^{-1}$ is piecewise linear and convex. □

Corollary 11.1. *If an endpoint of any curve segment in the support supp(μ) of μ is a hanging vertex (i. e., not shared by any other such segment), then this endpoint is either a zero or a pole of* $\frac{V(z)}{Q(z)}$.

Proof. If this were false, then take a point p which is a hanging vertex but not a zero of either $V(z)$ or $Q(z)$. The Cauchy transform $\mathcal{C}_\mu$ is supposed to satisfy an algebraic equation whose branching points are exactly the zeroes of $V(z)$ and $Q(z)$. In particular, it has no monodromy around p. This implies that the limits of $\mathcal{C}_\mu$ as we approach $\mathfrak{S}_\mu$ from both sides close to p are the same, which in turn implies that p is not in the support of μ. □

Proof of Theorem 11.3 starts with a series of additional observations. Notice that if we fix a local branch of $\sqrt[k]{\frac{V(z)}{Q(z)}}$ near some point, which is neither its root or its pole, and consider a multi-valued canonical coordinate

$$\Psi(z) = \int_{z_0}^{z} \sqrt[k]{\frac{V(t)}{Q(t)}}dt$$

globally, i. e., take its full analytic extension, then $\Psi(z)$ will be well defined and univalent only on the universal covering of $\mathbb{C} \setminus (\mathbb{Z}(V) \cup \mathcal{Z}(Q))$ where $\mathcal{Z}(V)$ (resp. $\mathbb{Z}(Q)$) is the set of all roots of $V(z)$ (resp. $Q(z)$). But due to the existence of a measure μ, we can choose an almost global representative of $\Psi(z)$ on $\mathbb{C}$ substantially reducing its multi-valuedness. Namely, let Ω be the complement to the support of μ, i. e., $\Omega = \mathbb{C} \setminus supp(\mu)$. Define the complex potential $\Psi^+(z)$ of μ in Ω by

$$\Psi_+(z) = \int_{\mathbb{C}} \log(z - \zeta)d\mu(\zeta).$$

Obviously, $\Psi_+(z)$ is a part of the complete multi-valued function $\Psi(z)$ since for any $z \in \Omega$ one has $\Psi_+'(z) = \int_{\mathbb{C}} \frac{d\mu(\zeta)}{z-\zeta} = \mathcal{C}_\mu(z)$ while $\mathcal{C}_\mu(z)$ satisfies (11.3). Since Ω is never simply-connected the function $\Psi_+(z)$ is still multi-valued. Namely, going once around some connected component K of *supp*(μ) in the clockwise direction, one increases $\Psi_+(z)$ by $2\pi i \times \mu(K)$, where $\mu(K) > 0$ is the mass of the measure μ concentrated on K. Nevertheless the real part $\Re(\Psi_+(z))$ is a well-defined single-valued function in Ω coinciding with the logarithmic potential $u(z)$ of μ, namely

$$u(z) = \Re(\Psi_+(z)) = \int_{\mathbb{C}} \log|z - \zeta|d\mu(\zeta).$$

Consider now the family Φ of curves in Ω defined by the condition

$$\mathfrak{I}(\Psi_+(z)) = \text{const}$$

which is obviously independent of the choice of the branch of $\Psi_+(z)$. One can easily show that the gradient grad $u(z)$ of $u(z)$ coincides with the vector field $\overline{\mathcal{C}}_\mu(z)\partial_z$. (Here $\overline{\mathcal{C}}_\mu(z)$ is the usual complex conjugate of $\mathcal{C}_\mu(z)$.) Thus the family Φ consists of the integral curves of the vector field $\overline{\mathcal{C}}_\mu(z)\partial_z$, which is well defined and non-vanishing in Ω. Moving along the trajectories of $\overline{\mathcal{C}}_\mu(z)\partial_z$ in positive time, we increase the value of $u(z)$. Finally, for sufficiently large $|z|$, one has

$$\overline{\mathcal{C}}_\mu(z) \simeq \frac{1}{\bar{z}} \quad \text{and} \quad u(z) \simeq \log|z|.$$

The next statement is crucial for the proof of Theorem 11.3.

Lemma 11.3.
1) *When $t \to +\infty$, any trajectory of $\overline{\mathcal{C}}_\mu(z)\partial_z$ tends either to ∞ in $\mathbb{C}P^1$ or to a root of $V(z)$.*
2) *Any trajectory of $-\overline{\mathcal{C}}_\mu(z)\partial_z$ tends to $\mathfrak{S}_\mu$ or, in other words, any trajectory of $\overline{\mathcal{C}}_\mu(z)\partial_z$ starts on $\mathfrak{S}_\mu$.*

Proof. According to Lemma 11.2, locally near any point $p \in supp(\mu)$ the logarithmic potential u is given as the maximum of a finite number of linear functions if one uses the canonical coordinate. This shows that locally on both sides of the complement to the support of μ, the gradient vectors everywhere point away from the support of μ, except possibly at a root of $V(z)$ where the gradient vanishes. At infinity we have $\overline{\mathcal{C}}_\mu(z) \approx \frac{1}{\bar{z}}$, which easily implies that the point at infinity is a sink for the flow defined by $\overline{\mathcal{C}}_\mu(z)\partial_z$. Denote by K_ϵ the Riemann sphere with an ϵ-neighborhood of $\{\infty\} \cup supp(\mu)$ removed. To obtain the proof we need to show that for any $\epsilon > 0$ a trajectory cannot stay in a subset of K_ϵ when $t \to +\infty$. By construction, any such K_ϵ does not contain singular points of $\overline{\mathcal{C}}_\mu(z)\partial_z$. Thus, we only need to rule out the occurrence of closed trajectories and recurrence in K_ϵ. But both of these possibilities are indeed forbidden by the fact that for any $\epsilon > 0$, $\overline{\mathcal{C}}_\mu(z)\partial_z$ is the gradient field of a function without singularities in K_ϵ. □

Definition 11.1. A trajectory of $\overline{\mathcal{C}}_\mu(z)\partial_z$ ending at a root of $V(z)$ is called *exceptional.*

Lemma 11.4. *There exists at most a finite number of trajectories of $\overline{\mathcal{C}}_\mu(z)\partial_z$ ending at a given root of V.*

Proof. Wlog assume that a root of $V(z)$ lies at the origin and has multiplicity p. Notice that

$$\mathfrak{I}(\Psi_+(z)) = \mathfrak{I}\int_0^z \sqrt[k]{\frac{V(t)}{Q(t)}}\,dt.$$

We now have $\sqrt[k]{\frac{V(z)}{Q(z)}} = z^{p/k}(a_p + a_{p+1}z + \cdots)^{1/k}$. Choosing a single-valued branch of $(a_p + a_{p+1}z + \cdots)^{1/k}$, say $b_0 + b_1 z + \cdots$, and taking $\xi = z(b_0 + b_1 z + \cdots)^{k/p}$, which can be chosen single-valued in a neighborhood of the origin, we get $\sqrt[k]{\frac{V(z)}{Q(z)}} = \xi^{p/k}$, which is valid in a neighborhood of the origin minus the slit defined by the support of our measure μ. Integration yields

$$\int_0^z \sqrt[k]{\frac{V(t)}{Q(t)}}dt = \left(\frac{p+k}{k}\right)\xi^{(p+k)/k}.$$

Since there is only a finite number of rays in the ξ-plane entering the origin along which $\Im(\xi^{(p+k)/k}) \equiv 0$, our claim follows. □

Remark 11.5. The proof of the previous lemma copies some of local studies of zeroes of quadratic differentials, see [179, Chapter 2]. It is indeed possible to interpret the support of μ as the union of several horizontal trajectories of a higher order differential, which in turn allows us to state more explicitly exactly how many exceptional trajectories enter a hanging vertex of the support of μ.

Definition 11.2. We denote the union of *supp*(μ) and all exceptional trajectories by Υ_μ and call it the *extended support* of μ.

The following statement is essential.

Lemma 11.5. *Topologically the extended support* Υ_μ *is an embedded tree in* $\mathbb{C}$.

Proof. Indeed, by Lemmas 11.3 and 11.4, the flow on $\mathbb{CP}^1 \setminus \Upsilon_\mu$ defined by the gradient vector field $\overline{C}_\mu(z)$ (which is non-vanishing there) contracts the whole domain to the point at infinity. The result follows. □

Let μ be a measure whose Cauchy transform satisfies (11.3) and denote $\Omega^* = \mathbb{C} \setminus \Upsilon_\mu$. Define now the following specialization of the function $\Psi_+(z)$ to Ω^*:

$$\Psi_{++}(z) = \int_{\mathbb{C}} \log(z-\zeta)d\mu(\zeta)$$

where $z \in \Omega^*$. Note that $\Psi_{++}(z)$ is multi-valued only up to addition of multiples of $2\pi i$.

Lemma 11.6. *The function* $\Psi_{++}(z)$ *determines a mapping of* Ω^* *onto a domain* $H = \{w \mid \Re w > h(\Im w)\}$ *where h is a piecewise-continuous function. Moreover,* $\Psi_{++}^{-1} : H \to \Omega^*$ *is a single-valued function, see Figure* 11.1.

Explanation to Figure 11.1

The left picture shows the support of a measure μ consisting of 2 components—one connecting three poles a, b, c and the other connecting the only zero α with the remaining pole d. The exceptional trajectory connects the point O on the first component with the zero α so that the extended support Υ_μ forms a tree. The map $\Psi_{++}(z)$ wraps out the complement $\mathbb{C} \setminus \Upsilon_\mu$ onto the strip bounded by two horizontal lines and the piecewise

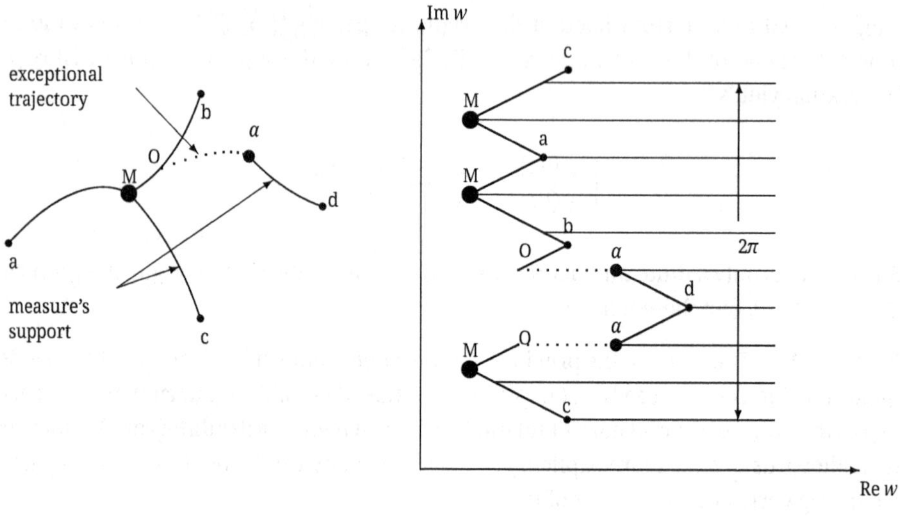

Figure 11.1: Extended support Υ_μ of a positive measure associated with a rational function $\frac{z-a}{(z-a)(z-b)(z-c)(z-d)}$ and its image under the mapping $\Psi_{++}(z)$.

linear curve shown on the right picture. This map has a period $2\pi i$. We go around the extended support in the anti-clockwise direction starting from the pole c. The piecewise linear curve on the right picture shows the images of the special points we meet while doing this: $c, M, O, a, d, a, O, b, M, a, M$ and returning back to c but with the imaginary part increased by 2π. Each time we meet one of the poles or the point M, we have to change our direction by $\pm 120°$.

Proof. We have already noticed that $\Psi_{++}(z)$ is defined up to a multiple of $2\pi i$ in Ω^* and that $\Psi'_{++}(z) = \mathcal{C}_\mu(z)$. If γ is a small curve segment of $supp(\mu)$ and if U is a small one sided neighborhood of γ, then Proposition 11.2 states that γ is mapped to a straight line segment by Ψ_{++}, which is not horizontal, and that U is mapped to the right of this straight segment. The latter follows if we observe that in the proof of Proposition 11.2 we have $\chi = 1$ in U and $\Re(\chi) \leq 1$ everywhere. If γ is a part of an exceptional trajectory, then it is mapped by Ψ_{++} to a straight horizontal line segment since exceptional trajectories are level curves of $\Im(\Psi_{++})$. Continuing Ψ_{++} around Υ_μ, we obtain a piecewise linear curve of the form $\{\Re(w) = h(\Im(w))\}$ bounding a domain H of the form $\{\Re(w) > h(\Im(w))\}$ where h is a piecewise continuous function. It is clear that Ψ_{++} maps Ω onto H with boundary to boundary. The function $\psi(z) = \exp(-\Psi_{++}(z))$ is single-valued on Ω^* and maps $\Omega^* \cup \{\infty\}$ to $D = \{\zeta : \log|\zeta| < -h(-\arg\zeta)\}$, does not vanish in Ω^* and has a simple zero at infinity. It follows that ψ is bijective on $\Omega^* \cup \{\infty\} \to D$ and hence $\Psi_{++}^{-1}(w) = \psi^{-1}(e^{-w})$ is single-valued. This concludes the proof of the lemma. □

We can finally prove the uniqueness of the required measure. Indeed, assume that there are two different probability measures μ_1 and μ_2 whose Cauchy transforms solve

(11.3) almost everywhere, and let $u_1(z)$ and $u_2(z)$ be their logarithmic potentials. Notice that there is only one branch of $\sqrt[k]{\frac{V(z)}{Q(z)}}$, which has $\frac{1}{z}$ as its asymptotics near ∞ in $\mathbb{C}P^1$. Therefore the Cauchy transforms and logarithmic potentials of μ_1 and μ_2 have to coincide in some neighborhood of ∞. We will show that $u_2(z) \geq u_1(z)$ in $\Omega_1^* = \mathbb{C} \setminus \Upsilon_{\mu,1}$ and $u_1(z) \geq u_2(z)$ in $\Omega_2^* = \mathbb{C} \setminus \Upsilon_{\mu,2}$. Indeed, $u_1(\Psi_{++}^{-1}(w)) = \Re(w)$ for all $w \in H_1$ and $u_2(\Psi_{++}^{-1}(w)) = \Re(w)$ for all sufficiently large $\Re(w)$. On the other hand, $u_2 \circ \Psi_{++}^{-1}$ is piecewise linear and convex on any ray $\Im(w) = \text{const}$ in H_1. Therefore, $u_2(\Psi_{++}^{-1}(w)) \geq \Re(w)$ for all $w \in H_1$. Changing place of $u_1(z)$ and $u_2(z)$ we get the second inequality. Since $u_1(z)$ and $u_2(z)$ are continuous in the whole $\mathbb{C}$, they should coincide. But the measures μ_1 and μ_2 are obtained as $\Delta u_1(z)$ and $\Delta u_2(z)$, therefore they coincide as well. □

We finish this section with some information about connected components of $supp(\mu)$.

Lemma 11.7. *Each connected component of $supp(\mu)$ contains either some number of roots of $Q(z)$ and equally many roots of $V(z)$ or it contains $k+j$ roots of $Q(z)$ and j roots of $V(z)$ for some $j = 0, \ldots, r$. (We count roots with multiplicities here.)*

Proof. Going around such a component should result in a trivial monodromy of the branches of $\sqrt[k]{\frac{V(z)}{Q(z)}}$. Notice that these branches are cyclicly ordered and a small clockwise oriented loop around a root of $V(z)$ results in the cyclic clockwise shift by $\frac{2\pi}{k}$ while a small clockwise oriented loop around a root of $Q(z)$ results in the cyclic counterclockwise shift by $\frac{2\pi}{k}$. □

11.3 Proving Theorem 11.2

Our scheme follows roughly the scheme suggested in Chapter 2, see [27]. We need to prove that under its assumptions the sequence $\{\mu_{n,i_n}\}$ of root-counting measures of the Stieltjes polynomials $\{S_{n,i_n}(z)\}$ converges weakly to a probability measure $\mu_{T,\widetilde{V}}$ whose Cauchy transform $C_{T,\widetilde{V}}(z)$ satisfies almost everywhere in $\mathbb{C}$ equation (11.1).

To simplify the notation, we denote by $\{\bar{S}_n(z)\}$ the chosen sequence $\{S_{n,i_n}(z)\}$ of Stieltjes polynomials whose normalized Van Vleck polynomials $\{\widetilde{V}_{n,i_n}(z)\}$ converge to $\widetilde{V}(z)$ and let $\{\bar{\mu}_n\}$ denote the sequence of the root-counting measures of $\{\widetilde{S}_n(z)\}$. Also let $\bar{\mu}_n^{(i)}$ be the root-counting measure of the ith derivative $\bar{S}_n^{(i)}(z)$. Assume now that NN is a subsequence of natural numbers such that

$$\bar{\mu}^{(j)} = \lim_{n\to\infty, n\in NN} \bar{\mu}_n^{(j)}$$

exists for all $j = 0, 1, \ldots, k$. The next lemma shows that the Cauchy transform of $\bar{\mu} = \bar{\mu}^{(0)}$ satisfies the required algebraic equation.

Lemma 11.8. *The measures $\bar{\mu}^j$ are all equal and the Cauchy transform $C(z)$ of their common limit satisfies equation* (11.1) *for almost every z.*

Proof. We have

$$\frac{\bar{S}_n^{(j+1)}(z)}{(n-j)\bar{S}_n^{(j)}(z)} \to \int \frac{d\bar{\mu}^{(j)}(\zeta)}{z-\zeta}$$

with convergence in L^1_{loc}, and by passing again to a subsequence we can assume that we have pointwise convergence almost everywhere. From the relation

$$T\bar{S}_n(z) + V_n(z)\bar{S}_n(z) = 0$$

it follows that

$$\frac{Q_k(z)\bar{S}_n^{(k)}(z)}{(n)_k\bar{S}_n(z)} + \frac{V_n(z)}{(n)_k} = -\sum_{l=0}^{k-1} \frac{Q_l(z)}{(n-l)_{k-l}} \left(\prod_{j=0}^{l-1} \frac{\bar{S}_n^{(j+1)}(z)}{(n-j)\bar{S}_n^{(j)}(z)} \right).$$

One can immediately check that $-\frac{V_n(z)}{(n)_k} \to \widetilde{V}(z)$, while the sum in the right-hand side converges pointwise to 0 almost everywhere in $\mathbb{C}$ due to presence of the factors $(n-l)_{k-l}$ in the denominators. Thus, for almost all $z \in \mathbb{C}$, one has

$$\frac{\bar{S}_n^{(k)}(z)}{(n)_k\bar{S}_n(z)} \to \frac{\widetilde{V}(z)}{Q_k(z)}$$

when $n \to \infty$ and $n \in NN$. If $u^{(j)}(z)$ denotes the logarithmic potential of $\bar{\mu}^{(j)}$, then one has

$$u^{(k)}(z) - u^{(0)}(z) = \lim_{n\to\infty} \frac{1}{n} \log\left| \frac{\bar{S}_n^{(k)}(z)}{(n)_k\bar{S}_n(z)} \right| = \lim_{n\to\infty} \frac{1}{n} (\log|\widetilde{V}(z)| - \log|Q_k(z)|) = 0.$$

On the other hand, we have that $u^{(0)}(z) \geq u^{(1)}(z) \geq \cdots \geq u^{(k)}(z)$, see Lemma 11.9 below. Hence all the potentials $u^{(j)}(z)$ are equal, and all $\mu_j = \Delta u^{(j)}/2\pi$ are equal as well. Finally, we get

$$C^k(z) = \lim_{n\to\infty} \Pi_{j=0}^{k-1} \frac{\bar{S}_n^{(j+1)}(z)}{(n-j)\bar{S}_n^{(j)}(z)} = \lim_{n\to\infty} \frac{\bar{S}_n^{(k)}(z)}{(n)_k\bar{S}_n(z)} = \frac{\widetilde{V}(z)}{Q_k(z)}$$

for almost all z. This completes the proof of the existence of a compactly supported probability measure whose Cauchy transform satisfies (11.1) in the case when $\widetilde{V}(z)$ is the limit of a sequence of normalized Van Vleck polynomials of a Heine–Stieltjes problem with the leading term of the operator T equal to $Q_k(z)\frac{d^k}{dz^k}$. The uniqueness of such a measure was obtained in § 11.2.

Lemma 11.9 (See Lemma 8 of [27]). *Let $\{p_m(z)\}$ be a sequence of polynomials, such that $n_m := \deg p_m \to \infty$ and there exists a compact set K containing the zeros of all $p_m(z)$ simultaneously. Finally, let μ_m and μ'_m be the root-counting measures of $p_m(z)$ and $p'_m(z)$*

resp. If $\mu_m \to \mu$ and $\mu'_m \to \mu'$ with compact support and $u(z)$ and $u'(z)$ are the logarithmic potentials of μ and μ', then $u'(z) \leq u(z)$ in the whole $\mathbb{C}$. Moreover, $u(z) = u'(z)$ in the unbounded component of $\mathbb{C} \setminus supp\,\mu$.

Proof. Assume wlog that all $p_m(z)$ are monic. Let K be a compact set containing the zeros of all $p_m(z)$. We have

$$u(z) = \lim_{m\to\infty} \frac{1}{n_m} \log|p_m(z)|$$

and

$$u'(z) = \lim_{m\to\infty} \frac{1}{n_m - 1} \log\left|\frac{p'_m(z)}{n_m}\right| = \lim_{m\to\infty} \frac{1}{n_m} \log\left|\frac{p'_m(z)}{n_m}\right|$$

with convergence in L^1_{loc}. Hence

$$u'(z) - u(z) = \lim_{m\to\infty} \frac{1}{n_m} \log\left|\frac{p'_m(z)}{n_m p_m(z)}\right| = \lim_{m\to\infty} \frac{1}{n_m} \log\left|\int \frac{d\mu_m(\zeta)}{z - \zeta}\right|.$$

Now, if $\phi(z)$ is a positive test function, it follows that

$$\begin{aligned}\int \phi(z)(u'(z) - u(z))d\lambda(z) &= \lim_{m\to\infty} \int \phi(z) \log\left|\int \frac{d\mu_m(\zeta)}{z - \zeta}\right| d\lambda(z) \\ &\leq \lim_{m\to\infty} \int \phi(z) \int \frac{d\mu_m(\zeta)}{|z - \zeta|} d\lambda(z) \leq \lim_{m\to\infty} \iint \frac{\phi(z)d\lambda(z)}{|z - \zeta|} d\mu_m(\zeta)\end{aligned}$$

where λ denotes Lebesgue measure in the complex plane. Since $\frac{1}{|z|}$ is locally integrable, the function $\int \phi(z)|z - \zeta|^{-1} d\lambda(z)$ is continuous, and hence bounded by a constant M for all z in K. Since $supp\,\mu_m \subset K$, the last expression in the above inequality is bounded by M/n_m. Hence the limit when $m \to \infty$ equals 0. This proves the inequality $u'(z) \leq u(z)$.

In the complement to $supp\,\mu$, $u(z)$ is harmonic and $u'(z)$ is subharmonic, hence $u'(z) - u(z)$ is a negative subharmonic function. Moreover, in the complement of K, $p'_n(z)/(n_m p_m(z))$ converges to the Cauchy transform $\mathcal{C}_\mu(z)$ of μ uniformly on compact sets. Since $\mathcal{C}_\mu(z)$ is a non-constant holomorphic function in the unbounded component of $\mathbb{C} \setminus K$, then by the above $u'(z) - u(z) = 0$ there. By the maximum principle for subharmonic functions it follows that $u'(z) - u(z) = 0$ holds in the unbounded component of $\mathbb{C} \setminus supp\,\mu$ as well. □

To accomplish the proof of Theorem 11.2, we need to show that we have the convergence for the whole sequence and not just for some subsequence. Assume now that the sequence $\bar{\mu}_n$ does not converge to $\bar{\mu}$. Then we can find a subsequence NN' such that for all $n \in NN'$, $\bar{\mu}_n$ stay away from some fixed neighborhood of $\bar{\mu}$ in the weak topology. Again by compactness, we can find a subsequence NN^* of NN' such that for all $n \in NN'$, all the limits for root-counting measures for derivatives exist. But then $\bar{\mu}^{(0)}$ must coincide with $\bar{\mu}$ by the uniqueness and the latter lemma. We get a contradiction to the assumption that μ_n stay away from $\bar{\mu}$ for all $n \in NN'$ and hence for all $n \in NN$. □

11.4 Open problems in Chapter 11

As an observant reader easily notices, the present chapter leads to many more questions than it provides answers to, some of those already mentioned in the introduction. We use this circumstance as an excuse for having this lengthy final section.

11.1. Call the differential monomial $Q_k(z)\frac{d^k}{dz^k}$ *the leading term* of a differential operator $T = \sum_{i=1}^{k} Q_i(z)\frac{d^i}{dz^i}$. Our first conjecture states that asymptotically Van Vleck and Stieltjes polynomials of equation (13.1) behave very similar to that of the equation

$$Q_k(z)\frac{d^k S(z)}{dz^k} + V(z)S(z) = 0. \tag{11.4}$$

Let Pol_r denote the space of all monic polynomials of degree r. Take a non-degenerate T. For some Van Vleck polynomial $V(z)$ of T denote by $\widetilde{V}(z)$ its monic scalar multiple. For a given positive integer n, denote by $\{V_{n,i}(z)\}$ the set of all Van Vleck polynomials $V(z)$ whose Stieltjes polynomials have degree exactly n. (Each $V(z)$ is repeated as many times as its multiplicity prescribes, see [164].) Notice that for sufficiently large n the set $\{V_{n,i}(z)\}$ belongs to Pol_r, i. e., each Van Vleck has degree exactly r. Transform now the set $\{V_{n,i}(z)\}$ into a finite measure $\sigma_n(T)$ in Pol_r by assigning to each element the finite mass equal to the inverse of the cardinality of $\{V_{n,i}(z)\}$.

Conjecture 11.1. *For any non-degenerate T,*
- *The sequence $\sigma_n(T)$ converges weakly to a measure $\Sigma(T)$ compactly supported in Pol_r.*
- *The measure $\Sigma(T)$ depends only on the leading monomial $Q_k(z)\frac{d^k}{dz^k}$ of T.*

Remark 11.6. One can define the natural 'projection' of the finite set of polynomials $\sigma_n(T)$ to the union of the zero loci $\mathfrak{Z}(\sigma_n(T))$ of these polynomials and then turn the latter set into a finite measure in the same way. Conjecture 11.1 implies that this set of finite measures converges to the standard measure supported inside $Conv_{Q_k}$, see two left pictures in Figure 11.3.

11.2. Let us present a quite surprising (at least to us) observation about the asymptotic distribution of Van Vleck versus Stieltjes polynomials obtained through numerical experiments. We state it first in the simplest case $r = 1$. Notice that for any non-degenerate Lamé operator $T = Q_k(z)\frac{d^{k-1}}{dz^{k-1}} + \cdots$ of order k with $r = 1$ and sufficiently large n there exist exactly $(n+1)$ Van Vleck polynomials all having degree 1, see Proposition 11.1. Denote by ξ_n the root-counting measure for the union of all zeros of these polynomials, and define $\xi = \lim_{n\to\infty} \xi_n$ (if it exists). Assuming that ξ exists, which is strongly supported by our numerics, we formulate the following fact illustrated in Figure 11.2 below.

Observation 1. The support of ξ has a very surprising topological and geometric resemblance with that for the limiting root-counting measure of the sequence of eigenpolynomials for the ES-operator $T = Q_k(z)\frac{d^k}{dz^k}$, see Theorem 3 of [27]. (Notice that in case

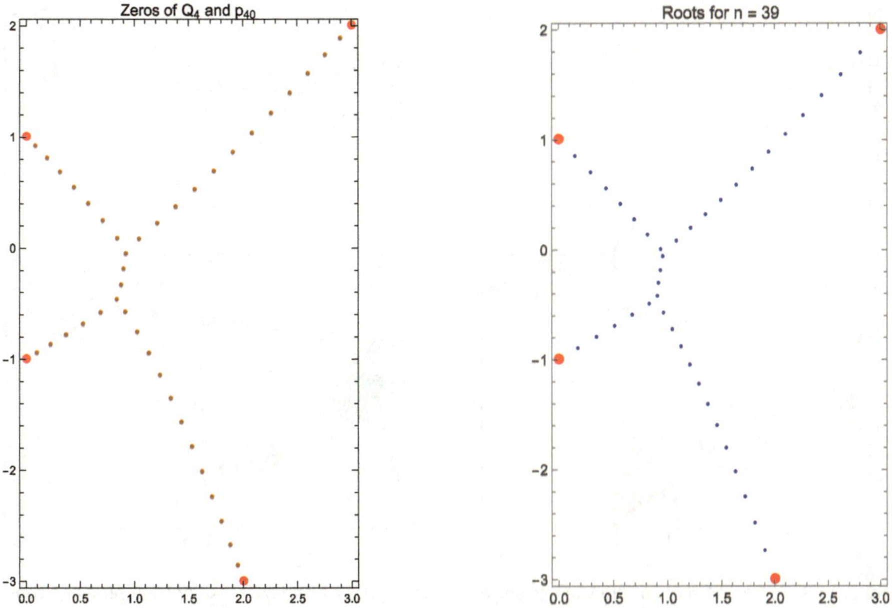

Figure 11.2: The zeros of the eigenpolynomial of degree 40 for the operator $Q_4(z)\frac{d^4}{dz^4}$ (left) and combined zeros of the 40 linear Van Vlecks having a Stieltjes polynomial of degree 39 for the operator $\frac{d^3}{dz^3}(Q_4(z)S(z)) + V(z)S(z) = 0$ with $Q_4(z) = (z^2+1)(z-3i-2)(z+2i-3)$.

$r = 0$ eigenpolynomials are natural analogs of Heine–Stieltjes polynomials for such an operator.)

This resemblance seems to persist for larger r. Namely, consider two higher Lamé operators T_1 and T_2 of the form

$$T_1 = Q_k(z)\frac{d^l}{dz^l} + \cdots \quad \text{and} \quad T_2 = Q_k(z)\frac{d^{l-1}}{dz^{l-1}} + \cdots,$$

where $\deg Q_k(z) \geq k$ and $l \leq k$. Let $\bar{V}_n(z)$ denote the product of all Van Vleck polynomials having Stieltjes polynomials of degree n for T_2 and let $\bar{S}_n(z)$ denote the product of all Stieltjes polynomials of degree n for T_1. Let ν_1 be the asymptotic root-counting measure for the sequence $\{\bar{S}_n(z)\}$ and ν_2 be the asymptotic root-counting measure for the sequence $\{\bar{V}_n(z)\}$ (if they exist).

Observation 2. In the above notation the supports of ν_1 and ν_2 have surprising geometric and topological similarities, see Figure 11.3.

Remark 11.7. In Figure 11.3 we use $Q_4(z)\frac{d^2}{dz^2}$ and $Q_4(z)\frac{d}{dz}$ with the same $Q_4(z) = (z^2 + 1)(z-3i-2)(z+2i-3)$ as described in Observation 2 above. Notice the resemblance of shapes!

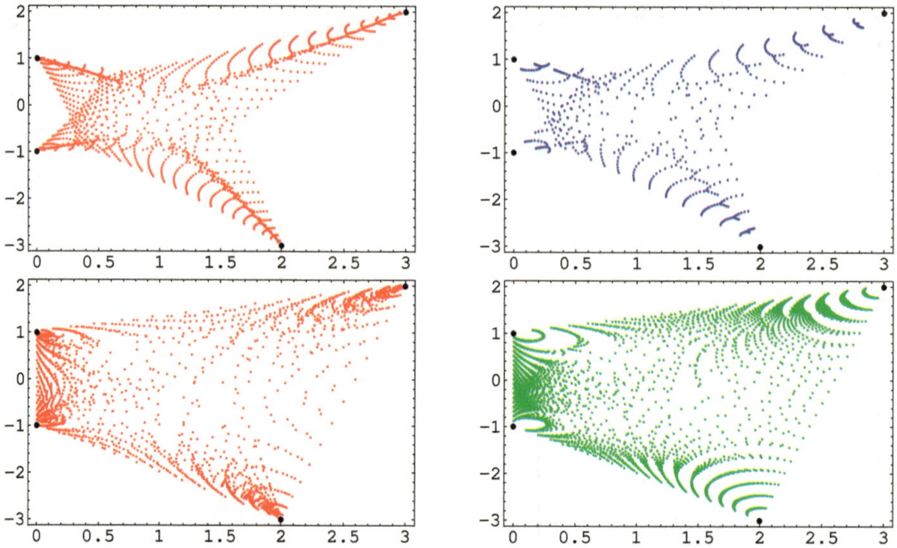

Figure 11.3: Combined zeros of the Stieltjes polynomials for $Q_4(z)\frac{d^3}{dz^3}$ and $Q_4(z)\frac{d^2}{dz^2}$ (two left pics) and combined zeros of the Van Vleck polynomials.

Problem 11.3. Find an explanation for the above observations.

11.3. Curves similar to that in the support of $\mu_{T,\tilde{V}}$ appeared earlier as Stokes lines of linear differential equations with polynomial coefficients.

Problem 11.4. What is the exact relation between the WKB-theory for the equation

$$y^{(k)} + \frac{\tilde{V}(z)}{Q(z)}y = 0 \tag{11.5}$$

and the asymptotic root-counting measure $\mu_{T,\tilde{V}}$ of the sequence of Stieltjes polynomials with the sequence of normalized Van Vleck polynomials converging to $\tilde{V}$?

In the case of a generalized Lamé equation, one at least has a strict definition of the global Stokes line for such an equation, see, e. g., [68] and the support of the corresponding $\mu_{T,\tilde{V}}$ coincides with a part of this global Stokes line. Even in this classical case there is an interesting problem of defining what part of the global Stokes line is covered by this support.

For the case of equations of order exceeding 2, the situation is much worse since even a good definition of the global Stokes line creates serious problems, see, e. g., [7, 8], references therein, and further publications of the same authors.

11.4. In [26] T. Bergkvist obtained a number of interesting results and conjectures in the case of degenerate ES-operators, i. e., $\deg Q_k < k + r$, see Chapter 3. Motivated by her results, we formulate the following conjecture and a question.

Conjecture 11.2. *For any degenerate Lamé operator and any positive integer N_0, the union of all the roots to polynomials V and S taken over* $\deg S \geq N_0$ *is always unbounded. Therefore, this property is a key distinction between non-degenerate and degenerate Lamé operators.*

Problem 11.5. Extend the results of this chapter to the case of degenerate higher Lamé operators.

11.5. Numerical experiments suggest that Stieltjes polynomials of consecutive degrees show a stable root-interlacing pattern along the curves in *supp*(μ) where μ is the corresponding asymptotic root-counting measure. This phenomenon is especially easy to explain and illustrate in the case of ES-operators, i. e., $r = 0$.

Conjecture 11.3. *For any ES-operator T, the family* $\{S_n(z)\}_{n\in\mathbb{N},n\geq n_0}$ *of its eigenpolynomials* ($\deg S_n(z) = n$ *for* $n \in \mathbb{N}$) *has the interlacing property along the support of its asymptotic root-counting measure. In other words, the zeros of any two consecutive polynomials* $S_{n+1}(z)$ *and* $S_n(z)$ *interlace along every curve segment in supp*(μ) *for all sufficiently high degrees n.*

Some caution is required when defining the notion of interlacing since a) the zeros of $S_n(z)$ do not lie exactly on *supp*(μ) and b) *supp*(μ) has a non-trivial topological structure. One has to remove sufficiently small neighborhoods of the singular points of *supp*(μ) where several smooth branches meet and to project the zeros of $S_n(z)$ onto *supp*(μ) along some fixed in advance normal bundle to the smooth part of *supp*(μ).

An example illustrating the interlacing phenomenon conjectured above is shown on Figure 11.4.

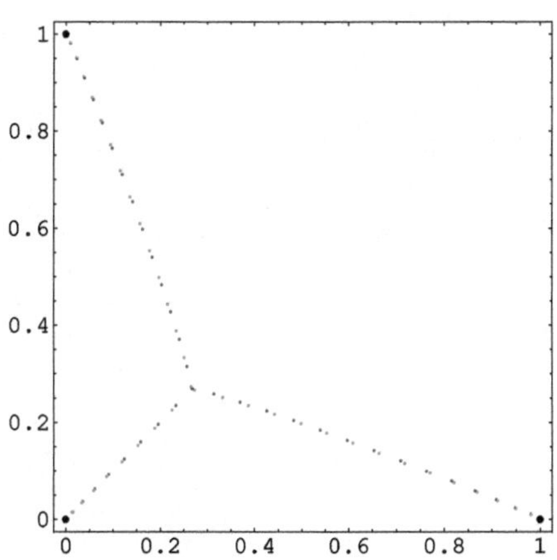

Figure 11.4: Interlacing pattern for the roots of $S_{40}(z)$ and $S_{41}(z)$, which are the eigenpolynomials of the operator $z(z-1)(z-i)\frac{d^3}{dz^3}$.

Similar interlacing property was observed for any family of Stieltjes polynomials of increasing degree as soon as the sequence of their normalized Van Vleck polynomials has a limit. Some results about interlacing can be found in [5].

11.6. In connection with the classical Bochner–Krall problem, which asks to describe all families of orthogonal polynomials appearing as the families of eigenpolynomials for linear differential operators with polynomial coefficients, it is natural to ask the following.

Problem 11.6. Describe all Lamé operators for which all their Van Vleck and Stieltjes polynomials have only real roots.

11.7. Finally, Theorem 11.2 can be reinterpreted as the statement that the level curves of the logarithmic potential $u_\mu(z)$ of the measure $\mu_{\widetilde{V}}$, which we construct in this theorem, are trajectories of the differential $(\sqrt{-1})^k \frac{\widetilde{V}(z)}{Q_k(z)} dz^k$ of order k. Since almost all level curves of $u_\mu(z)$ are closed curve, we are tempted to call the latter differential Strebel. This makes perfect sense when $k = 2$, see [168] and [131]. But for $k > 2$, even a definition of a Strebel differential is so far missing.

Problem 11.7. Define a notion of a Strebel (rational) differential of order $k > 2$.

Some initial steps related to Problem 11.7 can be found in Chapter 12.

The next question seems very natural in view of the latter theorem, but we do not have even a good guess about its possible answer except for some trivial cases.

Problem 11.8. Which rational functions $R(z)$ admit a probability measure as in Theorem 11.3?

12 Introducing quasi-Strebel differentials

(The material of this chapter is based on [166].)

12.1 Preliminaries

In the previous chapters we encountered kth differentials related to Heine–Stieltjes problem, which possess natural 1-parameter families of continuous piecewise broken trajectories covering the whole Riemann surface (up to measure 0). This property has a strong resemblance with the defining property of quadratic Strebel differentials. They are level curves of the logarithmic potential of the relevant asymptotic root-counting measure. In this chapter we attempt to develop a theory of high order differentials similar to that of Strebel differentials.

Definition 12.1. A (meromorphic) differential Ψ of order $k \geq 2$ on a compact orientable Riemann surface Y without boundary is a (meromorphic) section of the kth tensor power $(T^*_{\mathbb{C}}Y)^{\otimes k}$ of the holomorphic cotangent bundle $T^*_{\mathbb{C}}Y$. The zeros and the poles of Ψ constitute the set Cr_Ψ of *critical points* (alias, singularities) of Ψ.

In what follows, we will use the convention that the order of a zero is a positive integer while the order of a pole is a negative integer. For a differential Ψ of order k given locally by $f(z)dz^k$ in a neighborhood of a non-critical point, we define k locally distinct direction fields called *horizontal*, which are given by the condition that $f(z)dz^k$ is positive. Integral curves of these direction fields are called (horizontal) *trajectories*. Any two consecutive direction fields differ by the angle $\frac{2\pi}{k}$ and for even $k = 2\ell$, they form ℓ locally distinct line fields. However, for a general differential of order $k > 2$, it is impossible to globally distinguish these k direction fields because every two of them are obtained by the analytic continuation of each other. Multiplication of Ψ by the scalar factor $e^{\frac{\pi i}{k}}$ does not change these direction fields and their trajectories.

Extracting the kth root, one can globally interpret an arbitrary differential of order k as a multi-valued meromorphic abelian differential on Y. Outside of the poles and the zeros of Ψ, it is locally representable as the kth power of a holomorphic abelian differential ω. Integration of ω in a neighborhood of a point z_0 gives local coordinates whose transition maps are of the type $z \mapsto e^{i\theta}z + c$ where $\theta \in \{0, \frac{2\pi}{k}, \dots, (k-1)\frac{2\pi}{k}\}$. In such a way, the Riemann surface Y punctured at its singularities is locally isometric to the complex plane $\mathbb{C}$ endowed with the flat structure induced by dz. In other words, we obtain a flat metric on $Y \setminus Cr_\Psi$ induced by the integration of ω where the distance between two points p_1 and p_2 is $|\int_{p_1}^{p_2} \omega|$, see [17, 182]. (In particular, dz induces the usual Euclidean metric on $\mathbb{C}$.) The surface $Y \setminus Cr_\Psi$ with the latter flat metric will be called the *flat model* of Ψ.

This flat metric extends from $Y \setminus Cr_\Psi$ to all zeros and poles of order $\ell > -k$. Such singularities are called *conical of order* ℓ. Any conical singularity of order ℓ has a neigh-

https://doi.org/10.1515/9783112207741-014

borhood on Y isometric to a neighborhood of the apex of a flat cone with the total angle $\frac{(\ell+k)2\pi}{k}$. Analogously, a pole of order $\ell \leq -k$ corresponds to the point at ∞ of a flat cone with the total angle $-\frac{(\ell+k)2\pi}{k}$, see Proposition 12.1. By convention, flat cones of angle 0 (i. e., poles of order $-k$) are half-infinite cylinders.

In general, differentials of order $k > 2$ are much less studied than their quadratic counterpart. Nevertheless they sometimes appear in the literature, see, e. g., [180, 17, 25, 63, 91, 118, 149, 182, 49] and, as we already mentioned, are closely related to the theory of locally flat surfaces, which is used in the study of the moduli spaces of curves and dynamics of interval exchange maps, see, e. g., [201]. Moreover, the theory of translation surfaces (which correspond to the abelian differentials) provides new important insights into the dynamics of billiards by means of algebraic geometry and renormalization theory.

Moduli spaces of pairs (Y, Ψ) consisting of a compact Riemann surface Y and a k-differential on it are naturally stratified by the loci where the differentials have the same set of multiplicities of their singular points. For example, in the case of genus zero, a meromorphic differential of order k can be globally represented as $\Psi(z) = R(z)dz^k$, where $R(z)$ is a rational function. So, up to a non-vanishing constant factor, Ψ is completely determined by the multiplicities and the positions of its singularities. In particular, the strata of k-differentials can be identified with certain configuration spaces of divisors on $\mathbb{C}P^1$ and are connected, see [43].

A natural system of coordinates on the latter strata is given by the periods of $\sqrt[k]{\Psi}$ with respect to the relative homology of Y punctured at all poles of order smaller than or equal to $-k$ taken with respect to the set of all conical singularities (i. e., all zeroes together with all poles of order exceeding $-k$). In the genus zero case, projective automorphisms allow us to assign positions of three singularities to $(0, 1, \infty)$ and the strata of k-differentials with m singularities on the sphere are complex-analytic orbifolds of dimension of $m - 2$. More details can be found in [17].

Local classification of singularities of differentials of order $k > 2$ under the action of the group of local biholomorphisms is quite similar to that of quadratic differentials and can be found in Proposition 12.1 below.

The main motivation of this chapter is as follows. Recall that a quadratic differential on a compact Riemann surface Y is called *Strebel* if almost all its horizontal trajectories are closed. Such a phenomenon can never happen for a k-differential Ψ of order $k \geq 3$ unless it is a power of a 1-form or a quadratic differential, see Corollary 12.1 below. Thus, for $k \geq 3$, in order to define something similar to a Strebel differential, one should instead of using only smooth closed horizontal trajectories of Ψ (which might exist in some domains of Y) allow a certain type of closed broken trajectories and try to cover Y with those. However, to avoid trivialities, one should impose substantial restrictions on the set where broken trajectories can switch from one smooth piece to another. An attempt to develop an appropriate notion is carried out below by introducing quasi-Strebel structures, see Definition 12.10.

The purpose of this chapter is to develop elements of a general theory of k-differentials which as much as possible resembles that of quadratic differentials and to point out the connection of a certain class of k-differentials with the classical Heine–Stieltjes theory. It seems very plausible that special types of k-differentials should naturally appear in the asymptotic analysis of linear ODE of high order, comp., e. g., [7]. In what follows, theorems, conjectures, etc., labeled by letters, are borrowed from the existing literature, while those labeled by numbers are hopefully new.

Our main result Theorem 12.1 claims the existence of a quasi-Strebel structure for every meromorphic k-differential with admissible singularities if k is even. A similar statement for odd k currently requires an additional (and very restrictive) hypothesis of rationality of the periods of a differential. On the other hand, we strongly believe that for k odd, there are arithmetic obstructions to the existence of a quasi-Strebel structure. In general, it seems that our attempt to introduce a meaningful analog of higher order Strebel differential was not very successful.

The chapter is organized as follows. § 12.2 contains main definitions and results related to k-differentials and quasi-Strebel structures. § 12.3 introduces a special class of positive differentials and provides a short explanation of how such differentials appear in the Heine–Stieltjes theory. Finally, in § 12.4 we present a number of relevant open questions.

12.2 Basic notions and the main theorem

12.2.1 Trivium on k-differentials

The next definitions are straightforward generalizations of those for quadratic differentials.

Definition 12.2. The *canonical length element* on Y associated with a k-differential Ψ given locally as $\Psi = f(z)dz^k$ is defined by

$$|dw| = |f(z)|^{\frac{1}{k}}|dz|.$$

(Notice that the associated area element on Y equals $|f(z)|^{\frac{2}{k}}dxdy$ where $z = x + iy$. Both $|dw|$ and the area element are globally well defined.)

Definition 12.3. a) The *norm* of a k-differential $\Psi = f(z)dz^k$ is defined as

$$\iint_{\Gamma} |f(z)|^{\frac{2}{k}}dxdy.$$

(Again $|f(z)|^{\frac{2}{k}}dxdy$ is a globally well-defined 2-form on Y.)

b) If $\Psi = f(z)dz^k$ is meromorphic, then its singular point p is called *finite or conical* if p has a neighborhood with finite associated area.

Remark 12.1. Finite critical points of Ψ are exactly its zeros and poles of order greater than $-k$ (i. e., conical singularities), and a meromorphic k-differential Ψ has a finite norm if and only if all its critical points are finite.

Definition 12.4 (Comp. Definition 20.1 in [179]). A trajectory of Ψ is called *critical* if it starts or ends at a conical singularity.

Definition 12.5. The *distinguished* or *natural* parameter W associated with a k-differential $\Psi = f(z)dz^k$ is defined by

$$W = \int \sqrt[k]{f(z)}dz$$

for some branch of the kth root.

Remark 12.2. Obviously, a sufficiently small neighborhood of any regular point p is homeomorphically mapped by a branch of W onto an open set in the W-plane. Notice that $dW = \sqrt[k]{f(z)}dz$, which implies $(dW)^k = f(z)dz^k$.

A standard basic result about k-differentials, which can be found in several sources, is as follows.

Lemma 12.1. *The Euler characteristic of $(T^*_{\mathbb{C}}Y)^{\otimes k}$ equals $\chi(Y)k$, where $\chi(Y)$ is the Euler characteristic of the underlying curve Y. Therefore, the difference between the number of poles and zeros (counted with multiplicity) of a meromorphic differential Ψ of order k on Y equals $k \cdot \chi(Y)$. In particular, the number of poles minus the number of zeros of any rational k-differential Ψ equals $2k$.*

Next, following closely § 6 of Chapter 3 in [179], we provide normal forms for a k-differential near its critical points. (These results can be found in Proposition 3.1 of [17] and were independently obtained about 15 years ago by the author (unpublished).)

Proposition 12.1. *Given a k-differential Ψ on a sufficiently small disk V centered at 0, denote by m the order $\operatorname{ord}_0\Psi$. There exist a conformal map $\xi : (\Delta_R, 0) \to (V, 0)$ defined on a disk of sufficiently small radius R and a number $r \in \mathbb{C}$ such that*

$$\xi^*(\Psi) = \begin{cases} z^m dz^k & \text{if } m > -k \text{ or } k\nmid m, \\ \frac{r}{z^k} dz^k & \text{if } m = -k, \\ (z^{m/k} + \frac{s}{z})^k dz^k & \text{if } m < -k \text{ and } k|m, \end{cases}$$

where in the last case $s \in \mathbb{C}$ and $s^k = r$. The germ of ξ is unique up to multiplication by a $(m + k)$th root of unity when $m > -k$ or $k\nmid m$, and up to multiplication by a non-zero constant if $m = -k$, see Figure 12.1.

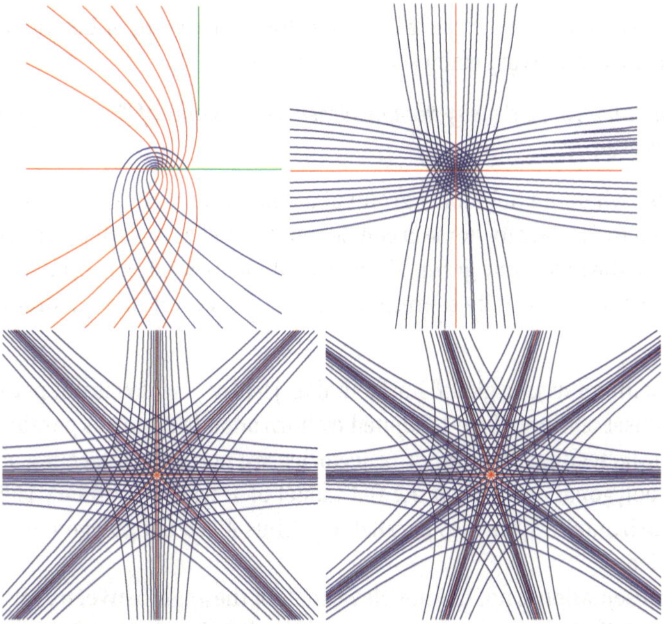

Figure 12.1: Local phase portraits of a cubic differential near a simple and double poles (top row) and simple and double zeros (bottom row).

The following statement describes possible behavior of non-critical global trajectories, see [182, Proposition 5.5].

Lemma A. Non-critical trajectories of a k-differential Ψ are of the following three types:
(i) Closed trajectories (with or without selfintersections).
(ii) Trajectories (with or without selfintersections) one or both ends of which approach a pole of order smaller than or equal to $-k$.
(iii) Trajectories (with or without selfintersections) that are dense in some polygonal domains in the flat model of Ψ.

Observe that the analytic continuation of a branch of $\sqrt[k]{\Psi}$ of a k-differential Ψ along some loop on Y may result in a different branch. The ratio between the values of these two branches is a kth root of unity called the *holonomy* of the loop. This ratio describes the failure of the identification of the circle of directions in the parallel transport along the loop. Observe that, in particular, the holonomy of a loop around a singularity of order n depends on the congruence class of n modulo k. To formulate our next result, we need the following notion.

Definition 12.6. In the above notation, the *holonomy group of* Y is the subgroup of $G_k \subset U(1)$ generated by the holonomies of all possible loops on Y where G_k denotes the group of kth roots of unity. Given a domain $\Omega \subseteq Y$, we define the holonomy of Ω as the subgroup of G_k generated by the holonomies of loops contained in Ω.

By a *density domain* we denote the closure of a trajectory in case (iii) of Lemma B without selfintersections. The following claim seems to be new.

Proposition 12.2. *The holonomy of a density domain either consists only of* $\{1\}$ *or is equal to* $\{\pm 1\}$ *in which case* k *is even.*

Remark 12.3. In the case when the holonomy group of a density domain is trivial, the k-differential Ψ restricted to the density domain can be represented as the kth power of a 1-form. In the case when the holonomy group of a density domain is $\{\pm 1\}$ k is even and Ψ restricted to the density domain can be represented as the $\frac{k}{2}$th power of a quadratic differential.

Proof. Consider a non-selfintersecting global horizontal trajectory γ, which is not periodic. Its closure A is a subset of the surface Y foliated by horizontal trajectories parallel to γ. Since γ is non-selfintersecting, these other trajectories are non-selfintersecting as well. Therefore, A is a polygonal domain on the flat model of Y whose boundary is a union of segments of horizontal trajectories parallel to γ. Thus A is a flat polygon with some of its sides identified.

Notice that such a presentation is not unique. Nevertheless, the angle between every side of a polygon and the foliation by the trajectory lines is well defined. Therefore, the holonomy of any simple loop that crosses exactly one side exactly once is either 1 or −1. Such loops generate the fundamental group of A, which finishes the proof. □

Corollary 12.1. *If a meromorphic k-differential* Ψ *on a compact Riemann surface* Y *with* $k \geq 3$ *has a family of smooth periodic trajectories covering* Y *almost everywhere, then there exists either a 1-form* ω *or a quadratic differential* q *such that either* $\Psi = \omega^k$ *or* $\Psi = q^{\frac{k}{2}}$.

Proof. If there is a family of smooth closed trajectories covering Y almost everywhere, then the direction field of Ψ tangent to this family is well defined on the whole surface Y. This implies that the holonomy of Ψ along any closed loop on Y is equal to ± 1. Assume first that the holonomy group of Ψ is trivial. Then there exists a 1-form ω on Y such that $\Psi = \omega^k$. (Indeed, in this situation every branch of $\sqrt[k]{\Psi}$ is a well defined 1-form on Y.) In the remaining case the holonomy group of Ψ is $\{\pm 1\}$, which implies that k branches split into $\frac{k}{2}$ pairs where the two branches in each pair are analytic continuations of each other. Moreover the two branches in each pair coincide with the two branches of $\sqrt{q}$ for some quadratic differential q. Additionally, $\Psi = q^{\frac{k}{2}}$ independently of the choice of a pair. □

Definition 12.7. For a k-differential $\Psi = f(z)dz^k$, we call by its *dual* the differential $\Psi^* = i \cdot f(z)dz^k$, $i = \sqrt{-1}$. Obviously, Ψ^* is globally well defined and attains purely imaginary values whenever Ψ attains real values and vice versa. Additionally, $(\Psi^*)^* = -\Psi$.

12.2.2 Introducing quasi-Strebel structures

The following definitions are inspired by the approach to Strebel quadratic differentials developed in [19] and by our previous results obtained in [122, 27, 84, 168].

Definition 12.8. Given a k-differential Ψ on Y, by its *finite broken trajectory* we mean a continuous curve on Y consisting of finitely many segments of trajectories of Ψ. A broken trajectory that is a non-selfintersecting closed curve on Y is called *closed.* Points where a trajectory is broken are called *switching points.* Usual periodic trajectories (which might exist for k-differentials, but cannot cover almost all Y unless it is a power of abelian or quadratic differential, see Corollary 12.1) is a particular case of closed broken trajectories.

Definition 12.9. Given a k-differential Ψ, we say that a continuous and piecewise smooth function $\Phi : Y \setminus Cr_\Psi \to \mathbb{R}$ is a *prelevel function* of Ψ if
(i) Φ is non-constant on any open subset of Y.
(ii) All connected components of its level curves are broken trajectories of Ψ.

We say that a prelevel function Φ is a *level function* of Ψ if its *switching set* $S_{\Psi,\Phi}$ (i. e., the set of all its switching points) consists of finitely many piecewise smooth compact curves on Y.

We say that two level functions Φ_1 and Φ_2 of Ψ are *level-equivalent* if every connected component of any level curve of Φ_1 is a connected component of some level curve of Φ_2 and conversely.

Definition 12.10. Given a meromorphic k-differential Ψ defined on a compact Riemann surface Y and a level function $\Phi : Y \setminus Cr_\Psi \to \mathbb{R}$, we say that Φ defines *a quasi-Strebel structure* for Ψ (or that a pair (Ψ, Φ) is quasi-Strebel) if
(i) Φ is a piecewise linear continuous function consisting of finitely many (local) linear functions in the charts of the flat model of Y.
(ii) At each point where Ψ is smooth its differential coincides with one of the k branches of $\operatorname{Im}(\sqrt[k]{\Psi})$.
(iii) Almost all connected components of almost all level curves of Φ are closed connected piecewise broken trajectories of Ψ and their canonical lengths are bounded from above.

Level functions appearing in Definition 12.10 are called *piecewise standard* or *ps-level* functions for short. Obviously the switching set of any ps-level function Φ coincides with the subset of Y where Φ is not smooth and it consists of a finite collection of straight segments on the flat model. More exactly, the following result holds.

Lemma 12.2. *For odd k, the switching set $S_{\Psi,\Phi}$ of a ps-level function Φ of a quasi-Strebel structure (Ψ, Φ) consists of finitely many segments of trajectories of Ψ only, while for even k, it consists of trajectories of Ψ together with trajectories of its dual Ψ^*.*

Proof. Indeed consider a generic point p of $S_{\Psi,\Phi}$. By the properties of the differential of Φ, in the flat model of Ψ, p lies on the line defined by the equation $l_1(W) = l_2(W)$, where W is a local canonical coordinate centered at p and $l_1(W) := \mathrm{Im}(\zeta_1 W)$ and $l_2(W) := \mathrm{Im}(\zeta_2 W)$ are the linear functions with $\zeta_1^k = \zeta_2^k = 1$. Any such line goes in the direction determined by the bisector of two kth roots of unity. Such bisectors coincide (up to $\pm$-sign) with kth roots of unity for odd k. For k even, bisectors either coincide (up to $\pm$-sign) with kth roots of unity or with kth roots of -1. These are exactly the directions tangent to the trajectories of either Ψ or Ψ^* at the point p. □

Remark 12.4. Recall that given a k-differential Ψ, its two ps-level functions Φ_1 and Φ_2 are called *level-equivalent* if they have coinciding connected components of their level sets. Since we are only interested in the behavior of the family of closed broken trajectories obtained as connected components of the level curves of Φ_1 (resp. Φ_2), we can identify the quasi-Strebel structures (Ψ, Φ_1) and (Ψ, Φ_2). Thus denoting by $[\Phi]$ the class of all ps-level functions equivalent to Φ, from now on we will say that a *quasi-Strebel structure* is a pair $(\Psi, [\Phi])$.

Definition 12.11. Given a meromorphic k-differential Ψ as above, we say that it has *admissible singularities* if
(i) Ψ has no poles of order smaller than $-k$.
(ii) At every pole of order $-k$ of Ψ the residue should be equal to $i^k \cdot a$ where $a \in \mathbb{R}$.

The motivation for the above definition is as follows.

Lemma 12.3. *If Ψ is a k-differential admitting a quasi-Strebel structure $(\Psi, [\Phi])$, then it has admissible singularities.*

Proof. Indeed, if Ψ has a pole of order smaller than or equal to $-k$, there exists a neighborhood V of this pole disjoint from the switching set $S_{\Psi,\Phi}$. If one of the level sets of Φ inside V is a periodic trajectory, then the pole has to be of order $-k$ with the residue of the form $i^k \cdot a$ where $a \in \mathbb{R}$. Otherwise there exists a level set whose intersection with V connects the boundary of V with the pole and therefore is of infinite length—contradiction. □

Definition 12.12. Given a ps-level function Φ for a differential Ψ, we say that a point p in the switching set Sw_Φ is its *node* if an arbitrary small neighborhood of p in Sw_Φ is not a straight segment without boundary. (In other words, a node is a vertex of Sw_Φ considered as a graph embedded in Y.) In the above notation, connected components of the level curves of Φ containing nodes will be called *special.*

One type of nodes is described by the following lemma.

Lemma 12.4. *For any quasi-Strebel structure (Ψ, Φ), each vertex of valency one of Sw_Φ (which is considered as an embedded graph on Y) is a singular point of Ψ whose order for*

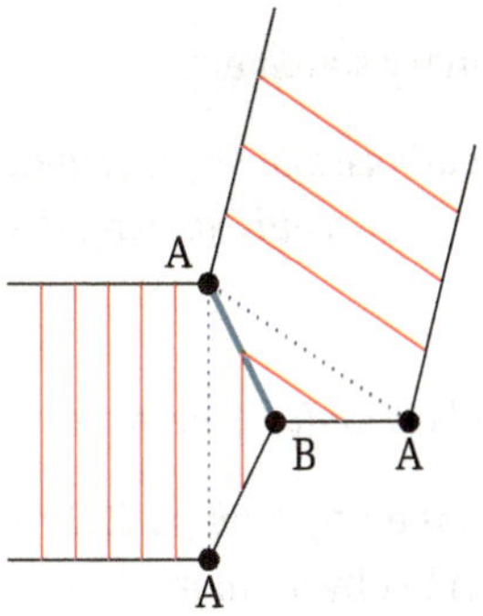

Figure 12.2: Cubic differential with one zero of order 2 (marked by *A*), one pole of order −2 (marked by *B*), and two poles of order −3 (at the ends of two infinite cylinders). The thick segment connecting *A* and *B* is the switching set of the quasi-Strebel structure whose broken trajectories are shown by the parallel lines. Finally, the dashed lines are the special broken trajectories separating the cylinders.

k is odd, is not divisible by k and for k even, not divisible by $\frac{k}{2}$. *Moreover, every singular point of* Ψ *whose order is not an integer multiple of* $\frac{k}{2}$ *is a node of* Sw_Φ.

(Obviously, vertices of valency one are nodes of Sw_Φ.)

Proof. Around a vertex of valency one of the switching set, the level curves are formed by some segments of smooth trajectories (possibly none) and one broken trajectory with a switching point of angle θ, where $0 < \theta < \pi$. Connecting these segments of trajectories by half circles, we get a loop around the vertex whose holonomy does not belong to $\{\pm 1\}$. Therefore, the vertex is a conical singularity whose total angle is not a multiple of π. Therefore its order is neither divisible by k nor by $\frac{k}{2}$ for k even.

Now if a conical singularity does not belong to Sw_Φ, then the level curves around it are formed only by some segments of smooth trajectories. Connecting as above these segments of trajectories by half circles, we get a loop around the vertex whose holonomy belongs to $\{\pm 1\}$. Thus the order of the conical singularity is an integer multiple of $\frac{k}{2}$. □

Remark 12.5. Notice that for $k > 2$, there usually exist nodes that are not the singular points of Ψ, see Figure 12.6. We will call them *secondary singularities*. (These nodes are similar to additional turning points appearing in the asymptotic theory of linear ODE of order exceeding 2, see [7].)

12.2.3 Main theorem

Theorem 12.1. (i) *For even* $k > 2$, *every k-differential* Ψ *with admissible singularities has a quasi-Strebel structure.*

(ii) *For odd* $k > 2$, *any k-differential* Ψ *with admissible singularities such that, up to a common factor, the period of* $\sqrt[k]{\Psi}$ *along every path connecting any two conical singularities of* Ψ *belongs to* $\mathbb{Q}[e^{\frac{2i\pi}{k}}]$ *has a quasi-Strebel structure.*

Remark 12.6. For k odd, differentials satisfying condition (ii) are dense in the space of all differentials with admissible singularities.

To prove Theorem 12.1, we will need the following preliminary statement.

Proposition 12.3. *For any $k \geq 3$ and any k-differential Ψ with admissible singularities, there exists a decomposition of the flat model of Ψ into triangles, trapezoids, and infinite cylinders such that*
(a) *Each singularity of Ψ is one of its vertices.*
(b) *Each boundary edge of the decomposition is a segment of a horizontal trajectory.*

Proof. To start with, observe that if Ψ has a pole of order $-k$, then by Lemma 12.3 its residue is such that the corresponding infinite cylinder is bounded by a union of some critical trajectories and therefore it can be cut off from Y along this union, see Figure 12.2. After cutting off all such cylinders, we get a new surface $\widetilde{Y}$ with finite area whose boundary consists of horizontal trajectories. (Recall that there exist only finitely many critical trajectories in Y since Ψ has a finite number of singularities.) Consider all the remaining critical trajectories satisfying the condition that they are nowhere dense in $\widetilde{Y}$ and remove their union from $\widetilde{Y}$. All the remaining critical trajectories are dense and some of them are self-intersecting while others are not. Let us additionally cut $\widetilde{Y}$ along a certain part of every remaining critical trajectory having self-intersections. Namely, we remove a fragment of such a trajectory starting at the conical singularity and ending at its first self-intersection. (If such a trajectory both starts and ends at conical singularities, we do the latter procedure for both of its endpoints.) Denote by CUT the constructed cut of Y, which includes every curve we have removed so far.

Denote by γ one of the remaining critical trajectories. Then γ is dense in some domain of Y and non-selfintersecting. At this point, there are two cases: either γ hits CUT or it is dense in some domain $D \subsetneq (Y \setminus CUT)$. In the first case, similarly to the above, we add to CUT the fragment of γ between the conical singularity and its first intersection with CUT.

Let us show by contradiction that the second case is impossible. Take some critical trajectory γ that is dense in some domain $D' \subsetneq (Y \setminus CUT)$. Then any other critical trajectory that starts in D' must be dense in some subdomain $D'' \subsetneq D'$.

The trajectory γ returns arbitrarily close to its starting point A (which is a singular point of Ψ). Therefore, there is an angular sector at A contained in D' and whose angle is at least π (see Figure 12.3). Since $k > 2$, there is still another non-selfintersecting critical trajectory α in D' starting at A transversally to γ. (Observe that this argument does not work for quadratic differentials, which explains why quadratic differentials are not always quasi-Strebel.) Notice that α should also be dense in some $D'' \subsetneq D'$ (because the geometry of D' is that of a quadratic differential, see Proposition 12.2). Notice that the case $D'' = D'$ is impossible since α and γ are transverse. Taking α instead of γ, we can iterate this step. However, we get a contradiction after a finite number of steps since there is no infinite sequence of density domains strictly included in each other.

In this way, the connected components of the complement to the constructed cut CUT have corners with angles smaller than or equal to π. Indeed, at each singularity

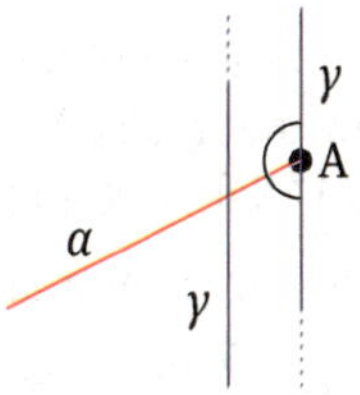

Figure 12.3: The local picture at a critical point lying in a density domain.

the angle between any two neighboring critical trajectories equals $2\pi/k < \pi$. Moreover, for any regular point where (at least) two segments of *CUT* intersect, there are either (at least) four angles smaller than π or there are three angles one of which is π. By the Gauss–Bonnet theorem, a flat surface such that every corner of its boundary has angle smaller than π is either a convex polygon (if at least one corner has an angle strictly smaller than π) or a cylinder (if every angle of every corner is π or, equivalently, if there are no corners). Cylinders can be cut out along some horizontal directions in order to get a parallelogram. Now, each convex polygon can be cut into a finite number of trapezoids and triangles by taking lines parallel to one side of the polygon. □

Proof of Theorem 12.1. To settle Claim (i), we use Proposition 12.3 to decompose the flat surface Y of Ψ into infinite cylinders, triangles, and trapezoids. Let us show that each of these tiles can be covered by a family of closed broken trajectories. Indeed, each infinite cylinder is covered by a family of smooth horizontal trajectories. Each triangle can be filled by the family of broken trajectories whose switching set consists of three segments of bisectors meeting at the incenter, i. e., the center of the inscribed circle, see the right-hand part of Figure 12.4. Finally, any trapezoid can be decomposed into a symmetric trapezoid and a triangle. Symmetric trapezoids can be covered by closed broken trajectories in a standard way shown in the left part of Figure 12.4.

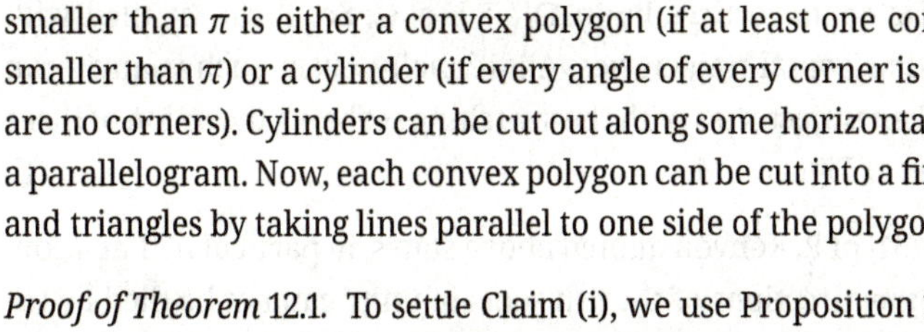

Figure 12.4: Tiling a general trapezoid by a symmetric trapezoid and a triangle and covering both tiles by closed broken trajectories.

To prove Claim (ii), we can without loss of generality assume that all the periods of $\sqrt[k]{\Psi}$ lie in $\mathbb{Q}[\zeta]$ where ζ is a kth root of unity. Firstly, define the subset $\mathcal{L} \subset Y$ consisting of all points p such that a period of Ψ over a path connecting p of one of the conical singularities is in $\mathbb{Q}[\zeta]$. (Observe that by our assumptions if one period satisfies this property, then all of them satisfy it as well.) We want to show that every vertex created in a decomposition from Proposition 12.3 belongs to $\mathcal{L}$.

Indeed, choose two points A and C in $\mathcal{L}$. Let B be a point in the intersection of the horizontal trajectories passing through A and C resp. We want to show that B also belongs to $\mathcal{L}$. By choosing the appropriate kth root, we can assume that $\int_A^B \sqrt[k]{\Psi} \in \mathbb{R}$ and $\int_B^C \sqrt[k]{\Psi} \in \zeta \cdot \mathbb{R}$. Since $\mathbb{Q}[\zeta]$ considered as a vector space over $\mathbb{Q}$ is the direct sum of $\mathbb{Q}[\zeta+\bar{\zeta}] \subset \mathbb{R}$ and $\zeta \cdot \mathbb{Q}[\zeta+\bar{\zeta}] \subset \zeta \cdot \mathbb{R}$, the decomposition of the number $\int_A^C \sqrt[k]{\Psi}$ into such summands is exactly $\int_A^C \sqrt[k]{\Psi} = \int_A^B \sqrt[k]{\Psi} + \int_B^C \sqrt[k]{\Psi}$. Thus $\int_A^B \sqrt[k]{\Psi}$ belongs to $\mathbb{Q}[\zeta+\bar{\zeta}] \subset \mathbb{Q}[\zeta]$ implying that $B \in \mathcal{L}$.

Therefore the decomposition provided by Proposition 12.3 consists of triangles and trapezoids whose vertices belong to $\mathcal{L}$. In Theorem 4 of [101] R. Kenyon proved that a convex polygon in $\mathbb{C}$ is tileable with triangles whose angles are multiples of $\frac{\pi}{k}$ if and only if up to a common factor its vertices lie in $\mathbb{Q}[e^{\frac{2\pi i}{k}}] \subset \mathbb{C}$. So, $\widetilde{Y}$ can be tiled with triangles whose sides are horizontal trajectories. Additionally, every such triangle can be filled by a family of broken trajectories with the switching set being the union of three bisectors, see the right part of Figure 12.4. □

Remark 12.7. The result [101] of R. Kenyon quoted above states, in particular, that a convex polygon with sides in the directions of the kth roots of unity can not be tiled by triangles with sides in the same set of directions if the lengths of the sides of the polygon do not belong to the field $\mathbb{Q}[\cos(2\pi/k)]$. Because of this result, we expect to find arithmetic obstructions to the existence of a quasi-Strebel structure for k-differentials with admissible singularities in the case of odd k.

12.3 Positive k-differentials and Heine–Stieltjes theory

The present section reconnects material of this chapter with the main body of the book and is somewhat similar to our treatment of gradient and positive gradient quadratic differentials presented in [19].

To connect quasi-Strebel structures with potential theory, let us recall some basic facts from complex analysis on Riemann surfaces, see, e. g., [74]. Let Y be a Riemann surface (open or closed) and h be a real- or complex-valued smooth function on Y.

Definition 12.13. The *Levy form* of h (with respect to a local coordinate z) is a $(1,1)$-form given by

$$\mu_h := 2i\frac{\partial^2 h}{\partial z \partial \bar{z}} dz \wedge d\bar{z}. \tag{12.1}$$

In terms of the real and imaginary parts (x, y) of z, μ_h is given by

$$\mu_h = \left(\frac{\partial^2 h}{\partial x^2} + \frac{\partial^2 h}{\partial y^2}\right) dx \wedge dy = \Delta h dx \wedge dy.$$

If h is a smooth real-valued function, then μ_h is a signed measure on Y with a smooth density. In potential theory h is usually referred to as the *(logarithmic) potential* of the measure μ_h, see, e. g., [74, Chapter 3].

Notice that (12.1) makes sense for an arbitrary complex-valued distribution h on Y if one interprets μ_h as a 2-current on Y, i. e., a linear functional on the space of smooth compactly supported functions on Y, see, e. g., [67]. Such a current is necessarily exact since the inclusion of smooth forms into currents induces the (co)homology isomorphism. Notice that if Y is compact and connected, then the exactness of μ_h is equivalent to the vanishing of the integral of μ_h over Y.

Now observe that by definition, any ps-level function Φ for a k-differential Ψ is piecewise-harmonic and continuous in the complement to the set of poles of order $-k$ where it tends to $\pm\infty$ depending on the residue of Ψ at such a pole, see Definition 12.11. Thus Φ defines a distribution on the Riemann surface Y and its Levy form μ_Φ is a measure supported on the union of the switching set $S_{\Psi,\Phi}$ and the set of poles of order $-k$.

The final notion motivated by our original study of the Heine–Stieltjes theory is as follows.

Definition 12.14. A k-differential Ψ with at least one pole of order $-k$ is called *positive* if there exists a ps-level function Φ whose Levy form μ_Φ is a real measure the negative part of which is supported only on the set of poles of order $-k$ of Ψ.

Let us now present some families of rational k-differentials that are positive in the above sense.

Theorem E. Given an arbitrary monic polynomial $Q_k(z) = z^k + a_1 z^{k-1} + \cdots + a_k$ of degree k, consider the rational k-differential $\Psi = \frac{i^k dz^k}{Q_k(z)}$, $i = \sqrt{-1}$. Then Ψ is a positive differential.

Theorem E is a special case of Corollary 12.2 below. Further, given a monic polynomial $Q(z)$ of degree ℓ and a positive integer $k \geq 2$, take the differential operator

$$T = Q(z)\frac{d^k}{dz^k}. \tag{12.2}$$

Assuming that $\ell \geq k$, consider the related Heine–Stieltjes problem which is as follows. For a non-negative integer n, find all pairs of monic polynomials $(V(z), S(z))$ such that $\deg V = \ell - k$ and $\deg S = n$, which solve the equation

$$TS_n(z) - (n)_k V(z)S(z) = Q(z)\frac{d^k S}{dz^k} - (n)_k V(z)S(z) = 0, \tag{12.3}$$

where $(n)_k := n(n-1)\cdots(n-k+1)$. ($V$ is classically known as the *Van-Vleck* polynomial and S as the *Stieltjes* polynomial for equation (12.3).)

The following can be found in Chapter 8, see [162, Theorem 4].

Theorem F. In the above notation, for every $n \geq k$, there exist exactly $\binom{n+\ell-k}{\ell-k}$ pairs $(V(z), S(z))$ counting multiplicities that solve (12.3). Moreover all roots of every $V(z)$ and every $S(z)$ are located in the convex hull $Conv(Q) \subset \mathbb{C}$ of the roots of $Q(z)$.

Given a non-negative integer n, denote by $\{V_{i,n}\}$, $i = 1, \ldots, \binom{n+\ell-k}{\ell-k}$ the set of all Van Vleck polynomials whose Stieltjes polynomials have degree n. In the above notation, assume that $\{V_{i_n,n}\}_{n=k}^{\infty}$ is a sequence of Van Vleck polynomials converging to some monic polynomial $\widetilde{V}(z)$. (Recall that all $V_{i,n}(z)$ have degree $k - \ell$ and their roots lie in $Conv(Q)$, which implies that there exist plenty of converging sequences $\{V_{i_n,n}\}_{n=N}^{\infty}$.) Denote by $S_{i,n}$ the Stieltjes polynomial corresponding to $V_{i,n}$. (Here $\deg S_{i,n} = n$.)

Theorem G. In the above notation, the sequence of root-counting measure $\{\mu_{i_n,n}\}$ of $\{S_{i_n,n}\}_{n=k}^{\infty}$ converges to the probability measure $\mu := \mu_{\widetilde{V},Q}$ satisfying the condition that

$$C_{\mu}^{k}(z) = \frac{\widetilde{V}(z)}{Q(z)}$$

a. e. in the complex plane $\mathbb{C}$, where $C_{\mu}(z) := \int_{\mathbb{C}} \frac{d\mu(\xi)}{z-\xi}$ is the Cauchy transform of μ.

Corollary 12.2. *The kth differential* $\Psi := \frac{i^k \cdot \widetilde{V}(z)}{Q(z)} dz^k$ *is positive. The corresponding ps-level function* Φ *is given by the logarithmic potential* $u_{\mu}(z)$ *of the probability measure* μ *where as before*

$$u_{\mu}(z) := \int_{\mathbb{C}} \log|z - \zeta| d\mu(\zeta).$$

Proof. (Sketch, cf. the final section of [84] and § 4 of [162].) Recall that

$$C_{\mu}(z) = \frac{\partial u_{\mu}(z)}{\partial z}; \quad \mu = \frac{1}{\pi} \frac{\partial C_{\mu}(z)}{\partial \bar{z}}$$

where the derivatives are understood in the distributional sense, which implies that μ is the Levy form of $u(z)$. The logarithmic potential $u_{\mu}(z)$ is continuous outside the set of point masses of μ and harmonic outside of its support. One can directly check that Theorem G implies the statement that the level curves of the logarithmic potential $u_{\mu}(z)$ of the measure μ constructed in this theorem consist of segments of horizontal trajectories of the differential $\Psi(z) = \frac{i^k \widetilde{V}(z)}{Q(z)} dz^k$. Indeed, the gradient of $u_{\mu}(z)$ equals $\bar{C}_{\mu}(z)$ where bar stands for the usual complex conjugation. Thus the tangent direction $\delta \in \mathbb{C}$ to the level curve of $u_{\mu}(z)$ at a point p is given by $\kappa \, i \, \bar{C}_{\mu}(p)$ for some real κ. Thus it will satisfy the condition $\frac{i^k \widetilde{V}(p)}{Q(p)} \delta^k > 0$. Additionally, $u_{\mu}(z)$ approaches $\log|z|$ when $|z| \to \infty$, which implies that almost all connected components of its level curves are closed non-selfintersecting curves in $\mathbb{C}$. The measure μ is positive in $\mathbb{C}$, and its extension to $\mathbb{C}P^1 \supset \mathbb{C}$ is the exact 2-current with the point mass -1 at ∞. Observe that ∞ is the pole of order $-k$ for the k-differential $\Psi(z) = \frac{i^k \widetilde{V}(p)}{Q(p)} dz^k$ with the residue $-i^k$. Finally, the support of μ consists of a finite number of curve segments globally forming an embedded forest in $\mathbb{C}$ where each curve is either a segment of a horizontal trajectory of $\Psi(z)$ or a segment of a trajectory of $\Psi^*(z) = \frac{i^{k+1} \widetilde{V}(z)}{Q(z)} dz^k$, see more details in [84]. □

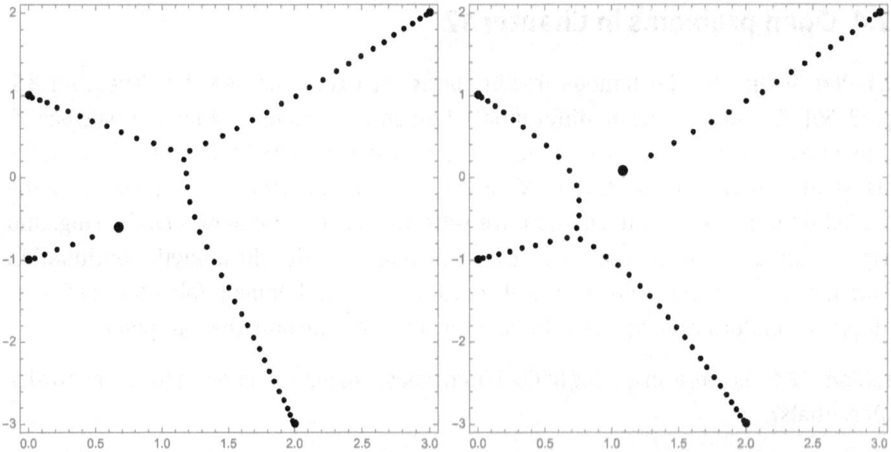

Figure 12.5: The switching sets for the logarithmic potential $u_\mu(z)$ in case of the cubic differentials $\Psi(z) = \frac{i^3\cdot(z-a)dz^3}{(z+i)(z-i)(z-2+3i)(z-3-2i)}$ with $a \simeq 0.7 - 0.6i$ (left) and $a \simeq 1.1 + 0.1i$ (right). (Possible choices of a are described in [84].)

Concrete illustrations of the switching sets for $u(z)$ of such cubic differentials with $\deg \widetilde{V} = 1$ and $\deg Q = 4$ are given in Figure 12.5, see more details and explanations in Figures 1 and 2 of [84].

The special case when $\deg Q = k$ is called *exactly solvable* and is considered in more detail in [122] and [27]. These papers contain substantial information about the switching sets of $u(z)$ for such differentials of order k, which are topologically planar trees in $\mathbb{C}$ with vertices of valency one coinciding with all roots of Q and special angles between the edges, see an illustration in Figure 12.6.

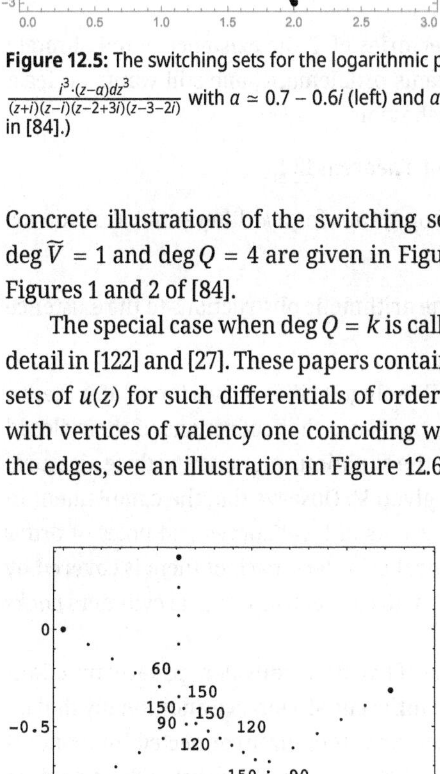

Figure 12.6: The switching set of $u(z)$ for the rational sextic differential $\Psi(z) = \frac{-dz^6}{z(z+i)(z-1-i)(z-2+2i)(z-2+3i)(z-3+2i)}$ shown in a chart of the corresponding flat surface. (The numbers are the angles between the neighboring edges.)

12.4 Open problems in Chapter 12

12.1. Recall that by the famous Jenkins basic structure theorem [95, Theorem 3.5, pp. 38–39], for any quadratic differential Ψ given on a compact Riemann surface Y, up to a few exceptions, the set $Y \setminus (K_\Psi \cup Cr_\Psi)$ consists of a finite number of the circle, ring, strip, and end domains. Here K_Ψ is the union of all critical trajectories. (For the detailed definitions and information, we refer to loc. cit.) The names circle, ring, and strip domain are reflecting the shapes of their images under the analytic continuation of the mapping given by the canonical coordinate; an end domain (also referred to as half-plane domain) is mapped by the canonical coordinate onto the half-plane.

Problem 12.1. Is there an analog of the Jenkins basic structure theorem for higher order differentials?

Although due to self-intersections of trajectories of Ψ the existence of only finitely many types of possible domains for $k > 2$ seems problematic, one still wants (at least crudely) to distinguish their possible global behavior.

12.2. The next question is related to Part (ii) of Theorem 12.1.

Problem 12.2. Find necessary and sufficient conditions for a k-differential Ψ of odd order k to admit a quasi-Strebel structure?

Hopefully it is at least possible to find some arithmetic obstructions to the existence of a quasi-Strebel structure for odd k.

12.3. Typically a k-differential Ψ with admissible singularities allows for an abundance of non-equivalent quasi-Strebel structures in the sense that collections of their closed broken trajectories are distinct. To streamline the situation, one can introduce a natural partial order on quasi-Strebel structures of a given Ψ. Observe that the complement in Y to the union of all special connected components of level curves and poles of order $-k$ is a disjoint union of finitely many topological cylinders. Each of them is covered by closed broken trajectories of the same type. We will call such topological cylinders *packs of broken trajectories.*

In Figure 12.7 we present a simple example of two non-equivalent ps-level functions (quasi-Strebel structures) for the same quartic differential. Observe additionally that on the right picture in Figure 12.7, the switching set consists of three connected components one of which is non-contractible. On the left picture of Figure 12.7, the switching set does not contain the zero of order 12. Another exmaple is given in Figure 12.8.

Given two ps-level functions Φ_1 and Φ_2 on Y, we say that Φ_1 gives a *coarser subdivision of* Y than Φ_2 if the closure of every pack of broken trajectories of Φ_1 is the union of the closures of some packs of broken trajectories of Φ_2. Notation, $\Phi_1 \succeq \Phi_2$. In this situation we will also say that Φ_2 provides a *finer subdivision of* Y than Φ_1. We say that a quasi-Strebel structure (Ψ, Φ) is *maximal* if there is no coarser ps-level function, i. e., no $\widetilde{\Phi}$ such that $\widetilde{\Phi} \succ \Phi$ exists.

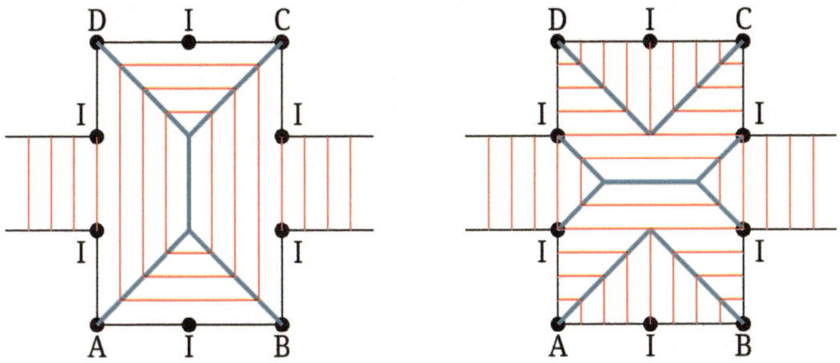

Figure 12.7: A quartic differential with zero of order 12 (marked by I), four poles of order -3 (marked by A, B, C and D), and two poles of order -4 together with two nonequivalent ps-level functions. (The broken trajectories are shown by the thin lines, while the switching set is shown by the thick ones.)

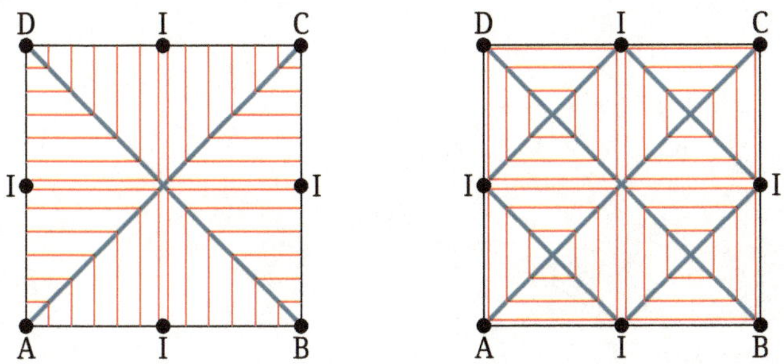

Figure 12.8: A quartic differential with a zero of order 4 (marked by I) and four poles of order -3 (marked by A, B, C, and D) together with two non-equivalent ps-level functions.

Problem 12.3. Given a k-differential Ψ with admissible singularities, describe all its maximal quasi-Strebel structures.

Part III: **Extras**

13 Outlook and appendices

In this chapter we make some concluding remarks about the material of the book and comment on the problems formulated at the end of each chapter.

13.1 Further directions in the study of ES-operators

Let us shortly summarize the results and problems formulated in the first part of the book.

- The asymptotic theory of eigenpolynomials for non-degenerate ES-operators seems to be in reasonable shape. One might still ask for a fixed positive integer k, how the space of monic polynomials Q_k is divided by non-generic polynomials, i. e., those for which the support of the measure μ_{Q_k} is not a trivalent tree.
- The asymptotic theory of eigenpolynomials for degenerate ES-operators has a complete conjecture that is settled only in special cases. The structure of the asymptotic root-counting measure is not quite clear.
- One would like to have more applications of the asymptotic theory to Bochner–Krall problem than has been obtained so far. In particular, one can ask which ES-operators have only real-rooted eigenpolynomials.
- The asymptotic theory of eigenpolynomials for homogenized spectral problem for ES-operators is in its initial stage. One would like to show that the asymptotic root-counting measures exist as well as understand their properties using the defining algebraic equation.
- one would like to have a more complete theory of algebraic equations admitting a positive measure as well as its applications since such equations often appear in asymptotic analysis.
- we present a complete conjectural solution to the generalized Bochner–Krall problem but solve it only for equations of order 3. More advances are desirable.
- WKB method so far is developed only for sequence of eigenpolynomials of a non-degenerate ES-operators. As a first step one needs to extend it to all solutions of a non-degenerate ES-operators and then to the homogenized spectral problem for such.

13.2 Further directions in Heine–Stieltjes theory

Let us shortly summarize the problems formulated in the second part of the book.

- A general theory of Heine–Stieltjes problem for higher order operators is in a reasonable shape. One might think about other multi-parameter problems having polynomial solutions as, for example, an appropriate homogenized version.
- The asymptotic theory for Stieltjes polynomials is also in reasonable shape. Asymptotic behavior of sequences of (generalized) Stieltjes polynomials for homogenized

https://doi.org/10.1515/9783112207741-016

version of Heine–Stieltjes problem has not been considered yet. It will be more difficult to study in analogy with the asymptotic of eigenpolynomials for the homogenized spectral problem for ES-operators. It will require the development of new tools and methods in potential theory.
- The asymptotic theory for (products of) Van Vleck polynomials is just in the beginning and seems very fascinating.
- Our attempts to develop a non-trivial theory of quasi-Strebel differentials of higher order have not been very successful so far. Probably one needs more connection of such differentials with positive measures.
- A very ambitious goal would be to try to develop elements of Heine–Stieltjes theory for linear differential operators on curves of positive genera than just on $\mathbb{C}P^1$ using instead of polynomials sections of natural linear bundles.

(The material of Appendix I is based on [165].)

Appendix I. On eigenvalues of rectangular matrices

In recent years there appeared a number of publications discussing the eigenvalues of pencils of non-square matrices and their approximations, see, e. g., [38, 51, 186] and the references therein. But to the best of our knowledge the following natural problem either has been overlooked by specialists in linear algebra or is deeply buried in the (enormous) literature on this topic.

Problem 13.1. Given a $(k+1)$-tuple of $(m\times n)$-matrices $A, B_1, \dots, B_k$, $m \leq n$, describe the set of all values of the parameters $\lambda_1, \dots \lambda_k$ for which the rank of the linear combination $A+\lambda_1 B_1+\dots+\lambda_k B_k$ is less than m or, in other words, when the linear system $v\cdot(A+\lambda_1 B_1+\dots+\lambda_k B_k) = 0$ has a non-trivial left solution $v \in \mathbb{C}^m \setminus 0$, which we call an *eigenvector*, where the symbol "$\cdot$" denotes the usual matrix/vector multiplication.

Let $\mathcal{M}(m,n)$, $m \leq n$, be the linear space of all $(m\times n)$-matrices with complex entries. In what follows we will consider k-tuples of $(m\times n)$-matrices $B_1, \dots, B_k$ that are linearly independent in $\mathcal{M}(m,n)$ and denote their linear span by $\mathcal{L} = \mathcal{L}(B_1, \dots, B_k)$. Given a matrix pencil $\mathcal{P} = A + \mathcal{L}$, where $A \in \mathcal{M}(m,n)$, let $\mathcal{E}_\mathcal{P} \subset \mathcal{P}$ be its *eigenvalue locus*, i. e., the set of all matrices in $\mathcal{P}$ whose rank is less than m. Elements of $\mathcal{E}_\mathcal{P}$ will be called *(generalized) eigenvalues.* Denote by $\mathcal{M}^1 \subset \mathcal{M}(m,n)$ the set of all $(m\times n)$-matrices with positive corank, i. e., whose rank is non-maximal. Its codimension equals $n-m+1$ and its degree as an algebraic variety equals $\binom{n}{m-1}$, see [47, Proposition 2.15]. Consider the natural left-right action of the group $GL_m \times GL_n$ on $\mathcal{M}(m,n)$, i. e., GL_m (respectively, GL_n) acts on $(m\times n)$-matrices by left (respectively, right) multiplication. This action on $\mathcal{M}(m,n)$ has finitely many orbits, each orbit being the set of all matrices of a given (co)rank, see, e. g., [14, Chap. I, § 2]. Note that by the well-known product formula for coranks the codimension of the set of matrices of rank r equals $(m-r)(n-r)$. Obviously, for any pencil $\mathcal{P}$, one has that the eigenvalue locus coincides with $\mathcal{E}_{\mathcal{P}} = \mathcal{M}^1 \cap \mathcal{P}$. Thus, for a generic pencil $\mathcal{P}$ of dimension k, the eigenvalue locus $\mathcal{E}_{\mathcal{P}}$ is a subvariety of $\mathcal{P}$ of codimension $n-m+1$ if $k \geq n-m+1$ and it is empty otherwise. The most interesting situation for applications occurs when $k = n-m+1$, in which case $\mathcal{E}_{\mathcal{P}}$ is generically a finite set. From now on we assume that $k = n-m+1$.

Denoting as above by $\mathcal{L}$ the linear span of $B_1, \ldots, B_{n-m+1}$, we say that $\mathcal{L}$ is *transversal to* $\mathcal{M}^1$ if the intersection $\mathcal{L} \cap \mathcal{M}^1$ is finite and *non-transversal to* $\mathcal{M}^{12}$ otherwise. Notice that due to the homogeneity of $\mathcal{M}^1$ any $(n-m+1)$-dimensional linear subspace $\mathcal{L}$ transversal to it intersects $\mathcal{M}^1$ only at 0 and that the multiplicity of this intersection at 0 equals $\binom{n}{m-1}$.

An important and most natural example of such a subspace $\mathcal{L}$ is motivated by the Heine–Stieltjes theory [82] and its higher order generalizations, see, e. g., [162]. Denote by J_s, $s = 1, \ldots, n-m+1$, the $(m \times n)$-matrix whose entries are given by $a_{i,j} = 0$ if $i - j \neq s$ and 1 otherwise. We call J_s the *sth unit matrix* or the *sth diagonal matrix*. Let us denote the linear span of $J_1, \ldots, J_{n-m+1}$ by $\mathfrak{L}$ and call $\mathfrak{L}$ the *standard diagonal subspace*. Note that $\mathfrak{L}$ is transversal to $\mathcal{M}^1$ since any matrix in $\mathfrak{L}$ different from 0 has full rank, as one can easily check.

We start with the following simple statement.

Lemma 13.1. *If $\mathcal{L} \subset \mathcal{M}(m,n)$ has dimension $(n-m+1)$ and is transversal to $\mathcal{M}^1$, then for any matrix $A \in \mathcal{M}(m,n)$, the eigenvalue locus $\mathcal{E}_{\mathcal{P}}$ of the pencil $\mathcal{P} = A + \mathcal{L}$ consists of exactly $\binom{n}{m-1}$ points counted with multiplicities.*

Remark 13.1. Notice that since $\mathcal{M}^1$ is an incomplete intersection the same holds for the eigenvalue locus $\mathcal{E}_{\mathcal{P}}$ of a generic pencil $\mathcal{P} = A + \mathcal{L}$, i. e., in order to find $\mathcal{E}_{\mathcal{P}}$ for a given generic matrix A and a given generic subspace $\mathcal{L}$, one has to solve an overdetermined system of determinantal equations.

However, as was essentially discovered by Heine [82], the situation is different if one considers the standard diagonal subspace $\mathfrak{L}$ and any $A = (a_{i,j}) \in \mathcal{M}(m,n)$ that is upper triangular—that is, such that $a_{i,j} = 0$ whenever $i > j$—and has additionally distinct elements on the first main diagonal.

Theorem 13.1. *For any upper triangular matrix $A = (a_{i,j}) \in \mathcal{M}(m,n)$ with all distinct entries $a_{i,i}$ on the first main diagonal, the eigenvalue locus $\mathcal{E}_{\mathcal{P}}$ of the pencil $\mathcal{P} = A + \mathfrak{L}$, where $\mathfrak{L}$ is the standard diagonal subspace, is the union of m complete intersections, indexed enumerated by the first component of the eigenvalue.*

Remark 13.2. An explicit defining system of $(n-m)$ algebraic equations in $(n-m)$ variables for each such complete intersection is presented in the proof of Theorem 13.1, see § I.1 below.

Given $\mathcal{L}$ as above, consider the natural projection map $\pi_{\mathcal{L}} : \mathcal{M}(m,n) \to \mathcal{L}^{\perp}$ along $\mathcal{L}$, where $\mathcal{L}^{\perp} = \mathcal{M}(m,n)/\mathcal{L}$. Noticing that $\dim \mathcal{M}^1 = \dim \mathcal{L}^{\perp}$, we define the *set of critical values* of $\pi_{\mathcal{L}}$ to be the set $\mathbf{C}_{\mathcal{L}}$ of all points in $\mathcal{M}^1$ where $\pi_{\mathcal{L}}$ is not a local diffeomorphism of $\mathcal{M}^1$ on its image $\pi_{\mathcal{L}}(\mathcal{M}^1)$. In other words, $\mathbf{C}_{\mathcal{L}}$ is the set of all points $p \in \mathcal{M}^1$ such that the sum of $\mathcal{L}$ and the tangent space to $\mathcal{M}^1$ at p does not coincide with the whole $\mathcal{M}(m,n)$. In particular, independently of $\mathcal{L}$ the critical value set $\mathbf{C}_{\mathcal{L}}$ always includes the set $\mathcal{M}^2$ of all $(m \times n)$-matrices with corank at least 2.

Recall that $\mathcal{M}^1 \subset \mathcal{M}(m,n)$ has the classical small resolution of singularities $\widetilde{\mathcal{M}^1} \subset \mathcal{M}(m,n) \times \mathbb{C}\mathrm{P}^{m-1}$. Here $\widetilde{\mathcal{M}^1}$ consists of all pairs $(A, pker(A))$, where $A \in \mathcal{M}^1$ and $pker(A)$ is the projectivization of the left kernel of A.

Using this construction, one can parameterize a Zariski open subset of $\mathcal{M}^1$ as follows. Consider the product $P(m,n) = \mathcal{M}(m-1,n) \times \mathbb{C}^{m-1}$. Take the map $\nu : P(m,n) \to \mathcal{M}^1 \subset \mathcal{M}(m,n)$ sending a pair $(\mathcal{A}; k_1, \dots, k_{m-1})$ to the matrix $A \in \mathcal{M}(m,n)$ obtained by appending to $\mathcal{A}$ the last row such that its sum with the linear combination with the coefficients $(k_1, \dots, k_{m-1})$ of the respective rows of $\mathcal{A}$ vanishes.

The main result of this chapter is a simple determinantal representation of $\mathbf{C}_{\mathcal{L}}$ in the above coordinates.

Theorem 13.2. *Let $\mathcal{L}$ be any $(n-m+1)$-dimensional linear subspace in $\mathcal{M}(m,n)$ transversal to $\mathcal{M}^1$ and denote by $L_1, \dots, L_{n-m+1}$ some basis of $\mathcal{L}$. Then in the coordinates of $P(m,n)$, the critical value set ${\mathbf{C}_{\mathcal{L}}}^2$ is given by the determinantal equation*

$$\det \begin{pmatrix} \mathcal{A} \\ V_1 \\ \vdots \\ V_{n-m+1} \end{pmatrix} = 0. \tag{13.1}$$

Here $\mathcal{A}$ is a $(m-1,n)$-matrix with undetermined entries and V_j, $j = 1, \dots, n-m+1$, are row vectors given by $V_j = \boldsymbol{\kappa} \cdot L_j$, where $\boldsymbol{\kappa} = (k_1, \dots, k_m)$.

Remark 13.3. If one expands equation (13.1) in the variables $(k_1, \dots, k_{n-m+1})$„ then the coefficient of each monomial in these variables becomes a linear combination of the maximal minors of $\mathcal{A}$ (i. e., the Plücker coordinates) with complex coefficients depending only on the choice of $\mathcal{L}$. Moreover, the above equation contains a lot of information of geometric nature.

Example 13.1. For $m = 2$, the homogeneous defining polynomial of $\mathbf{C}_{\mathfrak{L}}$ with respect to the standard diagonal subspace $\mathfrak{L}$ is given by

$$a_n k_1^n + a_{n-1} k_1^{n-1} k_2 + a_{n-2} k_1^{n-2} k_2^2 + \cdots + a_1 k_1 k_2^{n-1},$$

where $a_j = a_{1,j}, j = 1, \dots, n$.

Example 13.2. For the standard diagonal subspace $\mathfrak{L}$ in the case of $\mathcal{M}(3,4)$, the homogeneous defining polynomial of $\mathbf{C}_{\mathfrak{L}}$ may be written as

$$\begin{vmatrix} a_{1,1} & a_{1,2} & a_{1,3} & a_{1,4} \\ a_{2,1} & a_{2,2} & a_{2,3} & a_{2,4} \\ k_1 & k_2 & k_3 & 0 \\ 0 & k_1 & k_2 & k_3 \end{vmatrix} = \Delta_{3,4}k_1^2 + \Delta_{1,4}k_2^2 + \Delta_{1,2}k_3^2 - \Delta_{2,4}k_1k_2 + (\Delta_{2,3} - \Delta_{1,4})k_1k_3 - \Delta_{1,3}k_2k_3,$$

where $\Delta_{i,j}$ is the (2×2)-determinant of the upper part $\mathcal{A}$ including the ith and jth columns.

Remark 13.4. The multiplicity of an eigenvalue $A \in \mathcal{E}_{\mathcal{P}}$ can be expressed in terms of the dimension of the corresponding local algebra. More exactly, for an $(m \times n)$-matrix A, we define the ideal I_A in the algebra $\mathbb{C}[[t_1, \dots, t_k]]$ of formal power series as the ideal generated by all Plücker polynomials $\Delta_{i_1,\dots,i_m}(A + \sum_{l=1}^{m} t_l B_l)$, where $\Delta_{i_1,\dots,i_m}(X_{m\times n})$ is the determinant of the $(m \times m)$ submatrix of X formed by the columns with the indices $i_1, i_2, \dots, i_m$. Now define the local algebra $\mathcal{A}_{\text{loc}}$ as the quotient algebra $A_{\text{loc}} = \mathbb{C}[[t_1, \dots, t_k]]/I_A$. Then the multiplicity of the eigenvalue A in the pencil $A + \mathcal{L}$ equals $\dim_{\mathbb{C}} A_{\text{loc}}$.

The main result of this section (Theorem 13.2) gives a simple explicit determinantal formula for the critical value set $\mathbf{C}_{\mathcal{L}}$ (in coordinates on the resolution of singularities $P(m, n)$). Its inverse image $\pi_{\mathcal{L}}^{-1}(\mathbf{C}_{\mathcal{L}})$ is an important hypersurface consisting of all matrices in $\mathcal{M}(m, n)$ having a multiple eigenvalue. However, the problem of explicitly obtaining its defining polynomial in the matrix entries seems to be quite delicate in general. As an illustration, let us show how this can be done in the simplest case of (2×3)-matrices.

Example 13.3 (Discriminant equation). For $m = 2, n = 3$, we will write the defining equation for the hypersurface $\pi_{\mathcal{L}}^{-1}(\mathbf{C}_{\mathcal{L}})$ of matrices with multiple eigenvalues in the space $\mathcal{M}(2, 3)$ itself.

For any pair of positive integers $m < n$, consider the extended matrix space $\mathcal{M}(m, n) \times \mathbb{C}P^{m-1} \times \mathbb{C}^{n-m+1}$, where the m-tuple of homogeneous coordinates in $\mathbb{C}^m$ is denoted by $\kappa = (\kappa_1 : \dots : \kappa_m)$ and the coordinates in $\mathbb{C}^{n-m+1}$ are denoted by $\lambda = (\lambda_1, \dots, \lambda_{n-m+1})$.

Given a matrix $M \in \mathcal{M}(m, n)$, we will write a system of polynomial equations for $A + \sum_{i=1}^{n-m+1} \lambda_i J_i$ as

$$\sum_{s=1}^{m} \kappa_s \left[A + \sum_{i=1}^{n-m+1} \lambda_i J_i \right]_{s*} = 0,$$

expressing the fact that λ is an eigenvalue of M while the κ_is are the corresponding coefficients of a linear dependence between the rows of the matrix.

Using resultants we can get rid of the additional variables λ and κ. This elimination leads to the defining equation for the hypersurface in question.

Namely, consider a (2×3)-matrix $A = \left(\begin{smallmatrix} a_{11} & a_{12} & a_{13} \\ a_{21} & a_{22} & a_{23} \end{smallmatrix}\right)$ and let as before $J_1 = \left(\begin{smallmatrix} 1 & 0 & 0 \\ 0 & 1 & 0 \end{smallmatrix}\right)$ and $J_2 = \left(\begin{smallmatrix} 0 & 1 & 0 \\ 0 & 0 & 1 \end{smallmatrix}\right)$. A generic element of the pencil $\mathcal{P}$ is thus given by

$$A(\lambda_1, \lambda_2) := A - \lambda_1 J_1 - \lambda_2 J_2 = \begin{pmatrix} a_{11} - \lambda_1 & a_{12} - \lambda_2 & a_{13} \\ a_{21} & a_{22} - \lambda_1 & a_{23} - \lambda_2 \end{pmatrix}.$$

For a generic matrix A, the condition that the rank of $A(\lambda_1, \lambda_2)$ is less than 2 translates into two equations: the minor consisting of the second and third columns vanishes, and the minor consisting of the first and third columns vanishes. These equations have the form

$$(a_{12} - \lambda_2)(a_{23} - \lambda_2) - a_{13}(a_{22} - \lambda_1) = 0 \tag{13.2}$$

$$(a_{11} - \lambda_1)(a_{23} - \lambda_2) - a_{13}a_{21} = 0. \tag{13.3}$$

Note that

$$\kappa := \kappa_1 = \frac{a_{23} - \lambda_2}{a_{13}}. \tag{13.4}$$

Moreover, from the determinantal equation of Theorem 13.2 we obtain a third equation. Substituting expression (13.4) into the latter gives the equation

$$a_{13}^2 a_{11} - a_{13}^2 \lambda_1 + a_{23}a_{13}a_{12} - 3a_{23}a_{13}\lambda_2 + a_{13}a_{23}^2 - \lambda_2 a_{13}a_{12} + 2a_{13}\lambda_2^2 = 0. \tag{13.5}$$

Now equation (13.2) has bidegree $(1, 2)$ with respect to λ_1, λ_2. Analogously, (13.3) has bidegree $(1, 1)$ and (13.5) has bidegree $(1, 2)$ with respect to the same variables. Clearly, any solution s of the system of equations consisting of (13.2), (13.3), and (13.5) annihilates any polynomial in the ideal generated by these three equations. In particular, the following eight equations have s as a common solution: (13.2), (13.2) multiplied by λ_2, (13.2) multiplied by λ_2^2, (13.3), (13.3) multiplied by λ_2, (13.5), (13.5) multiplied by λ_2, (13.5) multiplied by λ_2^2. Therefore, the (8×8) determinant

$$D = \det \begin{pmatrix} \Delta_{23} & a_{13} & -a_{12} - a_{23} & 0 & 1 & 0 & 0 & 0 \\ 0 & 0 & \Delta_{23} & a_{13} & -a_{12} - a_{23} & 0 & 1 & 0 \\ 0 & 0 & 0 & 0 & \Delta_{23} & a_{13} & -a_{12} - a_{23} & 1 \\ \Delta_{13} & -a_{23} & -a_{11} & 1 & 0 & 0 & 0 & 0 \\ 0 & 0 & \Delta_{13} & -a_{23} & -a_{11} & 1 & 0 & 0 \\ \delta & -a_{13}^2 & \sigma & 0 & 2a_{13} & 0 & 0 & 0 \\ 0 & 0 & \delta & -a_{13}^2 & \sigma & 0 & 2a_{13} & 0 \\ 0 & 0 & 0 & 0 & \delta & -a_{13}^2 & \sigma & 2a_{13} \end{pmatrix}$$

vanishes when (13.2), (13.3), and (13.5) have a common root. Here we use the following notation: $\delta = a_{13}^2 a_{11} + a_{23}a_{13}a_{12} + a_{13}a_{23}^2$, $\Delta_{23} = a_{12}a_{23} - a_{13}a_{22}$, $\Delta_{13} = a_{11}a_{23} - a_{21}a_{13}$, and $\sigma = -a_{13}a_{12} - 3a_{13}a_{23}$.

This observation implies that the required defining polynomial for $\pi_{\mathcal{L}}^{-1}(\mathbf{C}_{\mathcal{L}})$ is the product of some (but not necessarily all) irreducible factors of the polynomial D. Factorizing D, we obtain $D = a_{11}^6 D_0$, where

$$\begin{aligned} D_0 = {} & -12a_{13}a_{22}^2 a_{11} + a_{22}^2 a_{12}^2 + 12a_{13}a_{22}a_{11}^2 + a_{11}^2 a_{23}^2 + 4a_{21}a_{12}^3 - 4a_{21}a_{23}^3 + a_{11}^2 a_{12}^2 \\ & + 12a_{12}a_{23}^2 a_{21} - 12a_{12}^2 a_{23}a_{21} - 2a_{12}a_{23}a_{22}^2 - 2a_{12}a_{23}a_{11}^2 - 2a_{22}a_{11}a_{23}^2 \\ & - 18a_{13}a_{22}a_{23}a_{21} - 2a_{22}a_{11}a_{12}^2 + 18a_{13}a_{22}a_{21}a_{12} + 18a_{11}a_{23}a_{13}a_{21} \\ & - 18a_{21}a_{13}a_{12}a_{11} + 4a_{13}a_{22}^3 - 27a_{21}^2 a_{13}^2 - 4a_{13}a_{11}^3 + 4a_{12}a_{23}a_{22}a_{11} + a_{22}^2 a_{23}^2 = 0. \end{aligned}$$

Note that D_0 is of second degree in the variable a_{13} and its discriminant (with respect to this variable) $W = 16(3a_{12}a_{21} - 3a_{21}a_{23} - 2a_{11}a_{22} + a_{11}^2 + a_{22}^2)^3$ is not a complete square. Thus, we conclude that D_0 is irreducible. Hence the variety given by $\{D = 0\}$ is the union of the variety given by $\{D_0 = 0\}$ and the hyperplane $\{a_{11} = 0\}$ taken with multiplicity 6.

Since the hyperplane $\{a_{11} = 0\}$ is obviously not contained in $\pi_{\mathcal{L}}^{-1}(\mathbf{C}_{\mathcal{L}}) \subset \mathcal{M}(2,3)$, we obtain that $\pi_{\mathcal{L}}^{-1}(\mathbf{C}_{\mathcal{L}})$ is given by $\{D_0 = 0\}$.

I.1 Proofs

Proof of Lemma 13.1. This follows almost directly from homogeneity of $\mathcal{M}^1$. Indeed, take any matrix $0 \neq A \in \mathcal{M}(m,n)$. Let $\tilde{l} \in \mathcal{L}$ be its eigenvalue, that is, a matrix from $\mathcal{L}$ such that $A + \tilde{l}$ belongs to $\mathcal{M}^1$. Notice that, for any $\epsilon \in (0,1]$, the matrix $\epsilon\tilde{l}$ is the eigenvalue of the matrix ϵA. Considering the family of matrices ϵA with $\epsilon \in [0,1]$, we conclude that the total multiplicity of eigenvalues of the pencil $A + \mathcal{L}$ coincides with that of the linear pencil $\mathcal{L}$ if the latter multiplicity is finite, which gives the required statement. □

Proof of Theorem 13.1. To get the defining system of algebraic equations for $\mathcal{E}_{\mathcal{P}}$ under the assumptions of Theorem 13.1 we proceed exactly as in [82]. For a given upper triangular matrix $A \in \mathcal{M}(m,n)$ with distinct entries on the main diagonal, we want to find all $(n-m+1)$-tuples $(\lambda_1, \dots, \lambda_{n-m+1})$ such that the matrix $A + \lambda_1 J_1 + \lambda_2 J_2 + \cdots + \lambda_{n-m+1} J_{n-m+1}$ has positive corank. Since A is upper triangular with distinct $a_{i,i}$, then in order to get a positive corank it is necessary to require $\lambda_1 + a_{i,i} = 0$ for some $i = 1, \dots, m$.

The next observation is that under the above assumptions on A for any given $i = 1, \dots, m$, the total number of eigenvalues with $\lambda_1 + a_{i,i} = 0$ equals $\binom{n-i}{m-i}$, which gives the following count of the eigenvalues of A noticed already by Heine: $\binom{n}{m-1} = \binom{n-1}{m-1} + \binom{n-2}{m-2} + \cdots + \binom{n-m}{0}$. Indeed, if $\lambda_1 + a_{i,i} = 0$, then $\lambda_1 + a_{j,j} \neq 0$ for all $j \neq i$ and, in particular, due to the assumptions on A, the first $i-1$ rows of $A - a_{i,i}J_1 + \lambda_2 J_2 + \cdots + \lambda_{n-m+1} J_{n-m+1}$ are linearly independent for all values of $\lambda_2, \dots, \lambda_{n-m+1}$. On the other hand, the remaining rows $i, i+1, \dots, m$ can become linearly dependent under an appropriate choice of $\lambda_2, \dots, \lambda_{n-m+1}$. Since the matrix $A - a_{1,1}J_1$ is upper triangular with the (i,i)th entry vanishing the condition that $A - a_{i,i}J_1 + \lambda_2 J_2 + \cdots + \lambda_{n-m+1} J_{n-m+1}$ has positive corank is equivalent to the condition that the matrix obtained by removing its first i rows and $i-1$ columns has positive corank. By Lemma 13.1 the total number of eigenvalues of the matrix of the size $(m-i+1) \times (n-i)$ equals $\binom{n-i}{m-i}$.

Let us now, for any given $i = 1, \dots, m$, derive a system of algebraic equations in the variables $\lambda_2, \dots, \lambda_{n-m+1}$ whose solutions are exactly all the eigenvalues of A with $\lambda_1 + a_{i,i} = 0$. We will concentrate on the case $i = 1$ since all other cases are covered in exactly the same way by working with a smaller matrix obtained from A by removing the first $(i-1)$ rows and $(i-1)$ columns. Using $(k_1, \dots, k_m)$ for the coordinates of the left kernel and $\lambda_1, \lambda_2, \dots, \lambda_{n-m+1}$ for the eigenvalues, we get the following system of equations:

$$\begin{cases} 0 = k_1(a_{1,1} + \lambda_1) \\ 0 = k_1(a_{1,2}\lambda_2) + k_2(a_{2,2} + \lambda_1) \\ \vdots \\ 0 = k_1(a_{1,m} + \lambda_m) + k_2(a_{2,m} + \lambda_{m-1}) + \cdots + k_m(a_{m,m} + \lambda_1) \\ 0 = k_1(a_{1,m+1} + \lambda_{m+1}) + k_2(a_{2,m+1} + \lambda_m) + \cdots + k_m(a_{m,m+1} + \lambda_2) \\ \vdots \\ 0 = k_1(a_{1,m+1} + \lambda_{m+1}) + k_2(a_{2,m+1} + \lambda_m) + \cdots + k_m(a_{m,m+1} + \lambda_2) \\ 0 = k_1(a_{1,n} + \lambda_n) + k_2(a_{2,n} + \lambda_{n-1}) + \cdots + k_m(a_{n,n} + \lambda_{n-m+1}) \end{cases}$$

expressing the existence of a non-trivial left kernel of $A + \lambda_1 J_1 + \lambda_2 J_2 + \cdots + \lambda_{n-m+1} J_{n-m+1}$. (To simplify notations, we assume here that $\lambda_j = 0$ for $j > n - m + 1$.) In order to get the required system of equations in $\lambda_1, \ldots, \lambda_{n-m+1}$, we have to eliminate from the above system the variables $k_1, \ldots, k_m$. Notice that under our assumptions on A the possible corank of $A + \lambda_1 J_1 + \lambda_2 J_2 + \cdots + \lambda_{n-m+1} J_{n-m+1}$ can be at most 1 and in the case of corank 1 the linear dependence must necessarily include the first row, i. e., $k_1 = 1$. Note also that the first m equations are triangular with respect to $k_1, \ldots, k_m$, which together with our assumptions on A allows us to successfully eliminate them. Namely, from the first equation we get $\lambda_1 = -a_{1,1}$ and $k_1 = 1$. Then, for any $i = 2, \ldots, m$, we solve the ith equation with respect to k_i and get

$$k_i = \frac{1}{a_{1,1} - a_{i,i}}(k_1(a_{1,i} + \lambda_i) + k_2(a_{2,i} + \lambda_{i-1}) + \cdots + k_{i-1}(a_{i-1,i} + \lambda_2)).$$

With the initial value $k_1 = 1$ and taking into account that the only possible denominators occurring in the above expressions for k_i are $a_{1,1} - a_{i,i}$, we recurrently find all k_i, $i = 1, \ldots, m$, as the functions of the matrix entries and λs. Substituting these found expressions in the remaining $n - m$ equations, we get the required system of algebraic equations to determine $\lambda_2, \ldots, \lambda_{n-m+1}$. (Notice that $\lambda_1 = -a_{1,1}$ was already obtained from the first equation.) □

Example 13.4. Any matrix $A \in \mathcal{M}(2,4)$ has four eigenvalues (counted with multiplicities) with respect to the standard diagonal subspace $\mathfrak{L}$. If A is upper triangular with distinct elements on the first main diagonal, then these eigenvalues split into two groups depending on the value of λ_1. Namely, there are three eigenvalues for which $\lambda_1 = -a_{1,1}$ and 1 eigenvalue for $\lambda_1 = -a_{2,2}$. For $\lambda_1 = -a_{1,1}$, the above system (before elimination) has the form

$$\begin{cases} 0 = k_1(a_{1,1} + \lambda_1) \\ 0 = k_1(a_{1,2} + \lambda_2) + k_2(a_{2,2} + \lambda_1) \\ 0 = k_1(a_{1,3} + \lambda_3) + k_2(a_{2,3} + \lambda_2) \\ 0 = k_1 a_{1,4} + k_2(a_{2,4} + \lambda_3). \end{cases}$$

From the first equation we get $k_1 = 1$ and $\lambda_1 = -a_{1,1}$. From the second equation we get $k_2 = \frac{a_{1,2}+\lambda_2}{a_{1,1}-a_{2,2}}$. Substituting in the remaining two equations, we get the next system to determine λ_2 and λ_3

$$\begin{cases} (\lambda_2 + a_{1,2})(\lambda_2 + a_{2,3}) + (a_{1,1} - a_{2,2})(\lambda_3 + a_{1,3}) = 0 \\ (\lambda_2 + a_{1,2})(\lambda_3 + a_{2,4}) + (a_{1,1} - a_{2,2})a_{1,4} = 0. \end{cases}$$

In the case $\lambda_1 + a_{2,2} = 0$, one gets a very simple linear system

$$k_2(a_{2,2} + \lambda_1) = k_2(a_{2,3} + \lambda_2) = k_2(a_{2,4} + \lambda_3) = 0,$$

which gives $k_2 = 1, \lambda_1 = -a_{2,2}, \lambda_2 = -a_{2,3}, \lambda_3 = -a_{2,4}$.

Proof of Theorem 13.2. As we already mentioned in the introduction, the set $\mathbf{C}_{\mathcal{L}}$ can be determined as the set of all matrices $M \in \mathcal{M}^1$ such that the sum of the tangent space to $\mathcal{M}^1$ at M and the linear space $\mathcal{L}$ does not coincide with the whole $\mathcal{M}(m,n)$. Let us describe a basis of the tangent space to $\mathcal{M}^1$ at a sufficiently generic matrix M. Since $GL_m \times GL_n$ acts on $\mathcal{M}(m,n)$ with finitely many orbits, the tangent space to the $GL_m \times GL_n$-orbit of M under this action coincides with the tangent space to $\mathcal{M}^1$ at M. Note that $GL_m \times GL_n$ acts on $\mathcal{M}(m,n)$ by elementary row and column operations. Thus, if we take for example the affine chart in which the determinant formed by the first $(m-1)$ rows and columns is non-vanishing, then the tangent space to $\mathcal{M}^1$ at any matrix M belonging to this chart is generated by the following two groups of operations: (i) add to each column of M one of its first $m-1$ columns and (ii) add to the last row of M one of its other rows. One has therefore a total of $n(m-1) + (m-1) = (n+1)(m-1) = \dim \mathcal{M}^1$ generators. Taking the wedge of these generators with the chosen basis of $\mathcal{L}$ and representing an $(m \times n)$-matrix as an mn-vector by patching together its rows, we obtain the following $(mn \times mn)$-matrix that has a block structure of an $(m \times m)$-matrix with $(n \times n)$-blocks of the form given below:

$$\mathfrak{D} = \begin{pmatrix} a_{1,1}I_n & a_{2,1}I_n & \cdots & a_{m-1,1}I_n & -\sum_{j=1}^{m-1} k_j a_{j,1} \cdot I_n \\ a_{1,2}I_n & a_{2,2}I_n & \cdots & a_{m-1,2}I_n & -\sum_{j=1}^{m-1} k_j a_{j,2} \cdot I_n \\ \vdots & \vdots & \ddots & \vdots & \vdots \\ a_{1,m-1}I_n & a_{2,m-1}I_n & \cdots & a_{m-1,m-1}I_n & -\sum_{j=1}^{m-1} k_j a_{j,m-1} \cdot I_n \\ 0_{m-1,n} & 0_{m-1,n} & \cdots & 0_{m-1,n} & \mathcal{A}_{m-1,n} \\ L_{1,1} & L_{2,1} & \cdots & L_{m-1,1} & L_{m,1} \\ L_{1,2} & L_{2,2} & \cdots & L_{m-1,2} & L_{m,2} \\ \vdots & \vdots & \ddots & \vdots & \vdots \\ L_{1,n-m+1} & L_{2,n-m+1} & \cdots & L_{m-1,n-m+1} & L_{m,n-m+1} \end{pmatrix}.$$

Here $\mathcal{A} = \mathcal{A}_{m-1,n} = (a_{i,j}), i = 1, \dots, m-1, j = 1, \dots, n$, I_n is the identity $(n \times n)$-matrix, $0_{m-1,n}$ is the $((m-1) \times n)$-matrix with all vanishing entries, and, finally, $L_{i,j}$ is the ith row of the matrix L_j, see Theorem 13.2. Notice that the determinant $\det(a_{i,j}I_n)$, $i = 1, \dots, m-1$, $j = 1, \dots, m-1$, of the upper-left block of $\mathfrak{D}$ equals Δ^{m-1}, where $\Delta = \det(a_{i,j})$, $i = 1, \dots, m-1$, $j = 1, \dots, m-1$, is the leftmost principal minor of $A_{m-1,n}$. By the above assumption the matrix $\mathcal{A}$ lies in the affine chart where $\Delta \neq 0$. Finally, we clear the low-left block $(L_{i,j})$, $i = 1, \dots, m-1$, $j = 1, \dots, n-m+1$, of $\mathfrak{D}$ by "killing" all its elements through row operations using the above upper-left block (which is a square and non-degenerate $((m-1)n \times (m-1)n)$-matrix) to obtain the low-right block coinciding exactly with the matrix in formula (13.1). Thus the determinant of the whole matrix $\mathfrak{D}$ equals the product between Δ^{m-1} and the determinant from Theorem 13.2. Since in the considered chart one has $\Delta \neq 0$, the result follows. □

I.2 Open problems in Chapter 13.2

13.1. By analogy with the above case, for a given triple n, m, r one can also consider $(m-r)(n-r)$-dimensional pencils of matrices in $\mathcal{M}(m, n)$ and study their intersections with the subvariety $\mathcal{M}^r$ of all matrices of corank at least r. In particular, a natural question is to find an analog of Theorem 13.2 in this situation.

13.2. It would be interesting to determine the equation for $\pi_{\mathcal{L}}^{-1}(\mathbf{C}_{\mathcal{L}})$ in general, see Example 3 in the introduction. Another important direction is to determine the local multiplicity of a given eigenvalue in terms of the defining polynomial for $\mathbf{C}_{\mathcal{L}}$. Is there any analog of the Jordan normal form allowing to determine the multiplicity of a given eigenvalue?

13.3. Notice that the left-right action of $GL_m \times GL_n$ extends from the space $\mathcal{M}(m, n)$ to every space of (in)complete flags in $\mathcal{M}(m, n)$. For simple dimensional reasons, in most cases this action cannot have finitely many orbits.

Problem 13.2. On which spaces of (in)complete flags the above left-right action of $GL_m \times GL_n$ has finitely many orbits?

Appendix II. E. Heine "On the existence and number of Lamé functions of higher degree"

II.1 Comments on Heine's result and history around it

As Heine himself mentions in [82], the requirement of algebraic independence of the coefficients of $Q_2(z)$ and $Q_1(z)$ being sufficient is too strong and restrictive for his purposes, but he fails to give any other explicit condition guaranteeing the same result. Heine's original motivation for the consideration (10.1) comes from the classical Lamé equation (1.1) in which case $Q_1(z)$ and $Q_2(z)$ are very much algebraically dependent, namely, $Q_1(z) = Q_2'(z)/2$. Also in order to prove that the upper bound $\binom{n+l-2}{n}$ is actually achieved for algebraically independent $Q_1(z)$ and $Q_2(z)$, Heine uses an inductive argument where he forces Q_2 and Q_1 to become algebraically dependent in a special way. Theorem 8.2 below due mainly to Stieltjes is another clear indication that the algebraic independence is apparently an inappropriate condition for the goal.

Interpretation of Heine's text written in a rather cumbersome 19th century German and the exact statements it contains seems to create difficulties for mathematicians starting from 1870s and up to now, see, e. g., [126]. The main classical sources, namely, [176] from 1885, [150] from 1912, and [181] from 1939 are rather unclear about what it is that Heine actually proved and under what assumptions on $Q_2(z)$ and $Q_1(z)$ one can guarantee that for a given positive n the number of possible polynomial pairs (V, S) such that the corresponding S has degree exactly n is finite and bounded by $\binom{n+l-2}{n}$. Being aware of the existence of a gap in his proof, Heine uses a common trick. He covers it by a reference to an authority, in his case a letter of his friend Leopold Kronecker who has (under unknown conditions) shown that for a given degree n the eliminant of the system of algebraic equations defining the coefficients of the polynomial $V(z)$ does not vanish identically. This is equivalent to the finiteness of the number of these polynomials. Heine mentions also a short note of Kronecker's on this topic presented in the January issue of Monatsbericht der Berliner Akademie from 1864. Unfortunately the track ends here. All one can find in this issue is the phrase that at the meeting on the section of physics and mathematics of the Prussian Academy of Science held on November 14, 1864 “Herr Kronecker las über die verschieden Faktoren des Discriminante von Eliminantions-Gleichungen”, i. e., “Herr Kronecker gave a lecture on different factors of the discriminant of the elimination equation.” In attempt to overbridge this gap, we undertook the tedious task of translating and (even more so) decoding Heine's arguments.

In what follows we give a phrase-by-phrase translation of portions of § 135 and § 136 from Heine's book [82], which are relevant for our consideration. Some of the phrases were difficult to understand literally, and we provided our interpretation and comments of the content placed within the slash signs. We allowed ourselves to correct several obvious misprints without a special mention, tried to keep (to certain extent) the language flavor, and preserved the enumeration of the formulas of the original text.

II.2 Translation of §§ 135–136

§ 135 . . .

Next we ask which conditions the polynomials $\chi(x)$ and $\theta(x)$ must satisfy in order for the differential equation

$$\psi(x)\frac{d^2W}{du^2} + \chi(x)\frac{dW}{du} + \theta(x)W = 0 \tag{88}$$

to have a solution, which is a polynomial of degree n in x assuming that ψ is of degree $p+1$ and χ and θ are of degrees at most p resp. $p-1$.

/*Comment:* In fact, $\chi(x)$ is fixed and $\theta(x)$ is a variable polynomial. It is not clear why Heine talks about both $\chi(x)$ and $\theta(x)$ here./

This is always the case when it is the question about, as in the case of Lamé functions, a differential equation whose general solution does not contain any higher transcendentals than a rational function of integrals of algebraic functions and which has a definite order at $x = \infty$.

/*Comment:* This and the next two phrases explain why one should assume that $\deg(\chi) < \deg(\psi)$./

We say about a function W that it is of the order α for the finite value $x = a$, if $(x-a)^{\alpha-\epsilon}W$ and $(x-a)^{\alpha+\epsilon}W$ will be 0 respectively ∞ for $x = a$, however, small ϵ is taken; we assign to it the order α at infinity if $x^{-\alpha-\epsilon}W$ and $x^{-\alpha+\epsilon}W$ for $x = \infty$ becomes 0 resp. ∞. Thus, for example, $\log x$ has a definite order, namely 0.

If y and z are two particular solutions of (88), then one has

$$\log(yz' - zy') = \int \frac{\chi}{\psi} dx.$$

If χ were not of smaller degree than ψ, then $yz' - zy'$ would at $x = \infty$ go to 0 or ∞ as an exponential function and would therefore have no order.

For the solution to have an order for every x where $\psi(x)$ vanishes, $\frac{\chi(x)}{\psi(x)}$, after possible cancelations can in the denominator only have different factors. This follows from the same equality between two particular solutions, which we used above.

The following theorem answers the question posed at the beginning:

If the two polynomials $\psi(x)$ and $\chi(x)$ are given, the first of degree $p+1$, the second of degree p, then for exactly $\frac{(n+1)(n+2)\cdots(n+p-1)}{1.2\cdots(p-1)}$ different functions $\theta(x)$, there exists a particular solution of (88), which is a polynomial of degree n in x.

For $p = 1$, we understand the number given above, which in general might be denoted by (n, p) as 1. It is assumed above that the coefficients of ψ and χ are mutually independent. I call the numbers a, b, etc. mutually independent when there is no algebraic equation with integer coefficients that they satisfy.

/*Comment:* As stated in the introduction, Heine actually proves his theorem under the assumption of the algebraic independence of the coefficients./

It may be immediately added here that we will say about the numbers $a, b, \dots$, which already satisfy one or more algebraic equations that "they are further specialized" when in addition to these equations they satisfy one or more additional equations—which of course are not allowed to contradict the earlier ones.

The just mentioned assumption for the validity of the theorem demands more than is necessary. One could say with the same right that a polynomial of degree n in x with mutually independent coefficients vanishes for n different values of x, while for this it would suffice that the coefficients did not satisfy a particular equation with integer coefficients, namely, the well-known one which describes the coincidence of roots. Also for the validity of our theorem, it suffices that the coefficients do not satisfy certain finite number of algebraic equations, which equations in every case can be found but not in a comprehensible way.

/*Comment:* As we mentioned in the introduction, Heine realizes that for a given fixed n the set of all pairs of polynomials $(\psi(x), \chi(x))$ of degrees at most $p+1$ and p respectively for which there are less than (n, p) functions $\theta(x)$ solving the problem under consideration is an algebraic hypersurface with integer coefficients. It is by no means clear to us how Heine could possibly conclude that all the coefficients of the discriminantal equation are integers in the basis of the coefficients of $\psi(x)$ and $\chi(x)$. The development of the corresponding theory can be traced back to Cayley, see Appendix in [73], but no general results were obtained until much later./

§ 136

To obtain the proof of the theorem, one substitutes in (88) polynomials of degree n and $p-1$ for W and θ, namely

$$W = x^n + g_1x^{n-1} + \cdots +,$$
$$\theta = k_0x^{p-1} + k_1x^{p-2} + k_2x^{p-3} + \cdots.$$

It is clear that the necessary and sufficient condition that W satisfies equation (88) is that certain $n+p$ equations, which are linear in both the g_is and k_js and in the coefficients of ψ and χ, are satisfied. To show the structure of these, without having to work with too clumsy formulas, I present them for the case $p = 3$.

Let the given functions be

$$\psi(x) = x(c_0x^3 + c_1x^2 + c_2x + c_3),$$
$$\chi(x) = b_0x^3 + b_1x^2 + b_2x + b_3$$

and the sought functions

$$W(x) = g_0x^n + g_1x^{n-1} + g_2x^{n-2} + \cdots + g_n,$$
$$\theta(x) = k_0x^2 + k_1x + k_2.$$

/*Comment:* Notice that Heine apparently realized that, in general, it might be impossible to find $W(x)$ as a polynomial of degree exactly n and introduced even the leading coefficient as a new variable without special explanations./

For W to satisfy the differential equation (88), the coefficients g_is and k_js must satisfy the system of equations

$$\begin{aligned}
0 &= g_0[k_0 + nb_0 + n(n-1)c_0];\\
0 &= g_1[k_0 + (n-1)b_0 + (n-1)(n-2)c_0] + g_0[k_1 + nb_1 + n(n-1)c_1];\\
0 &= g_2[k_0 + (n-2)b_0 + (n-2)(n-3)c_0] + g_1[k_1 + (n-1)b_1 + (n-1)(n-2)c_1]\\
&\quad + g_0[k_2 + nb_2 + n(n-1)c_2];\\
0 &= g_3[k_0 + (n-3)b_0 + (n-3)(n-4)c_0] + g_2[k_1 + (n-2)b_1 + (n-2)(n-3)c_1]\\
&\quad + g_1[k_2 + (n-1)b_2 + (n-1)(n-2)c_2] + g_0[nb_3 + n(n-1)c_3];\\
0 &= g_4[k_0 + (n-4)b_0 + (n-4)(n-5)c_0] + g_3[k_1 + (n-3)b_1 + (n-3)(n-4)c_1]\\
&\quad + g_2[k_2 + (n-2)b_2 + (n-2)(n-3)c_2] + g_1[(n-1)b_3 + (n-1)(n-2)c_3];\\
&\vdots
\end{aligned}$$

In this way the equations will continue to be formed so that the next one will give the relation between the four g_is with the indices 5, 4, 3, 2. The final equations will be

$$\begin{aligned}
0 &= g_{n-1}[k_0 + b_0] + g_{n-2}[k_1 + 2b_1 + 2\cdot 1c_1] + g_{n-3}[k_2 + 3b_1 + 3\cdot 2c_2] + g_{n-4}[4b_3 + 4\cdot 3c_3];\\
0 &= g_n[k_0] + g_{n-1}[k_1 + 1\cdot b_1] + g_{n-2}[k_2 + 2b_2 + 2\cdot 1c_2] + g_{n-3}[3b_3 + 3\cdot 2c_3];\\
0 &= g_n[k_0] + g_{n-1}[k_1 + 1\cdot b_1] + g_{n-2}[2b_3 + 2\cdot 1c_3];\\
0 &= g_n[k_0] + g_{n-1}[b_3].
\end{aligned}$$

From the first equation the coefficient k_0 is completely determined in terms of the given coefficients b_0 and c_0 of ψ and χ; the next n equations give all g_is expressed in terms of the same known coefficients b_is and c_js and the $(p-1)$ (in our example 2) unknowns $k_1, k_2, \ldots$. The values of g_is that are obtained from the second to the $(n+1)$th equations when substituted in the last $(p-1)$ equations, will then give $(p-1)$ equations of higher degrees between the unknowns $k_1, k_2, \ldots, k_{p-1}$ and the known coefficients of ψ and χ, which only appear rationally in these equations.

/*Comment:* Heine apparently means that g_is will be given by rational functions of c_ls and b_ms, and substituting these one gets a system of rational equations defining k_js./

Once the k_js have been determined from these $p-1$ equations, the substitution of the found values in the 2nd to the $(n+1)$st equations will give the g_is. Two systems of related k_js are called *different*, that is, the two systems $k_1, \ldots, k_{p-1}$ and $k_1', \ldots, k_{p-1}'$ are called *different* when they are not equal. One realizes with the full confidence from the form of the 2nd to the $(n+1)$th equation that every system k_js corresponds to a system of g_is and different systems of k_js correspond to different system of g_is.

/*Comment:* The latter statement of Heine is false as is. It requires that the diagonal entries in the upper triangular system are non-vanishing (see our non-resonant condition in the introduction). But it is certainly true if the coefficients of $\psi(x)$ and $\chi(x)$ are algebraically independent./

One obtains thus that

There exist as many different equations (88) and therefore as many different polynomials W of degree n as there are different systems of k_js.

Next it is realized that

The degree of the elimination equation is at most (n, p), that is, there can only be at most (n, p) different systems of k_js.

If one throws a glance at the $(n + p)$ equations, which, with the exception of the first one for k_0, one can find above for the special case that $p = 3$, one will perhaps not realize the truth of this statement immediately, and instead believe that the degree of the elimination equation is larger.

/*Comment:* One is supposed to disregard the 1st equation in the system above and study the remaining $(n + p)$ equations./

But if one instead of $k_2, k_3, \ldots$, substitutes $x_2^2, x_3^3, \ldots$ where the lower numbers are indices and the upper numbers are exponents, and for symmetry we use $x_i \leftrightarrow k_i$, one realizes immediately that $g_1, g_2, \ldots, g_n$ are polynomials in the x_is of degree $1, 2, \ldots, n$ resp. so that after the substitution the $(p-1)$ last equations will have degrees $n+1, n+2, \ldots, n+p+1$ in x_is. That means that the degree of the elimination equation will at most grow to $(n + 1)(n + 2) \cdots (n + p + 1)$. If one takes into account that every value of $k_1, k_2, \ldots$ corresponds to one value of x_1, two values of x_2, three values of x_3, then the above assertion is proved.

/*Comment:* This is an excellent passage! What Heine does is call in the modern language of algebraic geometry the weighted Bezout theorem, see, e. g., [61]. The author was unable to find a reliable proof of this result in the literature prior to the 1970s./

Under the assumptions that the elimination equation is not identically zero, it will indeed have the above mentioned degree and give (n, p) different systems of ks.

/*Comment:* Crucial claim but not completely proved below./

There exist as I will show below indeed (n, p) distinct systems of the coefficients if the coefficients of ψ and χ are specialized in a certain way.

/*Comment:* Heine will show by induction that for a special choice of ψ and χ one can obtain exactly (n, p) distinct solutions. But instead of algebraically independent coefficients of ψ and χ, he needs to make them dependent to get an example of (n, p) distinct solutions. This is OK as soon as one knows that even for the specialized situation the total number of solutions is finite. This finiteness probably follows from his specific choice of specialization, see below./

That this elimination resultant does not vanish identically follows from the next observation which I take from the letter of my friend Kronecker.

/*Comment:* One needs to check also that the system of equations has (under very unclear non-degeneracy assumptions on the coefficients of ψ and χ) only isolated solutions. This is equivalent to the non-vanishing of eliminant./

If in the mentioned final equation, which will determine the functions $\theta(x)$ and $W(x)$, all coefficients vanish, then according to the general principle of elimination will at least one of the roots of $W(x) = 0$ be unrestricted.

/*Comment:* This is an interesting and rather obvious observation. It would be nice to find it explicitly in the old literature on elimination theory./

If one assigns to this root all values for which $\psi(x)$ vanishes, then one gets through this (procedure) certain restrictions on the function $\chi(x)$ but the latter function does not satisfy them even after the mentioned specialization.

Herr Kronecker added in the mentioned message that these restrictions are actually satisfied and one of the roots of $W(x) = 0$ remains undetermined if both $\psi(x)$ and $\chi(x)$ have the properties that for the known function $\theta(x)$ both solutions of (88) are polynomials in x.[1]

Concerning the number of systems with specialized ψ and χ, I set such relations between the coefficients which give that ψ has a factor $(x - a)$ twice and χ has it once.

/*Comment:* Then the algebraic independence is lost here since ψ has a double root and therefore lies on a discriminantal surface./

Then all W satisfying (88) have the form

$$U(n); (x-a)U(n-1); \ldots; (x-a)^n U(0),$$

where U are as in § 123 polynomials coprime with $(x - a)$ are their degrees are given in parenthesis to the left of the letter U. Under the substitution of this expression in (88) one gets for every U an equation like (88) in which instead of ψ and χ appear polynomials with unrestricted coefficients, which are not of degree $p+1$ and p but instead of p and $p-1$.

1 In Monatsbericht der Berliner Akademie from January 1864 Herr Kronecker commented my message with the following:
Introducing the roots of $W(x) = 0$ as variables in the equations

$$\psi(x_k)W''(x_k) + \chi(x_k)W'(x_k) = 0, \quad k = 1, 2, \ldots, n,$$

which define them, and substituting the coefficients of W' and W'' through the symmetric functions of $x_1, \ldots, x_n$, one sees directly that one of the unknowns x remains arbitrary if the elimination equation vanishes. Through a simple transformation of this system of equations one can determine the degree of the final equation and at the same time prove that certain coefficients are different from 0 as long as there are no special conditions on the functions ψ and χ.

/*Comment:* Note quite comprehensible proof of the non-vanishing of the eliminant./

/*Comment:* Apparently the logics here is as follows. We can use our result to prove that the number (n, p) of simple solutions is obtained if we can show that for the above specialization the total number of solutions is finite./

If one assumes that the general statement that we are proving is settled if ψ is the product of p linear factors (and for the product of two linear factors this is easy to show), then one gets the situation when ψ consists of $p + 1$ factors of which two are coinciding, altogether

$$(n, p - 1) + (n - 1, p - 1) + (n - 2, p - 1) + \cdots + (0, p - 1)$$

which after summation gives (n, p) different W, i. e., (n, p) different θ just as many as different systems of k's.

/*Comment:* This accomplished the induction step. What one misses is mentioning that the total number of solutions for Heine's specialization is finite. But this follows from his representation of all solutions as $U(n); (x - a)U(n - 1); \ldots ; (x - a)^n U(0)$. In each of these cases we already know that the number of solutions is finite and all of them are simple. To be completely rigorous, one should use a double induction on p and n as we did in § 11.3. Additional simplification of the order 2 case compared to the general order k case considered above comes from the fact that during the above specialization one can assume that the polynomials $\tilde{\psi}$ and $\tilde{\chi}$ such that $\psi = (x - a)\tilde{\psi}$ and $\chi = (x - a)\tilde{\chi}$ have algebraically independent coefficients, which does not work for higher order case./

Index

https://doi.org/10.1515/9783112207741-017

Notation index

$T = \sum_{j=1}^{k} Q_j(z)\frac{d^j}{dz^j}$	Notation for an exactly solvable differential operator used throughout the book, p. 2
$Q_j(z) = \sum_{i=0}^{j} q_{ji}z^i$	The coefficients of the polynomial $Q_j(z)$ in the above notation for an ES-operator T, p. 3
$\mathbb{C}_n[z]$	The linear space of complex-valued polynomials of degree at most n, p. 3
λ_n	The n-th eigenvalue of an ES-operator T, p. 3
$(n)_j = n(n-1)\dots(n-j+1)$	The Pochhammer symbol, p. 4
$p_n(z)$	The n-th eigenpolynomial of an ES-operator T, p. 4
$r = \max_{i=1,\dots,k}(\deg Q_i(z) - i)$	The Fuchs index of an arbitrary linear differential operator T with polynomial coefficients, p. 6
$V(z)$	A Van Vleck polynomial of a (higher) Heine–Stieltjes problem, p. 7
$S(z)$	A Stieltjes polynomial of a (higher) Heine–Stieltjes problem, p. 7
μ_n	The root-counting measure of a polynomial $p_n(z)$, p. 11
$\mu = \lim_{n\to\infty}\mu_n$	The asymptotic root-counting measure (if its exists) of a polynomial sequence $\{p_n(z)\}$, p. 11
$\mathcal{S} = supp(\mu)$	The (root) accumulation curve of the sequence $\{p_n(z)\}$, i. e., the support of the asymptotic measure μ, p. 11
Conv_Q	The convex hull of roots of a polynomial Q, p. 13
$\Psi_Q(z) = \int_{z_0}^{z}\frac{d\xi}{\sqrt[k]{Q(\xi)}}$	Canonical coordinate/transformation associated to a polynomial Q of degree k, p. 14
$C_\mu(z) = \int\frac{d\mu(\zeta)}{z-\zeta}$	The Cauchy transform of a measure μ supported in $\mathbb{C}$, p. 26
$u_\mu(z) = \int\log\lvert z-\zeta\rvert d\mu(\zeta)$	The logarithmic potential of a measure μ supported in $\mathbb{C}$, p. 26
r_n	The maximal absolute value of the roots of a polynomial p_n, p. 46
$\mu_n^{(j)}$	The root-counting measure of the j-th derivative $p_n^{(j)}$ of s_n (or p_n), p. 49
$C_{n,j}$	The Cauchy transform of the j-th derivative of s_n (or p_n), p. 49
$T_\lambda = \sum_{j=0}^{k} Q_j(z,\lambda)\frac{d^j}{dz^j}$	Spectral pencil of ES-operators, p. 132
$m_k(\mu) = \int_{\mathbb{C}} z^k d\mu(z),\ k = 0, 1, \dots$	Harmonic moments of a measure μ, p. 166
$Sp_n(\lambda) = \prod_{j=1}^{n+1}(\lambda - t_{n,j})$	The spectral polynomial of Lamé equation, p. 212
$\mathbb{K}(\zeta) = \int_0^1\frac{dt}{(1-t^2)(1-\zeta^2t^2)}$	Complete elliptic integral, p. 212
$F(a,b,c,z)$	Gauss hypergeometric series, p. 212
$\omega_{[a_1,a_2]}$	The arcsine measure supported on the interval with (complex) endpoints a_1 and a_2, p. 230

https://doi.org/10.1515/9783112207741-018

Bibliography

[1] M. J. Ablowitz, A. S. Fokas, *Complex Variables: Introduction and Applications*. Cambridge University Press, 2003.

[2] M. Abramowitz, I. A. Stegun (eds.), *Handbook of Mathematical Functions with Formulas, Graphs, and Mathematical Tables*. National Bureau of Standards Appl. Math. Ser., 55. Dover Publications, New York, 1964, xiv+1046 pp.

[3] M. Alam, Zeros of Stieltjes and Van Vleck polynomials. *Trans. Am. Math. Soc.* 252 (1979), 197–204.

[4] M. Alam, N. Zaheer, On the zeros of Stieltjes and Van Vleck polynomials. *Trans. Am. Math. Soc.* 229 (1977), 279–288.

[5] A. Agnew, A. Bourget, T. McMillen, On the zeros of complex Van Vleck polynomials. *J. Comput. Appl. Math.* 223 (2009), 862–871.

[6] P. Appell, Sur une classe de polynômes. *Ann. Sci. Éc. Norm. Supér. (2)* 9 (1880), 119–144.

[7] T. Aoki, T. Kawai, Y. Takei, New turning points in the exact WKB analysis for higher order ordinary differential equations, in: *Analyse Algébrique des Perturbations Singuliéres. I.* Hermann, 1994, pp. 69–84.

[8] T. Aoki, T. Kawai, Y. Takei, On the exact WKB analysis for the third order ordinary differential equations with a large parameter. *Asian J. Math.* 2 (1998), 625–640.

[9] T. Aoki, T. Kawai, Y. Takei, The Bender–Wu analysis and the Voros theory, in: *ICM-90 Satellite Conference Proceedings*. Springer, 1991, pp. 1–29.

[10] T. Aoki, T. Kawai, Y. Takei, The Bender–Wu analysis and the Voros theory II. *Adv. Stud. Pure Math.* 54 (2009), 19–34.

[11] A. I. Aptekarev, A. B. J. Kuijlaars, Hermite–Padé approximations and multiple orthogonal polynomial ensembles. *Russ. Math. Surv.* 66(6) (2011), 1133–1199.

[12] V. Arnold, Hyperbolic polynomials and Vandermonde mappings. *Funct. Anal. Appl.* 20(2) (1986), 52–53.

[13] V. Arnold, The Newton potential of hyperbolic layers. *Tr. Tbil. Univ.* 232/233 (1982), 23–29.

[14] V. Arnold, A. Varchenko, S. Gusein-Zade, *Singularities of Differentiable Maps. Vol. I. The Classification of Critical Points, Caustics and Wave Fronts*. Monographs in Mathematics, 82. Birkhäuser Boston, Inc., Boston, MA, 1985, xi+382 pp. Translated from the Russian by Ian Porteous and, Reynolds, Mark.

[15] F. M. Arscott, *Periodic Differential Equations*. Pergamon Press, 1964, vii+283 pp.

[16] M. J. Atia, A. Martínez-Finkelshtein, P. Martínez-González, F. Thabet, Quadratic differentials and asymptotics of Laguerre polynomials with varying complex parameters. *J. Math. Anal. Appl.* 416(1) (1 August 2014), 52–80.

[17] M. Bainbridge, D. Chen, Q. Gendron, S. Grushevsky, M. Moeller, Strata of k-differentials. *Algebr. Geom.* 6(2) (2019), 196–233.

[18] B. Bakalov, E. Horozov, M. Yakimov, General methods for constructing bispectral operators. *Phys. Lett. A* 222(1–2) (1996), 59–66.

[19] Y. Baryshnikov, B. Shapiro, Quadratic differentials and signed measures. *J. Anal. Math.* 144(1) (2021), 1–19.

[20] O. B. Bekken, B. K. Oksendal, A. Stray (eds.), *Spaces of Analytic Functions*. Lecture Notes in Math., 512. Springer-Verlag, Berlin, 1976.

[21] D. Belokolos, V. Z. Enolskii, Reduction of Abelian functions and algebraically integrable systems. II. Complex analysis and representation theory. *J. Math. Sci. (N.Y.)* 108 (2002), 295–374.

[22] S. R. Bell, *The Cauchy Transform, Potential Theory, and Conformal Mapping*. Studies in Advanced Mathematics. CRC Press, Boca Raton, FL, 1992, x+149 pp.

[23] Y. Ben Cheikh, K. Douak, A generalized hypergeometric d-orthogonal polynomial set. *C. R. Acad. Sci. Paris* 331(Série I) (2000), 349–354.

[24] C. M. Bender, T. T. Wu, Anharmonic oscillator. *Phys. Rev.* 184(5) (1969), 1231–1260.

[25] Y. Benoist, D. Hulin, Cubic differentials and hyperbolic convex sets. *J. Differ. Geom.* 98(1) (2014), 1–19.

https://doi.org/10.1515/9783112207741-019

[26] T. Bergkvist, On asymptotics of polynomial eigenfunctions for exactly solvable differential operators. *J. Approx. Theory* 149(2) (2007), 151–187.

[27] T. Bergkvist, H. Rullgård, On polynomial eigenfunctions for a class of differential operators. *Math. Res. Lett.* 9 (2002), 153–171.

[28] T. Bergkvist, H. Rullgård, B. Shapiro, On Bochner–Krall orthogonal polynomial systems. *Math. Scand.* 94(1) (2004), 148–154.

[29] H. L. Berk, W. M. Nevins, K. V. Roberts, New Stokes lines in WKB-theory. *J. Math. Phys.* 23 (1982), 988–1002.

[30] F. Beukers, A. van der Waall, Lamé equations with algebraic solutions. *J. Differ. Equ.* 197(1) (2004), 1–25.

[31] G. D. Birkhoff, On the asymptotic character of the solutions of certain linear differential equations containing a parameter. *Trans. Am. Math. Soc.* 9(2) (1908), 219–231.

[32] J.-E. Björk, J. Borcea, R. Bøgvad, Subharmonic configurations and algebraic Cauchy transforms of probability measures, in: *Notions of Positivity and the Geometry of Polynomials Trends in Mathematics*, 2011, pp. 39–62.

[33] J. Bochnak, M. Coste, M.-F. Roy, *Real Algebraic Geometry*. Ergebnisse der Mathematik und ihrer Grenzgebiete, 36. Springer, Berlin, 1998, x+430 pp. Translated from the 1987 French original, Revised by the authors.

[34] S. Bodine, R. Schäfke, On the summability of formal solutions in Liouville–Green theory. *J. Dyn. Control Syst.* 8(3) (2002), 371–398.

[35] S. Bochner, Über Sturm-Liouvillesche Polynomsysteme. *Math. Z.* 29(1) (1929), 730–736.

[36] J. Borcea, R. Bøgvad, Piecewise harmonic subharmonic functions and positive Cauchy transforms. *Pac. J. Math.* 240(2) (2009), 231–265.

[37] J. Borcea, B. Shapiro, Root asymptotics of spectral polynomials for the Lamé operator. *Commun. Math. Phys.* 282 (2008), 323–337.

[38] G. Boutry, M. Elad, G. Golub, P. Milanfar, The generalized eigenvalue problem for nonsquare pencils using a minimal perturbation approach. *SIAM J. Matrix Anal. Appl.* 27 (2005), 582–601.

[39] J. Borcea, R. Bøgvad, B. Shapiro, Homogenized spectral pencils for exactly solvable operators: asymptotics of polynomial eigenfunctions. *Publ. Res. Inst. Math. Sci.* 45 (2009), 525–568. Corrigendum: "Homogenized spectral pencils for exactly solvable operators: asymptotics of polynomial eigenfunctions", Publ. RIMS **85** (2012) 229–233.

[40] J. Borrego-Morell, On orthogonal polynomials with respect to a class of differential operators, arXiv: 2210.16605.

[41] A. Bourget, Nodal statistics for the Van Vleck polynomials. *Commun. Math. Phys.* 230(3) (2002), 503–516.

[42] A. Bourget, D. Jakobson, M. Min-Oo, J. A. Toth, A law of large numbers for the zeroes of Heine–Stieltjes polynomials. *Lett. Math. Phys.* 64(2) (2003), 105–118.

[43] M. Bôcher, The roots of polynomials that satisfy certain differential equations of the second order. *Bull. Am. Math. Soc.* 4 (1897), 256–258.

[44] R. Bøgvad, B. Shapiro, On motherbody measures with algebraic Cauchy transform. *Enseign. Math.* 62 (2016), 117–142. doi: https://doi.org/10.4171/LEM/62-1/2-8

[45] L. Brillouin, La mécanique ondulatoire de Schrödinger: une méthode générale de resolution par approximations successives. *C. R. Acad. Sci.* 183 (1926), 24–26.

[46] M. A. Brodsky, V. N. Strakhov, On the uniqueness of the inverse logarithmic potential problem. *SIAM J. Appl. Math.* 46(2) (Apr. 1986), 324–344.

[47] W. Bruns, U. Vetter, *Determinantal Rings*. Lecture Notes in Mathematics, 1327. Springer-Verlag, Berlin, 1988, viii+236 pp.

[48] P. Bränden, A generalization of the Heine–Stieltjes theorem. *Constr. Approx.* 34 (2011), 135–148.

[49] A. Bud, The image in $\mathcal{M}_g$ of strata of meromorphic and quadratic differentials. *Math. Z.* 298 (2021), 975–988.

[50] T. Chihara, *An Introduction to Orthogonal Polynomials*. Ser. Math. & Appl, 13. Gordon & Breach, 1978.

[51] D. Chu, G. Golub, On a generalized eigenvalue problem for nonsquare pencils. *SIAM J. Matrix Anal. Appl.* 28 (2006), 770–787.

[52] J. A. Cima, A. L. Matheson, W. T. Ross, *The Cauchy Transform*. Mathematical Surveys and Monographs, 125. American Mathematical Society, Providence, RI, 2006, x+272 pp.

[53] E. A. Coddington, N. Levinson, *Theory of Ordinary Differential Equations*. McGraw-Hill Book Company, Inc., New York–Toronto–London, 1955, 429 pp.

[54] L. Comtet, *Advanced Combinatorics: The Art of Finite and Infinite Expansions*. D Reidel Publishing Company, Dordrecht–Holland/Boston–USA, 1974.

[55] E. Delabaere, F. Pham, Resurgent methods in semi-classical asymptotics. *Ann. IHP, Phys. Théor.* 71(1) (1999), 1–94.

[56] J. P. Dedieu, Obreschkoff's theorem revisited: What convex sets are contained in the set of hyperbolic polynomials?. *J. Pure Appl. Algebra* 81(3) (1992), 269–278.

[57] H. Dette, W. Studden, Some new asymptotic properties for the zeros of Jacobi, Laguerre and Hermite polynomials. *Constr. Approx.* 11 (1995), 227–238.

[58] M. Dettweiler, S. Reiter, Middle convolution of Fuchsian systems and the construction of rigid differential systems. *J. Algebra* 318 (2007), 1–24.

[59] D. Dimitrov, W. Van Assche, Lamé differential equations and electrostatics. *Proc. Am. Math. Soc.* 128(12) (2000), 3621–3628.

[60] R. B. Dingle, *Asymptotic Expansions: Their Derivation and Interpretation*. Academic Press, London–New York, 1973.

[61] I. Dolgachev, Weighted projective varieties, in: *Group Actions and Vector Fields*, Vancouver, B.C., 1981. Lecture Notes in Math., 956. Springer, Berlin, 1982, pp. 34–71.

[62] J. J. Duistermaat, F. A. Grünbaum, Differential equation in the spectral parameter. *Commun. Math. Phys.* 103 (1998), 177–240.

[63] D. Dumas, M. Wolf, Polynomial cubic differentials and convex polygons in the projective plane. *Geom. Funct. Anal.* 25 (2015), 1734–1798.

[64] A. Eremenko, A. Gabrielov, An elementary proof of the B. and M. Shapiro conjecture for rational functions, in: *Notions of Positivity and the Geometry of Polynomials*, pp. 167–178, 2011.

[65] P. I. Etingof, A. A. Kirillov, Representations of affine Lie algebras, parabolic differential equations, and Lamé functions. *Duke Math. J.* 74(3) (1994), 585–614.

[66] W. N. Everitt, K. H. Kwon, L. L. Littlejohn, R. Wellman, Orthogonal polynomial solutions of linear ordinary differential equations. *J. Comput. Appl. Math.* 133 (2001), 85–109.

[67] H. Federer, *Geometric Measure Theory*. Die Grundlehren der mathematischen Wissenschaften, 153. Springer-Verlag New York Inc., New York, 1969, xiv+676 pp.

[68] M. Fedoruyk, *Asymptotic Analysis. Linear Ordinary Differential Equations*. Springer-Verlag, Berlin, 1993, viii+363 pp. Translated from the Russian by Andrew Rodick.

[69] J. Garnett, *Analytic Capacity and Measure*. Lecture Notes in Mathematics, 297. Springer-Verlag, Berlin–New York, 1972, iv+138 pp.

[70] S. J. Gardiner, T. Sjödin, Convexity and the exterior inverse problem of potential theory. *Proc. Am. Math. Soc.* 136(5) (2008), 1699–1703.

[71] S. J. Gardiner, T. Sjödin, Partial balayage and the exterior inverse problem of potential theory, in: *Potential Theory and Stochastics in Albac*, Theta, Bucharest, 2009, pp. 111–123.

[72] W. Gawronski, On the asymptotic distribution of the zeros of Hermite, Laguerre and Jonquire polynomials. *J. Approx. Theory* 50 (1987), 214–231.

[73] I. Gelfand, M. Kapranov, A. Zelevinsky, *Discriminants, Resultants, and Multidimensional Determinants. Mathematics: Theory & Applications*. Birkhäuser Boston, Inc., Boston, MA, 1994, x+523 pp.

[74] H. W. Gould, A. T. Hopper, Operational formulas connected with two generalizations of Hermite polynomials. *Duke Math. J.* 29 (1962), 51–63.

[75] G. Green, On the motion of waves in a variable canal of small depth and width. *Trans. Camb. Philos. Soc.* 6 (1837), 457–462.
[76] M. P. Grosset, A. P. Veselov, Elliptic Faulhaber polynomials and Lamé densities of states. *Int. Math. Res. Not.* 2006 (2006), 62120.
[77] M. P. Grosset, A. P. Veselov, Lamé's equation, quantum top and elliptic Bernoulli polynomials. *Proc. Edinb. Math. Soc.* 51 (2008), 635–650.
[78] F. A. Grünbaum, L. Haine, E. Horozov, Some functions that generalize the Krall–Laguerre polynomials. *J. Comput. Appl. Math.* 106(2) (1999), 271–297.
[79] F. A. Grünbaum, M. Yakimov, Discrete bispectral Darboux transformations from Jacobi operators. *Pac. J. Math.* 204(2) (2002), 395–431.
[80] B. Gustafsson, On motherbodies of convex polyhedra. *SIAM J. Math. Anal.* 29(5) (1998), 1106–1117.
[81] W. Hahn, Über die Jacobischen Polynome und zwei verwandte Polynomklassen. *Math. Z.* 39 (1935), 634–638.
[82] E. Heine, *Handbuch der Kugelfunctionen*, 1, pp. 472–479. G. Reimer Verlag, Berlin, 1878.
[83] K. Henseland, G. Landsberg, *Theorie der algebraischen Funktionen einer Variablen und ihre Anwendung auf algebraische Kurven und Abelsche Integrale*. Chelsea, New York, 1965, xvi+707 pp.
[84] T. Holst, B. Shapiro, On higher Heine–Stieltjes polynomials. *Isr. J. Math.* 183 (2011), 321–347.
[85] N. Honda, T. Kawai, Y. Takei, *Virtual Turning Points*, vol. 4. Springer, 2015.
[86] R. A. Horn, C. R. Johnson, *Matrix Analysis*, 2nd edn. Cambridge Universtiy Press, New York, 2013.
[87] E. Horozov, Vector orthogonal polynomials with Bochner's property. *Constr. Approx.* 48 (2018), 201–234.
[88] E. Horozov, d-orthogonal analogs of classical orthogonal polynomials. *SIGMA* 14 (2018), 063, 27 pp.
[89] E. Horozov, B. Shapiro, M. Tater, In search of higher Bochner theorem, arXiv:1807.01558, 2018.
[90] P-F. Hsieh, Y. Sibuya, *Basic Theory of Ordinary Differential Equations*. Springer, Berlin, 1999.
[91] Zh. Huang, J. Loftin, M. Lucia, Holomorphic cubic differentials and minimal Lagrangian surfaces in CH2. *Math. Res. Lett.* 20(3) (2013), 501–520.
[92] M. Hukuhara, Sur la théorie des équations différentielles ordinaires. *J. Fac. Sci., Univ. Tokyo. Sect. I* 7 (1958), 483–510.
[93] E. L. Ince, Further investigations into periodic Lamé functions. *Proc. R. Soc. Edinb.* 60 (1940), 83–99.
[94] E. L. Ince, *Ordinary Differential Equations*. Dover Publications, New York, 1944, viii+558 pp.
[95] J. A. Jenkins, *Univalent functions and conformal mapping. Ergebnisse der Mathematik und ihrer Grenzgebiete. Neue Folge, Heft 18. Reihe: Moderne Funktionentheorie*. Springer-Verlag, Berlin–Göttingen–Heidelberg, 1958, vi+169 pp.
[96] I. Kaplansky, *An Introduction to Differential Algebra*, 2nd edn., Actualités Scientifiques et Industrielles, 1251. Publications de l'Institut de Mathématique de l'Université de Nancago, Hermann, Paris, 1976.
[97] N. Katz, *Rigid Local Systems*. Ann. of Math. Stud., 139. Princeton University Press, 1996.
[98] M. Kac, W. L. Murdock, G. Szegö, On the eigenvalues of certain Hermitian forms. *J. Ration. Mech. Anal.* 2 (1953), 767–800.
[99] T. Kawai, Y. Takei, *Algebraic Analysis of Singular Perturbation Theory*, 227. American Mathematical Soc., 2005.
[100] T. Kawai, T. Koike, Y. Takei, On the exact WKB analysis of higher order simple–pole type operators. *Adv. Math.* 228(1) (2011), 63–96.
[101] R. Kenyon, Tilings of convex polygons. *Ann. Inst. Fourier* 47(3) (1997), 929–944.
[102] A. N. Kolmogorov, S. V. Fomin, *Elements of Function Theory and Functional Analysis*.
[103] V. Kostov, On the geometric properties of Vandermonde's mapping and on the problem of moments. *Proc. R. Soc. Edinb.* 112(3–4) (1989), 203–211.
[104] V. Kostov, On the hyperbolicity domain of the polynomial $x^n + a_1x^{n-1} + \cdots + a_n$. *Serdica Math. J.* 25(1) (1999), 47–70.
[105] H. L. Krall, *On Orthogonal Polynomials Satisfying a Certain Fourth Order Differential Equation*. The Pennsylvania State College Studies, 6. The Pennsylvania State College, State College, PA, 1940.

[106] A. M. Krall, *Hilbert Space, Boundary Value Problems, and Orthogonal Polynomials*. Operator Theory, Advances and Applications, 133. Birkhauser, Berlin, 2002.

[107] H. A. Kramers, Wellenmechanik und halbzahlige Quantisierung. *Z. Phys.* 39(10–11) (1926), 828–840.

[108] I. Krasikov, On the zeros of polynomials and allied functions satisfying second order differential equations. *East J. Approx.* 9(1) (2003), 51–65.

[109] I. Krasikov, Bounds for zeros of the Laguerre polynomials. *J. Approx. Theory* 121(2) (2003), 287–291.

[110] I. Krichever, M. Mineev-Weinstein, P. Wiegmann, A. Zabrodin, Laplacian growth and Whitham equations of soliton theory. *Phys. D* 198(1–2) (2004), 1–28.

[111] A. B. J. Kujilaars, K. T.-R. McLaughlin, Asymptotic zero behaviour of Laguerre polynomials with negative parameter. *Constr. Approx.* 20(4) (2004), 497–523.

[112] A. B. J. Kuijlaars, W. Van Assche, The asymptotic zero distribution of orthogonal polynomials with varying recurrence coefficients. *J. Approx. Theory* 99 (1999), 167–197.

[113] K. H. Kwon, L. L. Littlejohn, J. K. Lee, B. H. Yoo, Characterization of classical type orthogonal polynomials. *Proc. Am. Math. Soc.* 120 (1994), 485–493.

[114] K. H. Kwon, D. W. Lee, Characterizations of Bochner–Krall orthogonal polynomials of Jacobi type. *Constr. Approx.* 19 (2003), 599–619.

[115] K. Kwon, L. Littlejohn, G. Yoon, Bochner–Krall orthogonal polynomials, in: C. Dunkl, M. Ismail, R. Wong (eds.), *Special Functions*. World Scientific, Singapore, 2000, pp. 181–193.

[116] K. H. Kwon, L. L. Littlejohn, G. J. Yoon, Orthogonal polynomial solutions to spectral type differential equations: Magnus's conjecture. *J. Approx. Theory* 112(2) (October 2001), 189–215.

[117] K. H. Kwon, B. H. Yoo, G. J. Yoon, A characterization of Hermite polynomials. *J. Comput. Appl. Math.* 78 (1997), 295–299.

[118] F. Labourie, Flat projective structures on surfaces and cubic holomorphic differentials. *Pure Appl. Math. Q.* 3(4) (2007), 1057–1099. Special Issue: In honor of Grigory Margulis. Part 1.

[119] J. Liouville, Sur le développement des fonctions et séries. *J. Math. Pures Appl.* 1 (1837), 16–37.

[120] R. S. Maier, Algebraic solutions of the Lamé equation, revisited. *J. Differ. Equ.* 198(1) (2004), 16–34.

[121] R. S. Maier, Lamé polynomials, hyperelliptic reductions and Lamé band structure. *Philos. Trans. R. Soc., Math. Phys. Eng. Sci.* 366(1867) (2008), 1115–1153.

[122] G. Másson, B. Shapiro, On polynomial eigenfunctions of a hypergeometric-type operator. *Exp. Math.* 10(4) (2001), 609–618.

[123] F. Marcellán, A. Martínez-Finkelshtein, P. Martínez-González, Electrostatic models for zeros of polynomials: old, new, and some open problems. *J. Comput. Appl. Math.* 207(2) (2007), 258–272.

[124] M. Marden, *Geometry of Polynomials*, 2nd edn., Mathematical Surveys, 3. American Mathematical Society, Providence, R.I., 1966, xiii+243 pp.

[125] M. Marden, On Stieltjes polynomials. *Trans. Am. Math. Soc.* 33(4) (1931), 934–944.

[126] A. I. Markushevich, *Theory of Functions of a Complex Variable*. Prentice-Hall, Englewood Cliffs, N.J., 1965. Translated from the Russian by R. A. Silverman.

[127] P. Maroni, L'orthogonalité et les récurrences de polynômes d'ordre supérieur á deux. *Ann. Fac. Sci. Toulouse* 10(1) (1989), 105–139.

[128] A. Martínez-Finkelshtein, P. Martinez-Gonzalez, R. Orive, On asymptotic zero distribution of Laguerre and generalized Bessel polynomials with varying parameters. *J. Comput. Appl. Math.* 133(1–2) (2001), 477–487. Proceedings of the Fifth International Symposium on Orthogonal Polynomials, Special Functions and their Applications (Patras 1999).

[129] A. Martínez-Finkelshtein, P. Martínez-González, R. Orive, Asymptotics of polynomial solutions of a class of generalized Lamé differential equations. *Electron. Trans. Numer. Anal.* 19 (2005), 18–28 (electronic).

[130] A. Martínez-Finkelshtein, P. Martinez-Gonzalez, A. Zarzo, WKB approach to zero distribution of solutions of linear second order differential equations. *J. Comput. Appl. Math.* 145 (2002), 167–182.

[131] A. Martínez-Finkelshtein, E. A. Rakhmanov, Critical measures, quadratic differentials, and weak limits of zeros of Stieltjes polynomials. *Commun. Math. Phys.* 302 (2011), 53–111.

[132] A. Martínez-Finkelshtein, E. A. Rakhmanov, Critical measures, quadratic differentials, and weak limits of zeros of Stieltjes polynomials. *Commun. Math. Phys.* 302 (2011), 53–111.

[133] A. Martínez-Finkelshtein, E. A. Rakhmanov, On asymptotic behavior of Heine–Stieltjes and Van Vleck polynomials. *Contemp. Math.* 507 (2010), 209–232.

[134] A. Martínez-Finkelshtein, E. B. Saff, Asymptotic properties of Heine–Stieltjes and Van Vleck polynomials. *J. Approx. Theory* 118(1) (2002), 131–151.

[135] A. Martínez-Finkelshtein, Open problems I: Heine–Stieltjes polynomials, in: L. Baratchart, A. Martínez-Finkelshtein, D. Jimenez, D. S. Lubinsky, H. N. Mhaskar, I. Pritsker, M. Putinar, N. Stylianopoulos, V. Totik, P. Varju, Y. Xu (eds.), *Open Problems in Constructive Function Theory*. Electron. Trans. Numer. Anal., 25, 2006, pp. 511–525.

[136] T. Moteki, Y. Takei, Stokes geometry of higher order linear ordinary differential equations and middle convolution. *Adv. Math.* 310 (2017), 327–376.

[137] E. Mukhin, V. Tarasov, A. Varchenko, Higher Lamé equations and critical points of master functions. *Mosc. Math. J.* 7(3) (2007), 533–542, 575.

[138] E. Mukhin, A. Varchenko, Critical points of master functions and flag varieties. *Commun. Contemp. Math.* 6(1) (2004), 111–163.

[139] T. Murai, *A Real Variable Method for the Cauchy Transform, and Analytic Capacity*. Lecture Notes in Mathematics, 1307. Springer-Verlag, Berlin, 1988, viii+133 pp.

[140] G. Nemes, On the Borel summability of WKB solutions of certain Schrödinger-type differential equations. *J. Approx. Theory* 265 (May 2021), 105562.

[141] P. Nevai, *Orthogonal Polynomials*. Memoirs Amer. Math Soc., 213, pp. 1–185, 1979.

[142] P. S. Novikoff, Sur le probléme inverse du potentiel. *C. R. (Dokl.) Acad. Sci. URSS (N.S.)* 18 (1938), 165–168.

[143] W. Nuij, A note on hyperbolic polynomials. *Math. Scand.* 23 (1968), 69–72.

[144] N. Obreschkoff, *Verteilung und Berechnung der Nullstellen reeller Polynome*. VEB Deutscher Verlag der Wissenschafter, 1963.

[145] F. W. J. Olver, *Asymptotics and Special Functions*. A K Peters, Ltd., 1974.

[146] T. Oshima, *Fractional Calculus of Weyl Algebra and Fuchsian Differential Equations*. MSJ Memoirs, 28. Math. Soc. Japan, 2012.

[147] R. B. Paris, A. D. Wood, Stokes phenomenon demystified. *Bull. Inst. Math. Appl.* 31(1–2) (1995), 21–28.

[148] D. Pasechnik, B. Shapiro, On polygonal measures with vanishing harmonic moments. *J. Anal. Math.* 123(1) (July 2014), 281–301.

[149] J. V. Pereira, L. Pirio, An invitation to web geometry. From Abel's addition theorem to the algebraization of codimension one webs, in: *IMPA Mathematical Publications, 27th Brazilian Mathematics Colloquium*. Instituto Nacional de Matemática Pura e Aplicada (IMPA), Rio de Janeiro, 2009, xii+245 pp.

[150] G. Pólya, Sur un théoreme de Stieltjes. *C. R. Acad. Sci. Paris* 155 (1912), 767–769.

[151] G. Pólya, G. Szegö, *Problems and Theorems in Analysis*, 2. Springer-Verlag, 1976.

[152] G. Pöschl, E. Teller, Bemerkungen zur Quantenmechanik des anharmonischen Oszillators. *Z. Phys.* 83 (1933), 143–151.

[153] E. G. C. Poole, *Introduction to the Theory of Linear Differential Equations*. Dover Publications, Inc., New York, 1960, viii+202 pp.

[154] I. Pritsker, How to find a measure from its potential. *Comput. Methods Funct. Theory* 8(2) (2008), 597–614.

[155] T. Ransford, *Potential Theory in the Complex Plane*. LMS Student Texts, 28. Cambridge University Press, Cambridge, 1995, ix+234 pp.

[156] N. Reshetikhin, A. Varchenko, Quasiclassical asymptotics of solutions to the KZ equations, in: *Geometry, Topology, & Physics*. Conf. Proc. Lecture Notes Geom. Topology, IV. Int. Press, Cambridge, MA, 1995, pp. 293–322 (English summary).

[157] A. Ronveaux (ed.), *Heun's Differential Equations*. Oxford University Press, Oxford, 1995.

[158] M. Sakai, A moment problem on Jordan domains. *Proc. Am. Math. Soc.* 70(1) (1978), 35–38.

[159] T. V. Savina, B. Yu. Sternin, V. E. Shatalov, On a minimal element for a family of bodies producing the same external gravitational field. *Appl. Anal.* 84(7) (2005), 649–668.

[160] I. Scherbak, Rational functions with prescribed critical points. *Geom. Funct. Anal.* 12(6) (2002), 1365–1380.

[161] I. Scherbak, A. Varchenko, Critical points of functions, sl_2-representations, and Fuchsian differential equations with only univalued solutions. *Mosc. Math. J.* 3(2) (2003), 621–645, 745. Dedicated to Vladimir I. Arnold on the occasion of his 65th birthday.

[162] G. M. Shah, On the zeros of Van Vleck polynomials. *Proc. Am. Math. Soc.* 19 (1968), 1421–1426.

[163] G. M. Shah, On the zeros of Stieltjes and Van Vleck polynomials. *Ill. J. Math.* 14 (1970), 522–528.

[164] B. Shapiro, Algebro-geometric aspects of Heine–Stieltjes theory. *J. Lond. Math. Soc.* 83(1) (2011), 36–56.

[165] B. Shapiro, M. Shapiro, On eigenvalues of rectangular matrices. *Proc. Steklov Inst. Math.* 267 (2009), 248–255.

[166] B. Shapiro, G. Tahar, On the existence of quasi-Strebel structures for meromorphic k-differentials. *Enseign. Math.* 67(1–2) (2021), 187–207.

[167] B. Shapiro, M. Tater, On spectral polynomials of the Heun equation. I. *J. Approx. Theory* 162(4) (April 2010), 766–781.

[168] B. Shapiro, K. Takemura, M. Tater, On spectral polynomials of the Heun equation. II. *Commun. Math. Phys.* 311(2) (2012), 277–300.

[169] B. Shapiro, M. Tater, On spectral asymptotics of quasi-exactly solvable quartic potential. *Anal. Math. Phys.* 12(1) (2022), 2.

[170] H. S. Shapiro, *The Schwarz Function and its Generalization to Higher Dimensions*. Univ. of Arkansas Lect. Notes Math., 9. Wiley, N.Y., 1992.

[171] I. M. Sheffer, On the properties of polynomials satisfying a linear differential equation: Part I. *Trans. Am. Math. Soc.* 35(1) (1933), 184–214.

[172] Y. Sibuya, Sur réduction analytique d'un systéme d'equations différentielles ordinaires linéaires contenant un paramétre. *J. Fac. Sci., Univ. Tokyo* 7(5) (1958), 527–540.

[173] T. Sjödin, Mother bodies of algebraic domains in the complex plane. *Complex Var. Elliptic Equ.* 51(4) (2006), 357–369.

[174] A. Yu. Solynin, Modules and extremal metric problems. *Algebra Anal.* 11(1) (1999), 3–86. English transl.: St. Petersburg Math. J. **11** (2000), 1–65.

[175] A.Yu. Solynin, Quadratic differentials and weighted graphs on compact surfaces, in: *Analysis and Mathematical Physics*. Trends Math., pp. 473–505. Birkhäuser, Basel, 2009.

[176] T. J. Stieltjes, Sur certains polynômes qui vérifient une équation différentielle linéaire du second ordre et sur la theorie des fonctions de Lamé. *Acta Math.* 6(1) (1885), 321–326.

[177] G. G. Stokes, On the discontinuity of arbitrary constants which appear in divergent developments. *Trans. Camb. Philos. Soc.* 10 (1864), 106–128.

[178] G. G. Stokes, *Mathematical and Physical Papers, vol. IV*. Cambridge University Press, New York, 1904.

[179] K. Strebel, *Quadratic Differentials*. Ergebnisse der Mathematik und ihrer Grenzgebiete, 5. Springer-Verlag, Berlin, 1984, xii+184 pp.

[180] K. Strebel, On the metric $|f(z)|^\lambda |dz|$ with holomorphic f, in: *Complex Analysis*, Warsaw, 1979. Banach Center Publ., 11. PWN, Warsaw, 1983, pp. 323–336.

[181] G. Szegö, *Orthogonal Polynomials*, 4th edn., American Mathematical Society, Colloquium Publications, XXIII. American Mathematical Society, Providence, R.I., 1975, xiii+432 pp.

[182] G. Tahar, Counting saddle connections in flat surfaces with poles of higher order. *Geom. Dedic.* 196 (2018), 145–186.

[183] Y. Takei, WKB analysis and Stokes geometry of differential equations, in: *Analytic, Algebraic and Geometric Aspects of Differential Equations*. Springer, 2017, pp. 263–304.

[184] K. Takemura, Analytic continuation of eigenvalues of the Lamé operator. *J. Differ. Equ.* 228 (2006), 1–16.

[185] K. Takemura, On eigenvalues of the Lamé operator, arXiv:math.CA/0409247, 2004.

[186] L. Trefethen, T. Wright, Pseudospectra of rectangular matrices. *IMA J. Numer. Anal.* 22 (2002), 501–519.

[187] A. Turbiner, On polynomial solutions of differential equations. *J. Math. Phys.* 33 (1992), 3989–3994.

[188] A. Turbiner, Quasi-exactly solvable differential equations, in: N. H. Ibragimov (ed.), *CRC Handbook of Lie Group Analysis of Differential Equations*, 3. CRC Press, Boca Raton, FL, 1996, xvi+536 pp.

[189] A. G. Ushveridze, *Quasi-Exactly Solvable Models in Quantum Mechanics*. CRC Press, Boca Baton, 1994, 480 pp.

[190] J. Van Iseghem, Vector orthogonal relations. Vector QD-algorithm. *J. Comput. Appl. Math.* 19(1) (1987), 141–150.

[191] E. B. van Vleck, On the polynomials of Stieltjes. *Bull. Am. Math. Soc.* 4(9) (1898), 426–438.

[192] H. Volkmer, *Multi-Parameter Eigenvalue Problems and Expansion Theorems*. Lecture Notes in Mathematics, 1356. Springer-Verlag, Berlin, 1988, vi+157 pp.

[193] H. Volkmer, Four remarks on eigenvalues of the Lamé equation. *Anal. Appl.* 2(2) (2004), 161–175.

[194] A. Voros, The return of the quartic oscillator. The complex WKB method. *Ann. Inst. Henri Poincaré A, Phys. Théor.* 39(3) (1983), 211–338.

[195] W. Wasow, *Linear Turning Point Theory*. Applied Mathematical Sciences, 54. Springer-Verlag, New York–Berlin, 1985, 246 pp.

[196] G. Wentzel, Eine Verallgemeinerung der Quantenbedingungen für die Zwecke der Wellenmechanik. *Z. Phys.* 38(6–7) (1926), 518–529.

[197] E. T. Whittaker, G. Watson, *A Course of Modern Analysis. An Introduction to the General Theory of Infinite Processes and of Analytic Functions with an Account of the Principal Transcendental Functions*. Cambridge Mathematical Library. Cambridge Univ. Press, Cambridge, UK, 1996, vi+608 pp. Reprint of the 4th (1927) edition.

[198] N. Zaheer, On Stieltjes and Van Vleck polynomials. *Proc. Am. Math. Soc.* 60 (1976), 169–174, 1977.

[199] L. Zalcmann, *Analytic Capacity and Rational Approximation*. Lecture Notes in Mathematics, 50. Springer-Verlag, Berlin–New York, 1968, vi+155 pp.

[200] D. Zidarov, *Inverse Gravimetric Problem in Geoprospecting and Geodesy*. Elsevier, Amsterdam, 1990, vii+283 pp.

[201] A. Zorich, Flat surfaces, in: *Frontiers in Number Theory, Physics, and Geometry. I*, pp. 437–583. Springer, Berlin, 2006.

De Gruyter Expositions in Mathematics

Volume 79
Victor V. Przyjalkowski, Constantin Shramov
Weighted Complete Intersections, 2026
ISBN 978-3-11-071882-9, e-ISBN 978-3-11-071904-8

Volume 78
Vladimir P. Kostov
Uni-variate Polynomials in Analysis. Hyperbolic Polynomials, Order of Moduli, and Descartes' Rule of Signs, 2025
ISBN 978-3-11-914355-4, e-ISBN 978-3-11-221885-3

Volume 77
Abdel O. Kenoufi, Michel Gondran, Alexandre Gondran
Tropical Mathematics and Applications to Theoretical Physics and Scientific Computing. A New Approach of Non-Linear Problems, 2025
ISBN 978-3-11-076909-8, e-ISBN 978-3-11-076926-5

Volume 76/1
Tushar Das, Giulio Tiozzo, Mariusz Urbański, Anna Zdunik
Open Dynamical Systems. Volume 1: Symbolic Thermodynamic Formalism and Statistics, 2026
ISBN 978-3-11-103415-7, e-ISBN 978-3-11-103427-0

Volume 75
Renaud Gauthier
Derived Algebraic Geometry. An Elemental Approach, 2024
ISBN 978-3-11-133366-3, e-ISBN 978-3-11-133407-3

Volume 74
Walter Roth
Integral Representation. Choquet Theory for Linear Operators on Function Spaces, 2023
ISBN 978-3-11-131450-1, e-ISBN 978-3-11-131547-8

Volume 59
Friedrich Haslinger
The d-bar Neumann Problem and Schrödinger Operators, 2nd Edition, 2023
ISBN 978-3-11-118290-2, e-ISBN 978-3-11-118292-6

www.degruyterbrill.com